Wild Pigs in the United States

Wild Pigs in the United States

Their History,
Comparative Morphology,
and Current Status

John J. Mayer and I. Lehr Brisbin, Jr.

The University of Georgia Press
Athens and London

©1991 by the University of Georgia Press
Athens, Georgia 30602
All rights reserved
The paper in this book meets the guidelines for
permanence and durability of the Committee on
Production Guidelines for Book Longevity of the
Council on Library Resources.

Printed in the United States of America
95   94   93   92   91   5   4   3   2   1

Library of Congress Cataloging in Publication Data

Mayer, John J.
    Wild pigs in the United States : their history, comparative morphology, and current
status / John J. Mayer, I. Lehr Brisbin, Jr.
        p.   cm.
    Includes bibliographical references and index.
    ISBN 0-8203-1239-8 (alk. paper)
    1. Wild boar—United States. 2. Feral swine—United States. 3. Anatomy, Comparative.
I. Brisbin, I. Lehr. II. Title.
QL737.U58M38   1991
599.73'4—dc20   90-10945
                                                                                     CIP

British Library Cataloging in Publication Data available

With our thanks to Sandy West and her island

Contents

Illustrations

x   *Illustrations*

Plates

Tables

Acknowledgments

We are especially grateful for the encouragement and guidance provided for this undertaking by the late Dr. Ralph M. Wetzel, former Professor of Biology at the University of Connecticut. We are also grateful to Dr. Kentwood D. Wells, Assistant Professor of Biology, at the same institution for his helpful criticism and encouragement of the senior author during the course of this study. Eleanor T. West, Director of the Ossabaw Foundation, and her staff were constant sources of insight, inspiration, and encouragement to us, and we were constantly challenged to match their tireless efforts on behalf of conserving the unique feral swine of Ossabaw Island. Deserving of special mention in this regard are the efforts of Roger Parker, Ossabaw Island herdsman, and Dr. J. H. McKenzie, D.V.M., of Savannah, Georgia. Dr. E. P. Odum, Director Emeritus of the Institute of Ecology at the University of Georgia, was instrumental in drawing our attention to the need to study and better understand the role of feral swine as long-term integral components of the Ossabaw Island environment.

The following people were also particularly helpful to us in various ways: Dr. Ray F. Smart, James R. Davis, Richard Gilbert, Daniel W. Baber, James Phelps, Kenneth Schwarz, Michael Ottea, Dr. Lyle K. Sowls, and Constance A. Rinaldo for collecting pig skulls; John W. Reiner for collecting numerous specimens from South Carolina and Georgia and for many insightful late-night discussions concerning the biology of wild pigs; Kenneth O. Butts, Assistant Manager of the Aransas National Wildlife Refuge, for helping to collect specimens and other data from that locality; Dr. Ray F. Smart, Catherine H. Carter, Paul W. Collins, James R. Davis, Sheila Kortlucke, and Aryan I. Roest for providing helpful discussion and correspondence; Jean B. Coleman for preparing the outline maps and figures; Jan Hinton, Karin Knight, and Edith Towns for typing and editing the manuscript; Robert E. Dubos for preparing a large portion of the specimens and for tolerating large numbers of smelly pig heads; Dr. Joshua Laerm and M. Elizabeth McGhee for help in specimen preparation and curation; Dr. Carl W. Rettenmeyer for aiding in administrative matters and for his constant encouragement; and the directors and staffs of the Savannah River Ecology Laboratory, Aransas National Wildlife Refuge, Ossabaw Foundation, and the Welder Wildlife

Refuge for their help in specimen collecting and other logistic matters. The encouragement of the officers and staff of the American Minor Breeds Conservancy, and their viewing and promoting the publication of this book as supportive of their efforts to survey and document all forms of feral livestock in the United States, is also gratefully acknowledged.

We also owe a debt of gratitude to the many people who responded to our letters and questionnaires, for the useful information they generously provided. Their names and addresses are listed in Appendix A.

We thank the following people for their efforts and assistance in field work: J. C. Agostine, K. C. Anderson, C. Bagshaw, J. F. Bergan, F. Bridges, F. A. Brooks, M. Caudell, P. Collins, A. W. Conger, R. E. Dubos, D. T. Elliott, T. E. Fox, G. Hogsed, R. T. Hoppe, T. W. Hughes, W. Jamtgaard, P. E. Johns, T. Jones, R. A. Kennamer, J. D. LeBlanc, B. H. Miller, J. W. Reiner, O. E. Rhodes, Jr., B. Schwartz, Dr. L. M. Smith, G. L. Sullivan, Dr. J. R. Sweeney, Dr. L. D. Vangilder, and M. Vargo.

We are also indebted to the following institutions, curators, and other staff members who generously allowed us to examine specimens in their care: The American Museum of Natural History, Sydney Anderson, Karl F. Koopman, and Richard G. Van Gelder; Philadelphia Academy of Natural Sciences, Charles Smart; Aransas National Wildlife Refuge Collection, Frank Johnson and Ken Butts; Bernice P. Bishop Museum, Alan C. Ziegler and Carla H. Kishinami; California Academy of Sciences, Luis F. Baptista and Jacqueline Schonewald; Carnegie Museum of Natural History, Hugh H. Genoways and Suzanne B. McLaren; California Polytechnic State University, Aryan I. Roest; California State University, David G. Huckaby; Delaware Museum of Natural History, David M. Niles; The Field Museum of Natural History, Patricia Freeman; Florida State Museum, Steven R. Humphrey; University of Iowa, George D. Schrimper; University of Kansas, Robert S. Hoffman and Sheila Kortlucke; Louisiana State University, George H. Lowrey, Jr.; Museum of Comparative Zoology, John A. W. Kirsch and M. Elizabeth Rutzmoser; Mississippi Museum of Natural Sciences, Catherine H. Carter; University of Missouri, William H. Elder; Museum of Southwestern Biology, William R. Barber; Michigan State University, Donald O. Straney; Museum of Vertebrate Zoology, James L. Patton; North Carolina State University, Roger A. Powell; New Mexico State University, Charles S. Thaeler, Jr.; Northeastern University Vertebrate Museum, Gwilym S. Jones; Oklahoma State University, Bryan P. Glass and Bob J. Bollinger; Royal Ontario Museum, Randolph L. Peterson; Santa Barbara Museum of Natural History, Paul W. Collins; San Diego Natural History Museum, A. M. Rea; Savannah River Ecology Laboratory, Michael H. Smith; Texas Cooperative Wildlife Collection, David J. Schmidly and Jo Boyd; UCLA-Dickey Collection of

Birds and Mammals, Nancy E. Simerly; University of Connecticut, Ralph M. Wetzel and Robert E. Dubos; University of Georgia Museum of Natural History, Joshua Laerm and M. Elizabeth McGhee; University of Georgia, School of Forestry, R. Larry Marchinton and Daniel L. Adams; University of Illinois Museum of Natural History, Donald F. Hoffmeister and Victor E. Diersing; University of Massachusetts at Amherst, David Klingener; University of Michigan Museum of Zoology, Philip Myers and Nancy Petersen; University of Puget Sound, Murray L. Johnson; University of Pennsylvania Veterinary School, Peter Dodson; University of Southwestern Louisiana Biology Museum, Marshall B. Eyster; The National Museum of Natural History, Charles O. Handley, Jr., James G. Mead, and Henry W. Setzer; University of Texas at Austin, Vertebrate Paleontology Laboratory, Melissa Winans; University of Texas at El Paso, Robert B. Wilhelm; Museum of Wildlife and Fisheries Biology, Ronald E. Cole; Welder Wildlife Refuge, Lynn Drawe; Yale-Peabody Museum, John A. W. Kirsch, V. Louise Roth, and Kathleen M. Scott.

We are also indebted to the following private collectors for access to their personal specimens: Dr. Daniel L. Adams, Daniel W. Baber, Dr. Richard Deiters, Dr. Daniel H. Janzen, Dr. Milton C. Hildebrand, Robert Nester, Lynn Nowotny, Michael Ottea, Peter W. Payne, Leroy Rittimann, Kenneth Schwarz, Dr. Ray F. Smart, and Lisa M. Staley.

Finally, Virginia R. Mayer, Mary E. Nadal-Mayer, and Brenda Brisbin provided motivation and moral support, without which the completion of this book would not have been possible. This study was also supported in part by University of Connecticut Research Foundation doctoral research grants nos. 1171-00-22-0202-35-373 and 5171-000-22-0202-35-272 in 1981, by a contract (DE-AC09-76SROO-819) between the U.S. Department of Energy and the University of Georgia, and by the Ossabaw Foundation.

Abbreviations

AMNH	The American Museum of Natural History, New York
ANSP	Philadelphia Academy of Natural Sciences, Philadelphia
ANWR	Aransas National Wildlife Refuge Collection, Austwell, Texas
BPBM	Bernice P. Bishop Museum, Honolulu, Hawaii
CAS	California Academy of Sciences, San Francisco
CM	Carnegie Museum of Natural History, Pittsburgh, Pennsylvania
CPSU	California Polytechnic State University, San Luis Obispo
CSULB	California State University, Long Beach
DHJ	Dr. Daniel H. Janzen, Philadelphia, Pennsylvania
DLA	Dr. Daniel L. Adams, Athens, Georgia
DMNH	Delaware Museum of Natural History, Greenville
DWB	Daniel W. Baber, Eugene, Oregon
FMNH	The Field Museum of Natural History, Chicago, Illinois
FSM	Florida State Museum, Gainesville
JJM	Dr. John J. Mayer, Savannah River Ecology Laboratory, Aiken, South Carolina
IOWA	University of Iowa, Iowa City
KS	Kenneth Schwarz, San Antonio, Texas
KU	University of Kansas, Lawrence
LMS	Lisa M. Staley, Storrs, Connecticut
LN	Lynn Nowotny, San Antonio, Texas
LR	Leroy Rittimann, Boerne, Texas
LSU	Louisiana State University, Baton Rouge
MCH	Dr. Milton Hildebrand, Davis, California
MCZ	Museum of Comparative Zoology, Cambridge, Massachusetts
MMNS	Mississippi Museum of Natural Sciences, Jackson
MO	Michael Ottea, San Antonio, Texas
MOU	University of Missouri, Columbia
MSB	Museum of Southwestern Biology, Albuquerque, New Mexico
MSU	Michigan State University, East Lansing
MVZ	Museum of Vertebrate Zoology, Berkeley, California
NCS	North Carolina State University, Raleigh
NMSU	New Mexico State University, Las Cruces

NUVM	Northeastern University Vertebrate Museum, Boston, Massachusetts
OSU	Oklahoma State University, Stillwater
PWP	Peter W. Payne, Boerne, Texas
RD	Dr. Richard Deiters, Storrs, Connecticut
RFS	Dr. Ray F. Smart, Boerne, Texas
RN	Robert Nester, San Antonio, Texas
ROM	Royal Ontario Museum, Toronto
SBMNH	Santa Barbara Museum of Natural History, Santa Barbara, California
SDNHM	San Diego Natural History Museum, San Diego, California
SREL	Savannah River Ecology Laboratory, Aiken, South Carolina
TCWC	Texas Cooperative Wildlife Collection, College Station
UCLA	UCLA-Dickey Collection of Birds and Mammals, Los Angeles, California
UCONN	University of Connecticut, Storrs
UGA	University of Georgia Museum of Natural History, Athens
UGSF	University of Georgia, School of Forestry
UIMNH	University of Illinois Museum of Natural History, Urbana
UMASS	University of Massachusetts at Amherst, Amherst
UMMZ	University of Michigan Museum of Zoology, Ann Arbor
UPS	University of Puget Sound, Tacoma, Washington
UPVS	University of Pennsylvania Veterinary School, Philadelphia
USLBM	University of Southwestern Louisiana Biology Museum, Lafayette
USNM	The National Museum of Natural History, Washington, D.C.
UTAVP	University of Texas at Austin, Vertebrate Paleontology Laboratory, Austin
UTEP	University of Texas at El Paso, El Paso
WFBM	Museum of Wildlife and Fisheries Biology, Davis, California
WWR	Welder Wildlife Refuge, Sinton, Texas
YPM	Yale-Peabody Museum, New Haven, Connecticut

Wild Pigs in the United States

1. Introduction

The study of exotic, or non-native, mammals introduced into new areas is essential for the effective management of the ecosystems in which these forms occur. Exotic species often become serious economic pests and exert an unbalancing force on their host environments (Cottam, 1956; De Vos et al., 1956). Examples of the introduced European rabbit, *Oryctolagus cuniculus,* in Australia and the Scottish red deer, *Cervus elaphus,* in New Zealand illustrate the magnitude of impact on the environment that such exotics can have (Wodzicki, 1950; Laycock, 1966).

More than 30 exotic mammal species have become established as free-ranging populations in the United States since the period of early European colonization (De Vos et al., 1956; McKnight, 1964; Roots, 1976). Some research has been done on the effects of these animals on the native flora and fauna, but comparatively little is known about the biology of the exotic forms themselves in their new habitats. It is especially important to learn more about those exotic mammals that have become established in the United States and are ecologically difficult to manage.

This study is an investigation of the history, morphology, and current status of the wild-living pig (*Sus scrofa* ssp.) populations in the United States. The term *wild-living pig* encompasses the introduced Eurasian wild boar, feral hogs (wild-living descendants of domestic swine), and hybrids between these two forms.

True wild pigs in the family Suidae (Mammalia: Artiodactyla) are not native to the New World. Fossils and subfossils of *Sus scrofa* are known only from the Palearctic, Oriental, and Ethiopian realms (Vereshchagin, 1967; Martin and Wrights, 1967; Kurten, 1968; Trumbull and Reed, 1974; Cook and Wilkinson, 1978). The recent natural range of this species, excluding domestic swine (*Sus scrofa domesticus*), includes the region south of 60°N in the steppe and broadleaf woodland zones of the Palearctic and Oriental realms from western Europe to the Amur River valley in eastern Siberia; eastern and southern Asia, including isolates in Japan, Formosa, Malaya, Sumatra, Java, and Sri Lanka; the Middle East north of approximately Saudi Arabia; and extreme northern Africa from Tunisia to Rio de Oro (Ellerman and Morrison-Scott, 1966; Corbet, 1978; Figure 1). A discussion of the native

1

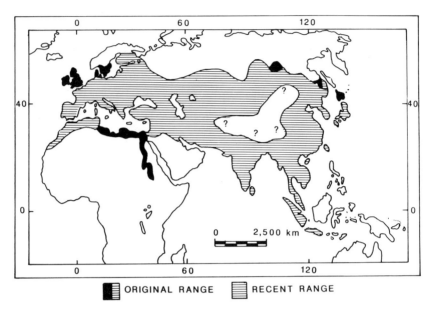

ORIGINAL RANGE   RECENT RANGE

Figure 1. Approximate original and recent native range of the Eurasian wild boar (*Sus scrofa* L.) based on Harrison (1968); Heptner, Nasimoviv, and Bannikov (1966); Van den Brink (1968); Ansell (1971); Lekagul and McNeely (1977); Genov (1981); and Sjarmidi and Gerard (1988).

subspecies of *Sus scrofa* is provided in Appendix B. The only piglike mammals native to the Nearctic and Neotropical realms are peccaries (family Tayassuidae), which are found from the southwestern United States south-ward to the Patagonian region of Argentina (Mayer and Brandt, 1982; Mayer and Wetzel, 1986, 1987). As a result of introductions by man, however, true pigs are currently distributed worldwide in either a wild, feral, or domesti-cated state (Anderson and Jones, 1967; Walker, 1975).

From as early as A.D. 750–1000, beginning with the settlement of Hawaii, there have been numerous introductions of swine into the United States (Towne and Wentworth, 1950; Simoons, 1961; Joesting, 1972). Both domestic swine and Eurasian wild boar have been subsequently released into the wild, either intentionally or accidentally, in at least 20 states (McKnight, 1964). Wild-living swine were documented around Spanish, French, and English colonies of the eastern and southern United States as early as the 16th and 17th centuries (Towne and Wentworth, 1950; Hanson and Karstad, 1959).

Free-living swine populations in this country currently consist of varying combinations of feral hogs, Eurasian wild boar, and hybrids between these two (Jones, 1959; Wood and Lynn, 1977; Rary et al., 1968; Figure 2). During

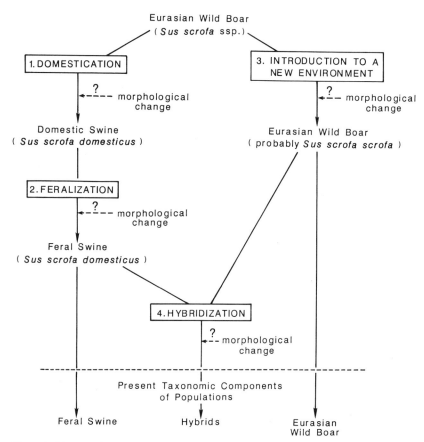

Figure 2. Lineages that led to the present taxonomic components of the wild pig populations in the United States, and the causal events (1-4) that may or may not have altered the morphology of these animals.

the last two decades, these scattered populations were estimated to number between 500,000 and 2 million animals (R. H. Conley, pers. comm.; R. L. Walker, pers. comm.; Springer, 1977; van der Leek, 1989; Nettles, 1989; Stewart, 1989), making this species the most abundant free-ranging introduced ungulate in the United States (McKnight, 1964; Decker, 1978). Wild pigs have mostly deleterious effects on their host environments and are very difficult to control or eradicate (Stegeman, 1938; Wahlenberg, 1946; Wakely, 1954); Hanson and Karstad, 1959; Jones, 1959; Bratton, 1975; Tisdell, 1979). Nonetheless, wild-living pigs are considered highly desirable big game animals by many individuals and state agencies (Ruch, 1966; Wooters, 1973,

1975; French, 1974). For these reasons, wild pigs have become important and controversial in many parts of the United States in recent years.

To date, most of the research on wild-living pigs in the United States has examined aspects of their biology related to their management and control as a big game species and/or a pest. No systematic work has been done on the morphological variation exhibited by these animals as a whole (Bratton, 1977). The few available studies cover less than a dozen populations in the United States. These studies either present morphological data as part of a general description of a specific population (Silver, 1957; Nichols, 1962a; Barrett, 1971; Pine and Gerdes, 1973; Wood and Brenneman, 1977), or analyze such data, mostly external body measurements, from only one or two populations (Henry, 1970; Brisbin et al., 1977a).

Because of the paucity of information on the morphological variation of these animals, the characteristics of the three types of populations are poorly understood. The lack of specific and identifiable definitions for wild boar versus feral hog versus hybrid animals has caused legal problems in the control of the hunting of wild pig populations in at least two states, Georgia and North Carolina (Lewis et al., 1965; R. B. Hamilton, pers. comm.). An understanding of the morphological characteristics of the three types of wild pigs would be useful in identifying these types for comparative studies, such as that of Brugh et al. (1964) comparing the susceptibility of wild boar and domestic swine to hog cholera. This would be especially true for populations for which the history is poorly known. In addition, morphological studies would aid in better understanding the adaptations and functional roles of the three types of wild-living and domestic swine. For example, the question could be addressed as to whether feral hogs diverge morphologically from their immediate domestic ancestor and come to more closely resemble their distant wild boar ancestor. The interpretation of morphological characteristics of these populations would be made more meaningful if a thorough history were available for each population unit under study.

The objectives of this research are thus (1) to compile a detailed history of the various wild-living populations of this introduced species in the United States; (2) to document the morphological variation exhibited by the various populations and taxonomic forms, using populations with known histories; (3) from these analyses, to determine identifying morphological characters for these taxa; (4) from these criteria, to determine the taxonomic status of wild pig populations of unknown or uncertain history; and (5) using all this information, to determine the current status of this species as a wild or feral animal in the United States. These objectives will be grouped into three major sections: (a) history of the introduction (objective 1 above); (b) comparative morphology (objectives 2, 3, and 4 above); and (c) current status (objective 5 above).

The following terms are defined here as they are used in the context of this study. Each term includes both male and female animals of all ages. Each definition is followed by the scientific name applied to the group. A more thorough discussion of the use of scientific nomenclature for these forms is provided in Appendix C.

Eurasian wild boar *(Sus scrofa* ssp.)—individuals from populations of wild-living *Sus scrofa* native to the Palearctic, Oriental, or Ethiopian realms which have no history of domestication in their ancestry; these include all subspecies of *Sus scrofa* except *Sus scrofa domesticus.*

Domestic swine *(Sus scrofa domesticus)*—individuals from populations of domesticated forms of *Sus scrofa* existing under some form of conscious artificial selection by man other than that exerted by hunting and trapping.

Feral swine or feral hogs *(Sus scrofa)*—individuals from populations of wild-living *Sus scrofa* with a domestic ancestry; these include recently escaped or released animals and animals from populations that have been wild for more than one generation.

Hybrids or wild boar × feral hog hybrids *(Sus scrofa)*—individuals from populations with some combination of both feral hogs and Eurasian wild boar in their ancestry; animals that are mostly Eurasian wild boar in their ancestry or that have a questionable presence of feral hog in their ancestry will be referred to as wild boar.

Wild or wild-living swine, hogs, or pigs *(Sus scrofa)*—any free-ranging form of *Sus scrofa*; can include Eurasian wild boar, feral hogs, or hybrids between these two.

Swine, hogs, or pigs *(Sus scrofa)*—any form of *Sus scrofa,* wild or domesticated.

2. History of the Introduction

The purpose of this chapter is to present a detailed history of the introduction of swine as wild-living animals into the United States. As mentioned earlier, true wild pigs are not native to the Nearctic realm, and all these populations are thus the result of direct introductions by man. No other comprehensive study or document currently exists detailing the origin and subsequent history of wild-living *Sus scrofa* in the United States.

The specific topics covered in this chapter pertain to (a) the present range of this species in each state where it occurs; (b) the types of pigs that were introduced into specific areas in the United States; (c) the location(s) from which these animals were originally obtained (especially in the case of the introduced Eurasian wild boar); (d) how, when, and by whom these animals were introduced; and (e) the fate and distribution of each population since the time of the introduction; that is, an account of how the range of these animals has changed, if at all, up to the present. This historical approach will then form a point of departure for determining the different types of wild pigs that current occur in different areas of the United States.

The information in this chapter was gathered through a search of past and recent literature and from letters and questionnaires sent to private individuals and federal and state personnel in 1981. A similar questionnaire (Appendix D) was again sent to appropriate personnel in all state agencies in 1988. Personal interviews were also conducted with certain private individuals and federal and state personnel.

The presentation of these data is divided into two subsections: (1) feral hog introductions and (2) wild boar introductions. The latter includes, in this case, hybrid populations. The resulting information was then compared with and/or supplemented by the results of a survey of feral and wild swine populations also conducted in 1988 by the Southeastern Cooperative Wildlife Disease Study (SCWDS, 1988; Appendix E). The histories of feral hog introductions are detailed on a state-by-state basis. The histories of wild boar introductions, however, are organized chronologically. Following the history of each initial introduction, the subsequent fate of these animals in that area is detailed, followed by descriptions of any other areas into which animals from the initial population may have been subsequently introduced.

Feral Hog Introductions

Feral hog populations have become established in at least 20 states in the United States. Numerous feral populations of this species were also present during colonial times in several other states where they no longer occur. This was especially true in the central East Coast and northeastern portions of this country (Towne and Wentworth, 1950). This study is concerned only with states having populations that have persisted into or were established during the 20th century.

Feral hog populations have originated through (1) the escape of domestic swine, (2) free-ranging of domestic swine as a method of husbandry, and (3) intentional release of domestic swine or wild-trapped feral hogs for the purpose of establishing wild-living populations (McKnight, 1964; Anon., 1970; Wood and Lynn, 1977). The original stocks used in these introductions varied greatly (McKnight, 1964). This variability has been complicated further in cases where feral individuals have interbred with introduced Eurasian wild boar.

Before detailing the feral hog introductions, it is necessary to discuss the origin of the first domestic swine in the United States, since these animals were the basis for the early feral hog populations. Historians generally agree that the first domestic swine introduced into the United States were those brought by the early Polynesian settlers to Hawaii during the first century A.D. The detailed history of this introduction will be covered in the subsequent discussion for that state. There is some disagreement, however, concerning the origin of domestic swine in the continental United States. Most authorities attribute the source of these animals to the early European explorers (Towne and Wentworth, 1950). At least one theory, discussed in the following paragraphs, would antedate the domestic swine of colonial European origin and possibly that of the Hawaiian introduction.

Despite the fact that swine are not native to the Nearctic realm, some skeletal remains of this species have been found in association with pre-Hispanic fossil and subfossil faunas. These specimens have been found in Florida (Webb, 1974), Texas, Arkansas (Quinn, 1970a, 1970b), Tennessee (Guilday, 1977), New Jersey (Cope, 1870), Maryland (F. C. Whitmore, pers. comm.), and South Carolina (R. E. Mancke, pers. comm.). Most of these records have been attributed to the incorporation of recent material into older assemblages, since these pig specimens either do not appear to be fossilized or were likely fossilized as more lately associated remains (Cope, 1870; Webb, 1974; F. C. Whitmore, pers. comm.). Quinn (1970a) has argued against this commonly accepted conclusion, claiming that some of these specimens suggest that pigs were hunted in North America by Archaic peoples, indicating a pre-Columbian introduction of this species into the New

World. One such specimen from near Paris, Texas, consists of a nearly complete skull with a large imbedded prehistoric-type projectile point (Quinn, 1970a). This specimen previously had been used to substantiate the recent employment of prehistoric-type points, thereby attributing the specimen's origin to the post-Columbian colonial introductions. Quinn (1970a) rejected this, suggesting that the animal, not the hunting point, had been misplaced in time.

Another pig specimen from an Indian midden in Arkansas was associated with late Archaic to early Woodland projectile points and with fragments of animal remains that apparently represented prey species hunted by these people (Quinn, 1970a). In Quinn's opinion, the presence of this specimen with its associated material implied that these people hunted pigs along with the native mammals of this region. Swine dental and cranial fragments recovered from an archaeological dig in northwestern Arkansas and aged using Carbon-14 dating were reportedly determined to be earlier than the 1490s (Quinn, 1970b). Quinn concluded that either domestic swine were introduced into North America by the Archaic people, or the feral animals themselves dispersed across the Bering land bridge during Altithermal times. If this hypothesis is correct, feral populations of swine could have been present in the Nearctic realm as early as perhaps 2,000 to 6,000 years B.P. Since the interpretation of this evidence is still debatable, few other paleontologists seem to have accepted Quinn's theory (F. C. Whitmore, pers. comm.).

It is also possible that the first domestic swine in the mainland United States were introduced as early as 1497, when John Cabot is said to have discovered the coast of New England. This is uncertain, and a more probable date for the introduction of swine to this region is during the 16th and early 17th centuries (Hoornbeek, 1980).

The earliest pigs brought to the continental United States were more likely the descendants of eight "selected" domestic swine that Admiral Christopher Columbus brought to Cuba in 1493 on his second voyage (Towne and Wentworth, 1950; Lewinsohn, 1964). These animals, taken on board with other livestock and supplies at the island of Gomera in the Canary Islands (Lewinsohn, 1964; Donkin, 1985), subsequently became the reputed progenitors of all the pigs that populated the islands of the Spanish West Indies. Swine were introduced into Puerto Rico between 1505 and 1508, into Jamaica as early as 1509, and into Cuba in 1511 (Donkin, 1985). On many of these islands, swine were turned loose to forage for themselves. On Hispaniola, Jamaica, and Cuba, they soon ran wild through the jungles and canebrakes (Towne and Wentworth, 1950). Thirteen years after Columbus introduced domestic swine into the West Indies, the Spanish colonists that settled on these islands found it necessary to hunt the now free-ranging descendants of these animals because they were killing cattle (Ensminger,

1961). These feral animals reportedly were very aggressive and often attacked Spanish soldiers hunting rebellious Indians or escaped slaves in these areas, especially when the animals were cornered (Towne and Wentworth, 1950). Droves of feral hogs also ravaged cultivated crops in the islands, notably maize and sugarcane (Donkin, 1985). When the Spanish explorers provisioned expeditions, they captured some of these feral hogs to take with them. From these sprang the immense droves that sustained such Spanish explorers as Cortés and De Soto on their journeys in the New World during the early 1500s (Towne and Wentworth, 1950). It was from the stock used by these initial expeditions that the first well-documented feral populations of this species originated in the continental United States.

Alabama

Feral hogs are found in scattered populations in the southwestern corner and eastern half of Alabama. A small population may persist on the Wheeler National Wildlife Refuge in the northern part of the state (T. Z. Atkeson, pers. comm.). These populations in Alabama are the result of free-range livestock practices dating back to the early settlements of the late 1770s. Some sources claim that descendants of domestic swine that escaped from De Soto's expedition in this area in 1541 have survived as feral populations up to the present. Although this is possible, there are no accounts to support the existence of any feral hog populations in this area before the American Revolution (J. R. Davis, pers. comm.). Some of these animals are of recent origin, although the majority of the feral hogs in the state have been wild for more than two generations (Lewis et al., 1965). One 157-kg feral boar, killed during a deer hunt on the Rob Boykin Wildlife Management Area in Washington County in the early 1950s, had been castrated, indicating a history of at least partial domestication or contact with man. This animal had been recognized and pursued for two years before it was killed by a hunter (McKnight, 1964).

Any feral hogs remaining on the Wheeler National Wildlife Refuge are descendants of escaped domestic stock from surrounding farms. This population probably never exceeded 25 individuals in number, and hunting by refuge staff and the general public continued to reduce the remaining numbers (T. Z. Atkeson, pers. comm.). All hogs were believed to have been removed from the refuge by the late 1970s (H. Fowler, pers. comm.).

A population in the central eastern portion of Cherokee County was reported to consist of feral hogs with some "Russian" wild boar ancestry (J. R. Davis, pers. comm.).

Starting in the late 1940s and extending through the 1960s, several attempts were made to develop strains of laboratory "miniature pigs" to serve as subjects for biomedical research. Because of their hardiness and relatively small body sizes, feral hogs from a number of populations in the United States were captured and used as foundation stock to produce miniature strains through crosses with various domestic breeds. Although Louisiana seems to have been the main source of feral animals used in most of these programs, the Hormel Institute of St. Paul, Minnesota, used "wild pigs from Alabama (Guinea hogs)." These animals were used along with feral hogs from Catalina Island, California, in crossing with Chester White domestic pigs to initiate their miniature pig development program in 1953 (Rempel and Dettmers, 1966). Later crosses in this program also involved "Piney Woods pigs" from Louisiana introduced in 1951 and 1953, although no further details are given concerning the numbers, sources, or characteristics of any of these feral animals.

Arizona

The largest feral hog population in Arizona is in Mohave County on the Havasu National Wildlife Refuge along the Colorado River below the Hoover Dam. Other smaller populations have been reported in the Dragoon Mountains, on the east slopes of the Chiricahua Mountains, in the Whetstone Mountains, in the southeast corner of Grand Canyon National Park, and in the San Pedro Valley of Cochise County (McKnight, 1964; R. Gilbert, pers. comm.). None of these populations reportedly were composed of wild boar or wild boar × feral hog hybrids.

The feral hogs on the Havasu National Wildlife Refuge are believed to have originated from domestic stock that escaped from the Soto Ranch north of Needles, California. Extensive flooding of the Colorado River apparently prompted their subsequent further dispersal throughout this area (R. Gilbert, pers. comm.). Reports of feral hogs on what is now the Havasu National Wildlife Refuge around Topock Marsh predate 1900. In the early 1900s, numerous feral hog hunting camps existed just outside the refuge boundary. Men from these camps processed large numbers of feral hogs illegally taken from land on and around the refuge (R. Gilbert, pers. comm.). These animals were estimated to number 300–400 in the mid 1970s (Decker, 1978). Today there are an estimated 200–300 feral hogs on the refuge (R. Gilbert, pers. comm.).

Arkansas

The Arkansas "razorback," or feral, hog is found in the southern half of the state and in the Ozark National Forest. All the wild pigs in Arkansas are strictly feral hogs (B. McAnally, pers. comm.). No Eurasian wild boar have ever been introduced into the state (L. Johnston, pers. comm.). Free-ranging domestic swine were known in Arkansas as early as the 1540s in Desha County, where a few pigs escaped from De Soto's expedition during the early part of that decade. In 1543, local Indians captured some of these animals and gave them to the men left in De Soto's returning party to supplement the expedition's food supplies (Lewis, 1907).

Free-ranging of domestic swine in the rural areas of Arkansas was a common practice among farmers and landowners from the early 1800s until the mid 1900s, when no livestock laws were in effect. In the early 1900s, this practice was most prevalent in areas that now form the present national forests and large tracts of paper company lands in the state (T. Taylor, pers. comm.). When the market was profitable, the owners of these animals would attempt to catch their stock for sale. Some of these domestic swine became feral when they escaped capture or were not claimed by their owners (B. Conley, pers. comm.; L. Owens, pers. comm.). Many of these feral populations have been in existence for several generations (F. Ward, pers. comm.).

In October 1966, the U.S. Forest Service closed all national forest land in Arkansas that had previously been open range for privately owned livestock. The Forest Service then initiated a trapping program designed to remove any remaining feral hogs on all government forest lands. During this period, the Arkansas state legislature enacted livestock laws to close all the remaining open range in the state (B. McAnally, pers. comm.; M. J. Rogers, pers. comm.). With the passage of these laws, the feral hog populations in Arkansas declined markedly (Sealander, 1979); however, in a few areas these livestock laws have been ignored, and some animals have been turned out to free-range during recent times (Lewis et al., 1965; B. Conley, pers. comm.). Today there are only isolated remnant populations of feral hogs in the more remote areas of the Ozark and Ouachita national forests (B. McAnally, pers. comm.) and in the bottomland forests and swamps of the state (Sealander, 1979). These populations seem to have stabilized in both numbers and range in these areas (T. Taylor, pers. comm.).

The wild pigs in the southwestern corner of Arkansas are feral hogs that originated from released domestic swine. Free-ranging of domestic swine was commonplace in this portion of the state before the closed-range livestock laws enacted in the mid 1960s. In the mid 1970s, feral hogs were found in the following counties (population size estimates in parentheses): Polk (50), Howard (50), Sevier (80), Lafayette (100), Miller (25–30), Nevada (40), Columbia (60), Pike (25–30), and Little River (40) (T. Taylor, pers. comm.).

Feral hogs that occurred in Philips, Lee, Cross, Crittenten, St. Francis, and Monroe counties were mostly the result of recently released domestic stock. Some, however, were from stock that had been feral for several generations.

A 1973 estimate of the number of feral hogs in all six counties was 300 (F. Ward, pers. comm.). No recent reports exist concerning the continued presence of these populations.

Free-ranging domestic swine existed as early as the late 1800s on the land that is now the Felsenthal National Wildlife Refuge in Bradley, Union, and Ashley counties. By the early 1900s, most residents in the surrounding area kept domestic swine on open range. The livestock laws passed in the mid 1960s were never enforced in the river bottoms of the Ouachita and Sabine rivers. Local individuals continued to keep swine in this area until the refuge was established in 1977, and people began removing those animals that could be recaptured. By 1981, however, a population of several hundred animals still remained on the refuge (L. L. King, pers. comm.).

Feral hogs recently reported to have occurred on the White River National Wildlife Refuge, in Desha, Arkansas, Phillips, and Monroe counties, originated from escaped domestic stock belonging to adjacent landowners. By 1981, however, these animals had apparently been completely removed from the refuge (R. R. McMaster, pers. comm.).

California

Populations of feral hogs are found throughout the oak-woodland and chaparral scrub areas of at least 33 of the 58 counties and on four of the six major channel islands of California. The range of these animals has increased through time by both natural dispersal and through numerous introductions by man.

Feral hogs have probably been in California since the arrival of the Spanish in the 1500s (C. Stanton, pers. comm.; Barrett, 1978). Many of the colonial domestic swine were allowed to forage freely in the surrounding oak-wooded hills near these early settlements (Barrett, 1970). The capacity of the forested areas of California to support large numbers of free-ranging domestic swine was noted as early as 1769 by Lieutenant Pedro Foges in an account of his travels through the state (Towne and Wentworth, 1950). In 1834, domestic swine were found around 21 of the Spanish missions in California (Towne and Wentworth, 1950). After 1850, as new settlements and homesteads spread throughout the state in the wake of the gold rush, domestic swine were commonly released to free-range in the hills and fatten during the acorn season (Barrett, 1978).

By 1880, the domestic swine population in California exceeded 600,000 by one estimate (Towne and Wentworth, 1950). In several localities throughout California, commercial swine operations liberated their stock to feed on acorns during the fall and winter. Often, these animals were allowed to forage in the hills until market prices were optimal. Thus numerous wild-living or semi-wild-living populations of this species have been long established throughout the state (Barrett, 1970). Many of these have since crossbred with introduced wild boar.

As the dates of origin are unknown for many of the present feral hog populations in California, their histories are presented in a county-by-county format, proceeding from north to south. Populations with known dates of origin are arranged chronologically within each county.

Siskiyou County—Feral hogs in Siskiyou County are descendants of hogs from Humboldt County that spread north along the Klamath River (Anon., 1970).

Humboldt County—Feral hog herds at Pilot Ridge, Rainbow Ridge, and Showers Pass all originated from stock that escaped from local ranches. The feral hogs along the Klamath River near Orleans originated from domestic stock that escaped from local Indians (Anon., 1970).

Shasta County—The populations of feral hogs on private lands in the eastern foothills of the county escaped from local ranches in the area (Anon., 1970).

Tehama County—In the 1840s, domestic swine were introduced on ranches along the Sacramento River in central Tehama County. Some were released to fatten during the season of acorn mast fall. As early as the 1880s, local settlers hunted feral hogs in the foothills of the county northeast of Red Bluff on what is now the Dye Creek Ranch. By 1910, feral hogs were common in the foothills north and south of Mill Creek, the southern border of the ranch. During the 1920s, most homesteaders had moved out of the roadless backcountry, reducing the hunting pressure on this population except in those areas along the edges of the Mill Creek valley. Hunting on the ranch was restricted in 1963, when it was purchased by William S. Keeler. The valley edge population began to increase, and in the summer of 1964 feral hogs invaded the irrigated pastures of neighboring landowners. A regulated hunting program was started in 1966 to control this population. Wild boar were released on the ranch at a later date (Barrett, 1978; R. H. Barrett, pers. comm.). Other feral populations in the eastern portion of the county originated from stock that escaped from local ranches. In the late 1960s, ranchers released feral hogs from the eastern Tehama herd into the Paskenta area (Anon., 1970).

Lake County—Feral hog populations in the Walker Ridge, Cow Mountain, and Middle Creek sections of the county all originated from stock that escaped from ranches in the surrounding areas (Anon., 1970).

Colusa County—The feral hogs on the Colusa National Wildlife Refuge originated from stock that escaped from neighboring ranches. Additional feral hogs have been introduced into the area of the refuge by parties interested in increasing the size of this population. By 1981, there were several hundred of these animals on the Colusa National Wildlife Refuge (L. Hill, pers. comm.). The 1988 SCWDS survey, however, does not indicate that any wild pig population currently persists in this county (Appendix E-7).

Sonoma County—Domestic swine were released on open range around Fort Ross by Russian settlers and fur traders perhaps as early as 1812 (Hutchinson, 1946). The feral hogs in the northwestern quarter of the county are the descendants of escaped domestic swine from early settlers in the area (Anon., 1970). By 1981, these animals were found along Elder Creek and other branches of the Eel River system (Lynn, 1978).

Santa Clara County—Domestic swine were released to start a feral population in the hills east of Los Altos (Anon., 1970).

San Benito County—Domestic swine first arrived in the county with the Spanish at the Mission San Juan Baptista in 1797. The first records of feral hogs in the county are from the 1870s. From 1900 until about 1955, all wild-living pigs in San Benito County were descendants of open-range domestic swine and were localized in the northeastern, central, and southern portions of the county. After 1955, these populations hybridized with introduced wild boar (Barrett and Pine, 1980).

Monterey County—In the 1880s, King City in the east-central part of this county was a major livestock shipping point. Herds of domestic cattle and swine were driven there from as far away as the coast to the west and Panocha to the east (Pine and Gerdes, 1969). Many pigs undoubtedly escaped during this process. King City, not incorporated under that name until 1911, was known as "Hog Town" until that time because of the large numbers of free-ranging domestic swine in and around the town. As market prices fluctuated, many hogs were abandoned and became established in the wilds of Monterey County during the late 1800s and early 1900s (Pine and Gerdes, 1969, 1973). Feral hogs have occurred on what is now the Fort Hunter Liggett area south of King City since the 1840s (P. A. Dubsky, pers. comm.). This population later hybridized with introduced wild boar (Pine and Gerdes, 1973).

San Luis Obispo County—On the land belonging to the Huasna Land & Cattle Company and in the Los Padres National Forest in the areas of the upper Huasna Creek, Lopez Canyon, and Stoney Creek, an expanding feral hog population originated from domestic swine from the R. Rust Ranch which had been released to forage for mast in the surrounding hills. Further north, in the area around Santa Margarita, feral hogs originated from stock that escaped from local ranchers. The feral hogs in the extreme northern end of the county, between San Simeon and Camp Roberts, are the descendants of an estimated 50 domestic swine left in the area by a rancher named Jane-way, who was unable to corral them when he moved out of the county (Anon., 1970).

Santa Barbara County—Feral hogs on Vandenberg Air Force Base are the descendants of escaped domestic stock from the Spanish explorers. In 1978, there were approximately 225 on the base (Schoning, 1978).

Feral hogs exist on only two of the Channel Islands in this county, Santa Cruz and Santa Rosa islands. Domestic swine formerly occurred on San Miguel Island, but neither feral nor domestic swine remained there in 1981 (P. W. Collins, pers. comm.). The feral hogs on Santa Cruz Island probably originated from several sources. The island was first discovered in October 1542 by a Spanish expedition led by Juan Rodríguez Cabrillo (Bolton, 1916). In 1582, Spain used the island as a penal colony, and about 600 prisoners were brought by ships and abandoned there with a few head of domestic cattle, horses, and swine. A short time later, the prisoners killer the cattle and horses and used the hides to cover crudely made boats to escape from the island. Before leaving, however, they turned the swine loose to wander the island. These animals multiplied and became free-ranging (Hogg, 1920). In 1830, a short-lived Mexican penal colony on the island may have introduced additional swine. In the 1850s, a poacher named Captain Box is said to have lived on the island and raised domestic swine for fresh meat (C. Stanton, pers. comm.). Although the details of these accounts are uncertain, the feral hogs on the island are unquestionably the result of domestic swine releases by one or more of the island's early residents (Wheeler, 1944). In the 1940s and 1950s, Edwin L. Stanton, owner of the island, introduced hog cholera into the Santa Cruz feral hog population. This was done on the advice of experts from the state universities in an effort to control this population (Wheeler, 1944). On the basis of interviews with former island residents, Nettles et al. (1989) report that the first virus introduction occurred before 1944, with at least two subsequent reintroductions in the early–mid 1950s. Each introduction of the virus was followed by observations of large numbers of dead or dying swine. A survey of sera and tissues from 31 hogs captured on the island in 1987, however, led these same authors to the conclusion that the hog cholera virus was no longer active on the island. Since the mid 1960s, a commercial sport hunting program has controlled the size of the population (P. W. Collins, pers. comm.). In the late 1960s, the island was estimated to have about 1,000 feral hogs (Anon., 1970).

In the late 1970s, the Nature Conservancy purchased nearly 90% of the land on Santa Cruz Island. By the early 1980s plans were under way to extirpate the island's feral livestock from the lands under their control. Through an agreement with the American Minor Breeds Conservancy of Pittsboro, North Carolina, an effort was made to determine whether any of the island's feral stocks represented unique genetic resources from which representative individuals should be obtained for captive propagation (D. Bixby, pers. comm.; M. E. Stanley, pers. comm.). In conjunction with this agreement, a number of feral sheep were captured and removed from the island; however, since the island's swine had been subject to a number of introductions of domestic stock over the years, they were judged not to represent a genetic resource of any unique value. Accordingly, the capture and removal of feral hogs for captive propagation was not recommended.

The feral hogs on Santa Rosa Island are of uncertain origin. The Cabrillo expedition also discovered Santa Rosa in late 1542 (Bolton, 1916). Between 1595 and the late 1700s, European explorers visited the island several times. Some sources believe that the Santa Rosa feral hogs were introduced by one of the early Spanish explorers who visited the island (Holland, 1962). It was a common practice among these early Spanish explorers in the Pacific to stock domestic goats and swine on islands they visited. This was done to provide later visitors and shipwreck victims with an available supply of fresh meat (C. Stanton, pers. comm.). It is unknown if any of these expeditions actually released any stock on Santa Rosa. Holland (1962) discounted this theory of an early Spanish origin of Santa Rosa's feral hogs, stating that no early records indicated wild pigs on the island according to either the early Spanish explorers or such people as George Nidever, who hunted on the island before 1843. Archaeological specimens of skulls (SBMNH 985, 988, and 989) of this species, however, are known from early Indian middens on the island that date no later than 1820–40 (P. W. Collins pers. comm.). These specimens consisted of one yearling and two adults. One of the adults was a large boar with large, well-developed, sharp canines. Because swine are usually slaughtered at younger ages and it is somewhat dangerous to maintain large male swine with intact canines under primitive husbandry conditions, it is likely that the Santa Rosa Indians were hunting free-ranging animals rather than raising them. Later analyses showed that these three animals were morphologically indistinguishable from the present-day feral hogs on the island. It is possible that the Canalino Indians living on Santa Rosa in the late 1700s or early 1800s could have brought some of these animals from Santa Cruz Island, where they had been introduced earlier. According to Holland (1962), these Indian fishermen traveled easily between the islands in their plank canoes during this period. In addition, North American and Russian fur hunters visited the island during the late 1700s and early 1800s (Holland, 1962) and may have been another means by which feral hogs from Santa Cruz could have been introduced on Santa Rosa.

In September 1844, Alpheus B. Thompson, owner of one-half of Santa Rosa at that time, released domestic cattle, horses, and sheep on the island (Holland, 1962). He later introduced domestic swine in 1853 (Holland, 1962; P. W. Collins, pers. comm.). The exact number of swine released is unknown, but it has been described as "a lot of hogs" (Holland, 1962). In 1859 and 1865, Thomas W. and Alexander P. More were able to purchase both halves of the island. Between then and the early 1900s, the island was used by the

More family for sheep ranching (Holland, 1962). In the late 1800s, wild-living representatives of the introduced domestic cattle and swine were found on the island, with the feral hogs numbering about 400 animals (P. W. Collins, pers. comm.).

In 1902, Santa Rosa was purchased by the Vail and Vickers Company (Holland, 1962), which owned the island until the mid-late 1980s, when the island was purchased by the National Park Service. While the island was owned by the Vail and Vickers Company, feral hogs remained common, and their rooting activities caused a great deal of damage to the island's rangeland, prompting the company to institute a shoot-on-sight policy in an effort to control them (P. W. Collins, pers. comm.). As in the case of Santa Cruz Island, Nettles et al. (1989) have also documented the introduction of hog cholera virus on Santa Rosa, in this case in 1949 and again in the early 1950s. The first introduction produced about 80% mortality in the island swine; the second introduction was less effective. Again, however, as was the case with Santa Cruz, evidence of the disease soon disappeared from the population, and a survey of sera and tissues from 61 hogs collected on Santa Rosa in 1987 indicated that the virus was no longer active on the island (Nettles et al., 1989). After assuming ownership of the island in the 1980s, the National Park Service instituted its own program of intensive fencing, trapping, and shooting to control feral hogs (D. K. Garcelon, pers. comm.).

Los Angeles County—Populations of feral hogs in this county occur only on Santa Catalina and San Clemente islands. Although these islands were visited by the early explorers and fur hunters, feral hogs were not introduced there until the early 1900s (Bolton, 1916; P. W. Collins, pers. comm.). The feral hogs on Santa Catalina Island were first introduced in the mid 1930s. These animals were already feral at the time of introduction and were obtained from the residents of Santa Rosa Island in trade for some captive Catalina Island California quail (*Lophortyx californica catalinensis*). The exact details of the introduction are unknown. The hogs were introduced by Jack White, a resident of the island, for the sole purpose of establishing a population of predators to eat the island's rattlesnakes (*Crotalus* sp.), which, however, are still common there. A subsequent introduction of one male and two female syndactylous, or mule-foot, hogs (Figure 3) was made on the island, supposedly in the 1950s. Two of these animals had four mule feet and one had only three mule feet. Although it is thought by some of the island residents that these animals came from somewhere in the southeastern United States, possibly Georgia or Florida, the specific details of this second introduction are unknown. The last animal seen with this morph was killed in 1972, and it had only one syndactylous foot. In spite of the fact that they were subjected to intense sport hunting, the population on the island in the early 1980s numbered between 2,000 and 2,500 animals (D. W. Baber, pers. comm.).

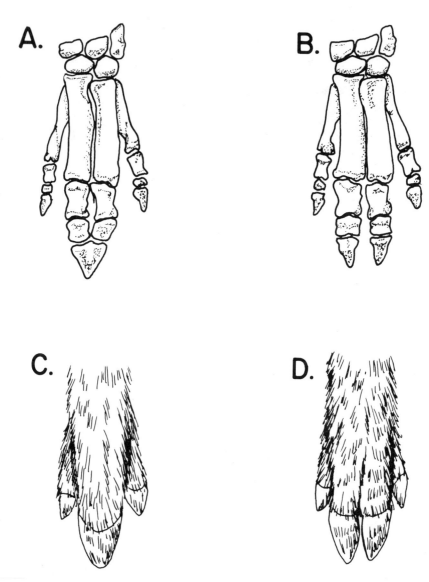

Figure 3. Skeletons and external appearances of the left manus of a syndactylous, or mule-foot, feral male *Sus scrofa* (A and C) and a dyadactylous, or normal, cloven-hoof hybrid male *Sus scrofa* (B and D). The specimens figured are mule-foot feral male—JJM 430; hybrid male—JJM 347.

A "wild boar" (undoubtedly a feral hog) from "Catalina Island, California," was used along with feral hogs from Alabama and Louisiana as foundation stock in developing two strains of miniature swine for biomedical research: the Hormel line (Rempel and Dettmers, 1966) and the Nebraska line (Welch and Twiehaus, 1966). This single male "wild boar" was introduced into the Hormel foundation herd in 1949 (Rempel and Dettmers, 1966), and subsequent progeny were transferred from the Hormel herd to the Nebraska herd in 1958 (Welch and Twiehaus, 1966). No description of this animal is given, nor is the reason given for its use in developing these strains of miniature swine.

Feral hogs were introduced on San Clemente Island in the 1950s, presumably from Santa Catalina Island, apparently for sport hunting purposes (P. W. Collins, pers. comm.). In the late 1960s, there were about 100 feral hogs on the island (Anon., 1970). Since then, their number has been as high as 1,000 (P. W. Collins, pers. comm.). In January 1980, the U.S. Navy began working on a plan to remove all feral animals from San Clemente Island (Anon., 1980a). An intensive removal-by-shooting effort began in the mid 1980s, with as many as 300 hogs per year removed under contract with the Institute for Wildlife Studies of Arcata, California (D. K. Garcelon, pers. comm.). Although there is some feeling that this periodic hunting has kept the number of hogs in check (P. W. Collins, pers. comm.), other estimates indicate that the number of feral hogs on the island is still not under control (D. K. Garcelon, pers. comm.).

Florida

In 1981, the Florida feral hog was found in 66 of the 67 counties in the state. Only Pinellas County reported no feral hog populations. In 1989, however, a report of a wildlife officer with the Florida Game and Fresh Water Fish Commission confirmed the presence of approximately 100 feral hogs in the northeast section of Pinellas County, thus establishing the occurrence of these animals in every county in the state (R. Wright, pers. comm.). This state along with Texas contains two-thirds of the feral hog range in the southeastern United States (Wood and Lynn, 1977). In three areas of Florida, wild boar from Tennessee have been introduced into the local feral populations with varying results (this will be discussed later). Florida was one of the earliest states, and possibly the first state, in the continental United States to have a feral hog population. Records of the existence of these animals in the wild date from the early colonial period in the 1500s up to the present (Hanson and Karstad, 1959).

The introduction of swine to Florida, and quite possibly to North America, is usually considered to have occurred in 1539, when Governor Hernando de Soto established an encampment at Charlotte Harbor, Lee County, Florida, as will be described below. A possibility exists, however, that swine may have actually been introduced 18 years earlier at this same site by the explorer Juan Ponce de León on his second journey to the Florida coast, in 1521. Although Ponce de León's first journey to Florida is both well known and well documented, less information is available concerning his second voyage, which was made for the clearly stated purpose of establishing a self-sustaining settlement (Spellman, 1948). From this settlement, expeditions were to be made to the west in order to determine whether the region of Florida he had discovered was an island or shared a common coastline with Mexico (Davis, 1935).

It seems that no detailed official report was ever made of this second voyage—quite possibly because it ended quickly in failure. As a result, details of the second voyage have been pieced together from accounts of Spanish historians of the time—notably that of Gonzalo Fernández de Oviedo, whose account, apparently obtained first-hand from the survivors of the expedition, has been reported by Davis (1935). According to this account, Ponce de León brought with him "mares and heifers and swine and sheep and goats and all kinds of domestic animals . . . useful in the service of mankind." The expedition arrived at what is assumed to have been Port Charlotte at a date that must be inferred as having been not more than a month after 20 February 1521, the known departure date from Puerto Rico. With considerable numbers of livestock on board, Ponce de León would almost certainly have made as direct a course as possible to the site of his proposed colony, and he was known to have been familiar with the Port Charlotte site, having visited there on his first expedition in the spring of 1513 (Davis, 1935). Unfortunately, Oviedo's account, as well as that of the historian Herrera (also reported by Davis, 1935), indicates that Ponce de León's party was quickly attacked by hostile Indians "when he had set foot on land in Florida." A number of Ponce de León's men were killed, and he was wounded in the thigh and forced to retreat immediately to Cuba, where he died of his wounds. Thus there is no indication that any of the swine brought to Florida by Ponce de León were ever unloaded, and it is quite possible that these animals were returned directly to Cuba in the course of the retreat. Regardless of whether any of this expedition's swine ever landed in Florida, Ponce de León is to be credited for having first brought this species to any part of what is now the continental United States (Davis, 1935; Spellman, 1948).

Before embarking for Florida from Santiago de Cuba in 1539, De Soto decided to follow the example of Cortés's march to Honduras and brought a herd of swine on his expedition to the mainland to feed his party and to leave animals at settlements established along the route (Maynard, 1930;

Doran, 1988). De Soto was determined not to suffer the fate of Pánfilo de Narváez, whose expedition had almost starved to death 11 years before while exploring Florida. The members of that ill-fated group had to eat their horses and leather items to stay alive (Doran, 1988). Apparently, many or all of De Soto's swine were a gift from Vasco Porcallo de Figueroa, who, upon being named by De Soto as the captain-general of the expedition, contributed "a great many" of his swine to help provision the armada (Lewis, 1907; Varner and Varner, 1951). These animals later saved the expedition from starvation numerous times (Maynard, 1930). As De Soto's expedition crossed Cuba to Porcallo's estate, his men hunted feral hogs to keep the forces supplied with pork (Varner and Varner, 1983). Departing from Havana on 18 May 1539, the fleet sighted the coast of Florida seven days later. On 30 May, the expedition put ashore at Charlotte Harbor, on the Florida mainland (Schell, 1966). According to the "Gentleman of Elvas," an anonymous chronicler of the expedition, De Soto initially brought only "thirteen sows" to Florida with the fleet (Lewis, 1907), and this is the number usually cited for the size of the herd of swine that De Soto brought to the United States (Belden and Frankenberger, 1977). This seems to contradict Elvas's earlier description of Vasco Porcallo's gift to the expedition of "a great many" swine. The 13 sows were referred to by Elvas in a description of how numerous the herd of swine later became. It is possible that he did not include boars and young swine in the count, since the sows were the ones responsible for farrowing the offspring that increased the size of the herd. None of De Soto's other chroniclers gave any estimates of the initial number of swine brought to Florida.

The expedition stayed in camp at Charlotte Harbor, making short trips into the interior of the peninsula, until July 1539. On the 15th of that month, the entire expedition started inland, taking with them "a large drove of pigs." De Soto's herd of swine must have grown quite numerous, since on the first day's march the swine were so difficult to control that the expedition succeeded in covering only three and one-half miles (Schell, 1966). The Spanish continued north up the peninsula and eventually traveled throughout much of the southeastern United States. In three years' time, the expedition covered 3,100 miles and traveled through eight states in addition to Florida (Towne and Wentworth, 1950). During this period, the herd of swine increased to 700 (Lewis, 1907). Throughout these travels, numerous swine either escaped or were stolen by Indians. Many of these animals became wild-living (Hanson and Karstad, 1959).

From 1562 to 1565, the French tried to compete with the Spanish and maintained two short-lived colonies in Florida (Lowery, 1959). Pork was supplied to them by local Indians who hunted the already established feral herds (Hanson and Karstad, 1959).

In August 1565, Admiral Pedro Menéndez de Avilés landed a company of several hundred colonists in Florida in an effort to expel the French (Lowery, 1959). The livestock on this expedition included 400 domestic swine. A century later, eight large towns, 72 missions, and two royal haciendas possessed pigs. These swine eventually came into the hands of Indians. As with most of the swine either given to or stolen by the Indians in the 16th and 17th centuries, no attempt was made to domesticate them. Rather they were permitted to roam the woods freely and were hunted there when the Indians needed meat (Towne and Wentworth, 1950).

During the early settlement of Florida, the European colonists adopted the Indians' practice of raising their domestic swine under semiwild conditions. Their animals roamed at large most of the time and were rounded up at infrequent intervals. Many of these animals became feral and flourished in populations in the broad central savannah and coastal areas of the state (Towne and Wentworth, 1950). In the late 1600s, Jonathan Dickinson found "wild hogs" in the woods of Florida north of St. Augustine (Towne and Wentworth, 1950).

During a survey of the mammals of Florida in the latter half of the 19th century, Maynard (1872) noted the presence of feral hogs in some sections of the state. He went on to mention that they were not found "in any great numbers" in these areas.

The wild pigs of the Big Cypress National Preserve are feral animals descended from the Spanish domestic stock released during the 1500s and more recent domestic animals introduced for open-range hog-raising operations (Duever et al., 1986). These animals range throughout the Big Cypress and have been hunted as game animals up to the present (Stone, 1987; Duever et al., 1986). On the basis of field survey data (USDI, 1969), the total population of wild pigs in the preserve was estimated at 300–500 animals. From 1974 to 1977, approximately 240 wild pigs live-trapped elsewhere in the state were released into the preserve by the Florida Game and Fresh Water Fish Commission (Duever et al., 1986). No current population estimates exist for this group.

Feral hogs were thought to have been on St. Vincent Island, Franklin County, since European settlers first arrived in the area. This population was upgraded around 1905–10, when R. V. Pierce released domestic stock there. A second such stocking to upgrade the quality of this feral herd occurred in the early 1940s when purebred domestic Brown Russian and Poland China swine were introduced on the island by local ranchers, the Loomis brothers. The island later became the St. Vincent National Wildlife Refuge. This population has been isolated since the establishment of the refuge. In 1979, a track-count study estimated that about 1,151 feral hogs were present on the refuge (M. D. Perry, pers. comm.).

Before 1900, it was common for farmers, ranchers, and others to free-range domestic swine around what is now the St. Marks National Wildlife Refuge in Wakulla and Jefferson counties. These animals were rounded up in the fall, merchantable animals were shipped to market, and the remainder were marked and released back into the woods. Purebred domestic swine were commonly released into the area to maintain the quality of this population. After the refuge was established in 1931, and until 1964, various people who claimed "hog rights" on refuge lands were issued permits that enabled them to continue to free-range domestic stock in the area. The progeny of these animals plus other hogs that wandered into the refuge from surrounding lands constitute the present St. Marks feral hog population. In the late 1970s, these animals numbered about 1,000 (C. S. Gidden, pers. comm.).

The feral hogs on the Chassahowitzka National Wildlife Refuge, Citrus and Hernando counties, originated from domestic stock that were released or escaped from local farms from around 1900 up to the 1940s. These animals were abundant on the refuge and adjacent lands until the mid 1960s, when they became a preferred game animal and a source of meat for local residents. Because of this, the refuge population was recently estimated to number less than 100 animals (E. Collingsworth, pers. comm.).

The feral hog population of the Lake Woodruff National Wildlife Refuge resulted from domestic stock released into open range in that part of Volusia County after the turn of the century. A control program reduced their number to considerably less than 100 individuals by 1981 (B. Lihoude, pers. comm.).

The feral hogs in what is now the Myakka River State Park, Manatee and Sarasota counties, originated from escaped domestic swine when the present park lands and adjacent areas were used for ranching in the late 1800s. In addition, hogs were introduced to enhance the hunting opportunities around one or two of the former ranches in the 1920s and early 1930s. Open range existed in the area until the park was authorized in 1933. These animals have increased in the park since that time, and despite control efforts by park personnel, feral hogs were still abundant in 1981 (R. Dye, pers. comm.).

Elsewhere in Manatee, Sarasota, Lee, DeSoto, Hardee, Charlotte and other counties in the vicinity of the Peace and Caloosahatchee rivers, several unique morphological traits have been identified in the resident feral hog populations. These include mule-footed hogs, many of which show neck wattles, and hogs with slaty-gray adult coat colorations and striped coat colors in the juveniles (the so-called watermelon pattern). The similarity of these latter coat color patterns to those of the Mangalitza, a European breed of domestic pig, suggests that the presence of a number of these traits in this particular region may be a result of the earliest Spanish introductions of swine into this area—particularly at Port Charlotte, Lee County (R. M. Wright, pers. comm.).

The feral hogs in the Osceola National Forest, Columbia and Baker counties, are the result of free-ranged domestic swine released into the area's forests by local farmers. The descendants of these animals were common in this area in 1981 (J. Craft, pers. comm.).

The Kissimmee River valley had an active homestead history from about 1840 up to the mid 1900s. Free-ranged domestic swine were a constant component of these homesteads. Some wild descendants of these animals have survived in the area and formed the feral hog population that was present in 1981 (R. L. Barker, pers. comm.). The portion of this population on the Avon Park Air Force Range later hybridized with introduced wild boar (R. C. Belden, pers. comm.).

Wild pigs found during the early 1940s on the University of Florida's Conservation Reserve in Welaka, Putnam County, were escaped domestic swine. The largest numbers of these animals in the preserve were present in the portion adjacent to the St. Johns River. Efforts to eliminate them on the reserve at that time succeeded only in thinning their numbers (Moore, 1946).

Before 1960, Eglin Air Force Base had a population of feral hogs that originated from domestic stock. These animals were being reared under free-range conditions and were not recaptured by their owners when the federal government purchased the land to establish an air base. There apparently had been no feral hogs in the area before this time (C. Maloy, pers. comm.).

Live-trapped feral hogs from Cumberland Island, Camden County, Georgia, were released in the late 1970s and early 1980s on a farm owned by Lucy Fergusson about 15 miles north of Hilliard, Nassau County, Florida. Like their progenitors from Cumberland Island, these animals are said to occasionally produce striped piglets.

The wild pig population inhabiting the Crystal River Hunt Club near Red Level in Citrus County, Florida, is of mixed feral ancestry. The feral hogs present on this privately owned club before 1988 exhibited typical long-term feral characteristics, that is, long rostrum, long bristles, proportionately long legs, and mostly solid black color (D. G. DuPont, pers. comm.). In an effort to increase the harvestable wild pig population on the property, the club members released domestic stock onto the hunt club lands. These animals, typical of modern domestic swine, had short rostrums, short bristles, proportionately short legs, and a variety of solid and spotted coat coloration patterns (D. G. DuPont, pers. comm.). The impact on the morphology of the resident feral population is unknown at this time.

In 1949, the Florida state legislature enacted regulations that effectively closed all open range for domestic livestock in Florida. All private ownership of unfenced domestic swine was legally eliminated, and animals found in violation of these laws were subject to public sale. By 1981, only a few local county authorities still honored ownership claims of unfenced swine (Belden and Frankenberger, 1977).

Since the 1950s, the Florida Game and Fresh Water Fish Commission has conducted a feral hog–stocking program throughout the state. Under this program nuisance feral hogs, mostly from state parks, are relocated to public hunting areas in the state. Between 1960 and 1976, for example, 2,848 feral hogs were stocked into the J. W. Corbett Wildlife Management Area, Palm Beach County, from other localities around the state (Belden and Frankenberger, 1977). In an earlier stocking in 1956, about 150 live-trapped feral hogs from the Myakka River State Park were relocated to the Everglades and J. W. Corbett wildlife management areas in an effort to increase numbers in these two areas (Anon., 1956). The purchase of these animals was partially financed by the Wildlife League of Palm Beach (Anon., 1956).

During the 1960s and into the early 1970s, several private organizations were also conducting feral hog–stocking programs in Florida. In the early 1960s, for example, 23 wild-trapped feral hogs from open range near Palmdale, Glades County, were purchased by the Everglades Conservation Club. These animals were released in an area south of the Tamiami Trail in the vicinity of the Collier-Monroe county line to increase the wild pig population in this area (Hunn, 1961).

In 1966, the U.S. Forest Service declared that all feral hogs or free-ranging domestic swine in Florida were privately owned livestock that did not belong on national forest lands. All suspected owners of these animals were notified and told to remove their livestock from the government lands. After a reasonable amount of time, the U.S. Forest attempted to capture all remaining feral hogs. The suspected owners had to buy them back, or the hogs were put up for public sale or destroyed. By the early 1980s, the Forest Service was working toward the elimination of all feral hogs on national forest lands in Florida (Belden and Frankenberger, 1977).

Feral hog populations in Florida declined or remained stable through the early 1980s, mostly due to a change in land use practices and the overharvest of the more vulnerable populations of these animals (L. E. Williams, Jr., pers. comm.; Frankenberger and Belden, 1976).

Georgia

Feral hogs in Georgia exist in scattered populations throughout the coastal plain, piedmont, and mountain sections of the state. In the Appalachian Mountains in northern Georgia, wild boar from Tennessee and North Carolina have dispersed into the state and interbred with the feral populations there (J. Scharnagel, pers. comm.). Feral hogs have probably been present in the state since the colonization of the sea islands by the Spanish in the 1540s (Golley, 1964; Johnson et al., 1974).

In 1697, while traveling up the coast of Georgia, Jonathan Dickinson reported that Indians in his expedition killed several wild pigs on what was probably St. Simons Island, Glynn County. Such reports of hunting wild pigs are uncommon in the early records of the state. In spite of the many reports of colonial domestic swine roaming freely in the woods of Georgia, only a few may have spread far from settlements and become feral. If feral hogs did exist in any large numbers, they were apparently not considered wild game. F. Moore, who visited the coast of Georgia in 1744, and F. A. Kemble, who spent the winter of 1838–39 on Butler's and St. Simons islands, did not mention feral hogs in their accounts of the game animals in these areas (Hanson and Karstad, 1959). Carl Bauer, a Hessian army officer preparing to support the British forces in their attack on Charleston, landed on Tybee Island, Chatham County, on 2 February 1780 and reported finding both feral horses and swine. He described these animals as "very wild . . . from the many attempts to capture them" (Jones, 1987). In 1981, feral hogs were common on several of the coastal islands of the state (Johnson et al., 1974).

Many of the recent feral hogs in Georgia originated from domestic swine allowed to free-range in the state during the 19th and early 20th centuries. These animals were rounded up or killed whenever the owners were ready to eat or sell them. Brands and tattoos placed in the ears were used to verify hog claims in the state during this time (Tye, 1971).

During the early 1940s, the state legislature closed the open range in the state (Lewis et al., 1965). Shortly after this time, the numbers of hogs throughout the coastal plain were very high. In 1958, for example, the density of the feral hogs on the Fort Stewart Army Reservation, Liberty County, was estimated to be 30–40 animals per square kilometer, with a total population on the base estimated to be roughly 20,000 animals (Hanson and Karstad, 1959). Despite the range laws, the practice of free-ranging domestic swine was still common in some areas of the state in the 1960s (Golley, 1964).

Some individuals, particularly in south Georgia, have continued to release domestic swine into unfenced areas to fend for themselves. Because the feral hog is considered a domestic animal, the Georgia Department of Natural Resources has no authority to regulate the hunter harvest of these animals except on state lands (J. Kurz, pers. comm.). In general, feral hogs have been considered the legal possessions of the landowners on whose land they exist (Golley, 1964; J. Scharnagel, pers. comm.; C. Allen, pers. comm.).

The feral hogs on Ossabaw Island, Chatham County, are one population of the Georgia sea islands that has remained relatively free of human interference since their introduction by the Spanish in the 16th or 17th century. Only a limited number of domestic swine may have been released on the island during the subsequent several hundred years (Brisbin et al., 1977a). In 1962, feral hogs were abundant on the island, and their rooting was doing considerable damage to the forest floor. At this time, Arthur Graves, the island manager, was live-trapping and selling Ossabaw Island feral hogs to Tennessee, where they were turned loose for hunts (Doutt, 1962). In the early 1970s, an estimated 1,500 feral hogs were on the island (Prestwood et al., 1975). One registered domestic Hampshire boar was introduced into this feral herd during the mid 1970s by Ossabaw's chief herdsman, Roger Parker, to improve the quality of this stock. Although some belted progeny later appeared in the immediate area where this boar was introduced, this domestic animal did not remain in the population long, and few if any belted progeny still persist. It is thus unlikely that this single introduction has had much impact on the genetic makeup of the island's feral population.

From the 1970s to the present, Parker has operated a live-trap-and-removal program. Large boars are trapped and sold to shooting preserves, one of which is near Englewood, Tennessee. Large numbers of these animals, purchased from Parker, have been released onto Seven Mile Bend Island in the Ogeechee River by Jack Douglas and his brother. Other animals are sold to local slaughterhouses on the mainland. About 10 years ago, between 50 and 100 Ossabaw Island feral hogs, many of them gravid sows, were purchased by the Long Lake Hunt Club, which released them in habitat along the Oconee River in Baldwin County, Georgia (J. Phelps, pers. comm.). At present, feral hogs are still abundant on the island.

Since the 1970s, a number of studies have documented unique biochemical characteristics in the Ossabaw Island feral swine. As will be discussed in detail later, many of these characteristics were presumably related to selection for rapid and effective storage and mobilization of fat energy reserves during periods of food abundance or scarcity. Stribling et al. (1984), for example, showed that at certain times of the year, the Ossabaw Island feral hogs have higher levels of total body fat than any other wild mammals that have been studied.

In 1938, 20–30 domestic swine were introduced onto St. Catherines Island, Liberty County, by the father of the present caretaker and foreman of the island, John Toby Woods, Jr. There were no pigs, wild or domestic, on the island before that time. During the 1950s, the Noble family, the owners of the island, released a small number of registered domestic Hampshire boars to improve the quality of the feral population there. In 1968, the board of directors of the Edward John Noble Foundation decided to undertake a control program to remove the pigs from the island because they were a non-native introduced species. A trapping program was initiated in 1971 and continued until 1976, with approximately 500 pigs trapped each year. An undetermined number of feral hogs were shot during that same period. In 1974, it was estimated that more than 1,000 feral hogs existed on the island (J. T. Woods, Jr., pers. comm.). Between 1975 and 1976, more than 1,200 feral hogs were trapped and removed from St. Catherines (Hudson, 1978);

however, a population of several hundred feral hogs still remained on the island in the early 1980s (J. T. Woods, Jr., pers. comm.). During December 1988, a Christmas Bird Count expedition to the island reported abundant evidence of hog rooting and the casual sighting of at least a dozen animals, indications that the resident hogs are now again increasing in number (V. Waters, pers. comm.).

Domestic swine were first brought to Cumberland Island, Camden County, by Spanish colonists and missionaries in about 1562. The free-ranging of these animals on the island for fattening probably dated back to the early settlements and was the apparent origin of the island's feral hogs (Singer and Stoneburner, 1977). Domestic swine have been maintained on the island from the early colonial period to the present century (McCrary, 1979). Periodic escapes and releases of domestic stock have further contributed to the gene pool of this feral population (Singer and Stoneburner, 1977). In the 1940s and 1950s, Lucy Fergusson, a lifetime resident and owner of a large portion of the island, released additional domestic swine into the island's feral herd (Z. T. Kirkland, pers. comm.). A few European wild boar also were allegedly introduced to improve the hunting characteristics of this population (Singer and Stoneburner, 1977). In May 1959, feral hogs were abundant on the island (Doutt, 1959). When the Cumberland Island National Seashore was created in 1972, the National Park Service became concerned with the potential damage that these feral hogs could do to the island's "natural resources." In late 1975, a live-trap-and-removal program was initiated to reduce the number of hogs on Cumberland Island (Singer and Stoneburner, 1977). Those animals captured were turned over to Fergusson's staff, who removed them for sale to farms and slaughterhouses on the mainland (Day, 1979). That fall, between 250 and 600 feral hogs were removed from the island. By 1977, only 420–600 feral hogs were estimated to remain (Singer and Stoneburner, 1977). By 1979, approximately 1,100 feral hogs had been live-trapped and removed (McCrary, 1979), and only 300–400 feral hogs were thought to be left in the population (Day, 1979). Two years later, this number had been reduced to about 250 animals (Singer, 1981). It is the ultimate plan of the National Park Service to remove all these feral hogs from Cumberland Island (Day, 1979; K. O. Morgan, pers. comm.).

In 1976, feral hogs were stocked in Crawford County by private individuals. The origin and present status of these animals are unknown (R. W. Whittington, pers. comm.).

In 1980, six captive hogs of feral ancestry were purchased by a man in Savannah, Georgia, and were released along the Ocmulgee River north of Douglas, Georgia, in an attempt to increase the size of the huntable feral hog population in that area. At least one of these animals was castrated before being released (R. Davis, pers. comm.).

Hawaii

At present, the Hawaiian feral hog is restricted to the islands of Niihau, Kauai, Oahu, Molokai, Maui, and Hawaii. Free-ranging domestic swine or feral hogs formerly occurred on all eight main islands of the Hawaiian chain, Laysan Island, and Sand Island in the Midways (Kramer, 1971). As determined by numbers harvested each year, feral hogs are the most commonly hunted game mammal in the state (Nichols, 1962b). Their distribution in Hawaii ranges from the coastal areas bordering the sea to the above-timberline lava flows of the highest volcanic peaks on the islands (Nichols, 1962a).

As mentioned earlier, the first domestic swine to be brought to the United States were introduced into Hawaii. Polynesian immigrants were in Hawaii by A.D. 750, and all the major islands were well populated by A.D. 1000 (Joesting, 1972). Presumably, all eight main islands that supported Polynesian settlements also had domestic swine (Tomich, 1969). This species was abundant in Hawaii at the time of the first European contact (Kramer, 1971). These early domestic swine originated in the East Indies and were carried through Melanesia from New Guinea to the Solomon Islands and finally to the Fiji chain (Sharp, 1956). The species was subsequently introduced by sailing canoes from Fiji across Polynesia as far as Samoa in the southeast and eventually Hawaii in the north (Simoons, 1961; Heyerdahl, 1979). It is thought that the Hawaiian islands were populated by early Polynesian voyagers from the Marquesas and/or Tahiti (Alexander, 1967). A further early addition of domestic swine into the Hawaiian islands may have occurred when Spanish explorers arrived there in the 16th century (Bryan, 1937). Lamb (1938) stated that some writers considered it likely that domestic swine were not brought to the islands before the Spanish did so.

The Hawaiians' practice of permitting their domestic swine to roam freely in and around their settlements probably occurred during the early history of the population of the islands, but it was first noted on Kauai by the famous navigator Captain James Cook when he visited the islands in 1778 (Tomich, 1969). On 1 February of that year, Cook presented the inhabitants of Kauai with a pair of pigs of "English breed" (Bryan, 1937). According to Tomich's (1969) citation of Cook's own account of the voyage, a domestic boar and sow, again of "English breed," were landed on Niihau on 2 February 1778. These animals were soon bred into the native domestic stock on both islands (Bryan, 1937; Tomich, 1969). Later settlers from other lands also brought domestic swine, representing a wide variety of breeds, and these have since interbred with the original Polynesian pigs in most areas where they are presently found (Nichols, 1962a; Baker, 1976). The continued introductions and reproduction of European domestic swine since Cook's time, together with other land use changes, have resulted in feral hog populations in most habitats and on most of the Hawaiian islands (Stone, 1985).

Early records of the existence of wild pigs in the Hawaiian islands are scattered. In 1823, W. Ellis noted the presence of "ferocious" free-ranging swine in the mountains of Hawaii (Tomich, 1969). In 1841, C. Wilkes found many wild pig tracks on the island of Kahoolawe, but saw only one animal (Kramer, 1971). These animals disappeared at some unknown date and were no longer present on the island when C. N. Forbes visited there in 1913 (Tomich, 1969). Their absence on the island was again verified by J. C. McAllister in 1931, during an archaeological survey of Kahoolawe (Kramer, 1971). In 1891, "semi-wild pigs" were present on Laysan Island, but these disappeared before 1910 (Kramer, 1971). The last wild pig on Lanai was shot sometime in the 1930s (Kramer, 1971). Although they have probably been present on Niihau since the early colonization of the island, wild pigs were not reported there until 1936 (Tomich, 1969). Feral hogs formerly inhabited the coastal flats and adjacent sloping lands of the west end of Molokai. The last surviving animals there were shot in 1946 or 1947. More recently, feral hogs were found only on the eastern half of that island (Kramer, 1971).

As in other states, the free-ranging of domestic swine by European settlers introduced more variable stock into the feral hog populations of Hawaii. In the mid 1880s, in a memorandum to the domestic swine farmers of Hawaii, R. S. Hollinsworth encouraged free-ranging as an economical method for rearing domestic swine (Tomich, 1969). In the late 1800s or early 1900s, ranchers on the island of Hawaii released imported domestic boars of selected breeds on the slopes of Mauna Kea to improve the quality of the feral hogs there (Tomich, 1969). These animals became quite numerous and began doing considerable damage to forest growth and pasturelands on Hawaii. In 1936, in an effort to control these animals, the Hawaii Division of Forestry issued 124 permits to hunters, who reportedly killed 2,000 feral hogs (Bryan, 1937).

No introductions of stock have been made into the feral hog populations of Hawaii in recent years except domestic strains that have commingled with the feral stock in areas where they both occur. In 1981, feral hogs were common on most of the islands where they occurred, and a rough estimate of the population size was about 80,000 (R. L. Walker, pers. comm.).

Iowa

McKnight (1964) reported the occurrence of a feral hog population in eastern Iowa, in what was probably Delaware, Linn, or Jones county. No other details were given about these animals. According to the Iowa Department of Natural Resources, there is no evidence or record of a persisting wild pig population in that part of the state (R. Bishop, pers. comm.).

Within the last decade, free-ranging domestic swine along the riparian area of the Raccoon River valley in Polk County were isolated from their farms by spring floods. These animals included pregnant sows that later farrowed in the wild. In time, these sows and their offspring became wild-living (J. Dearinger, pers. comm.). The population remained small and was restricted to Polk and possibly part of Dallas counties. These feral hogs were hunted locally because of the damage done to evergreens in the hills surrounding the valley (J. Dearinger, pers. comm.). By 1981, in fact, there was no longer any evidence that a feral population still persisted in this area.

Kentucky

Kentucky has had a small population of feral hogs in the southeastern corner of the state. In the early 1960s, this population was estimated to be less than 250. Their numbers have remained low, and there is some doubt as to whether any of the descendants of these animals from the 1960s still survive. No wild boar have ever been released or stocked into unfenced areas of the state (Lewis et al., 1965; J. Bruna, pers. comm.).

More recently, the feral hogs in this area of Kentucky have been transitional in nature, the result of free-range practices by local farmers. Some were not recaptured by their owners and became feral. By the early 1980s, this was a moderately rare occurrence. Few of these hogs lasted long, and not many litters were produced annually in the wild (J. Durell, pers. comm.).

Louisiana

The feral hog of Louisiana is found in populations throughout the bayous, forest swamps, and river bottomlands of almost the entire state. Most of these animals are of recent origin and are either escapees or the products of the free-range livestock practices. In the early 1960s, there were estimated to be more than 7,500 feral hogs in the state. Open range still exists in at least some locations in Louisiana, and except for animals on certain game management areas, feral hogs there are still considered the private property of the individual on whose land they exist (Lewis et al., 1965; D. John, pers. comm.; J. D. Newsom, pers. comm.).

Feral hogs on the Sabine National Wildlife Refuge, Cameron Parish, are the descendants of escaped or abandoned domestic stock. The original animals were brought in by the early trappers, and subsequent local landowners have continued the practice up to the present day. An increase of feral hogs on the refuge in the early 1980s resulted from a hurricane that flooded the areas surrounding the refuge and carried animals into the marsh from adjacent private lands. In 1981 there were fewer than 100 animals on the refuge (J. R. Walther, pers. comm.).

Wild pigs found in Caldwell Parish are feral hogs, the descendants of domestic stock that either escaped or were "turned out" into open range by local farmers (Cockerham, 1985).

The area around the Catahoula National Wildlife Refuge, La Salle and Catahoula parishes, has been open range from the time of the first colonists. The practice of free-ranging domestic swine around Catahoula Lake dates back to the early establishment of settlements in this area. The refuge was purchased by the federal government in 1958, and sections of it were fenced with net wire in the mid 1960s. The nearby Sabine Wildlife Management Area was similarly fenced by the state (J. W. Farrar, pers. comm.). The resident feral hogs were not removed from inside these fenced-in areas in the Catahoula refuge until 1973. In 1979, 4,300 of the refuge's 5,308 acres were enclosed by fences. In the recent past, unknown individuals either released pigs into the fenced-in sections or cut holes in the wire to allow loose pigs to wander into these exclosures. These practices were subsequently discouraged, and through an ongoing extermination program, both the refuge and the Sabine Wildlife Management Area were practically free of feral hogs by the late 1970s. Only one remaining feral hog, an adult, was known to be present on the Catahoula refuge in 1979 (S. K. Joyner, pers. comm.).

As noted earlier, feral hogs from Louisiana seem to have been more widely used than those of any other state to form captive parent stock for the development of miniature strains of swine for biomedical research (see section on Alabama). A "light-colored 'swamp hog' from Louisiana" was bred with offspring from a Palouse gilt–Pitman-Moore boar cross to initiate the Hanford strain of miniature swine (Bustad et al., 1966). Rempel and Dettmers (1966) and Welch and Twiehaus (1966) also note the inclusion of "Piney Woods pigs" from Louisiana as foundation stock for the development of the Hormel (in 1951) and Nebraska (in 1958) lines of miniature pigs, respectively. The latter authors also describe a later addition to the foundation stock: one gilt and three boars that were captured in the Piney Woods area of Louisiana in early 1961. No further description is given of what is meant by the "Piney Woods area of Louisiana." None of these authors indicate why feral hogs from Louisiana were particularly selected for these breeding programs; feral hogs of equal if not smaller body sizes are commonly found in a number of other states. Presumably, the "blonde" (probably light tan) body color was selected in an effort to produce white or light-skinned progeny for use in biomedical research on burns and other skin injuries.

Mississippi

Feral hogs in Mississippi are found in scattered populations along the Mississippi River, in the southeastern corner, and in two isolated localities in the northern and central portions of the state. In the early 1960s, Mississippi feral hog populations were estimated to number in excess of several thousand individuals, most of them the offspring of hogs that had been wild for generations (Lewis et al., 1965).

The feral hog populations in Mississippi date from the 16th and 17th centuries. De Soto's expedition with its herd of swine traveled through the state from 1539 to 1541. In January 1699, the French explorer Pierre le Moyne, Sieur d'Iberville, brought domestic swine along with other livestock to Biloxi. Descendants of d'Iberville's swine later spread throughout parts of Mississippi and Louisiana (Towne and Wentworth, 1950). Because of the free-range livestock practices of the Indians and early European settlers in this area, some of these swine probably became feral.

Feral hogs on the Gulf Islands National Seashore, Harrison and Jackson counties, originated from domestic stock released onto the islands during the mid 1800s (Singer, 1981). On Horn Island, the first documented domestic hogs were brought to the island by the Waters family in 1845 (Baron, 1979). These hogs were turned loose and hunted only when needed for meat. The Waters family were the last farmers on the island known to have domestic swine. These animals were at least part of the ancestral stock that gave rise to the recent feral hog population on the island (J. A. Jackson, pers. comm.). Additional hogs may have been released by hunters during the period from the Civil War to the present (Baron, 1979). Portions of the island became a national wildlife refuge in 1958. Most of the island was later included in the Gulf Islands National Seashore in 1972. Approximately one-third of it is still under private ownership (Baron, 1979). In 1981, these hogs numbered more than 60 individuals (Singer, 1981). Because of rooting damage, an eradication program was implemented to remove all these animals from the island (Baron, 1979; J. A. Jackson, pers. comm.).

Missouri

At one time, Missouri had a considerable population of feral hogs. These animals originated from domestic stock released onto open range, especially during years of heavy acorn production in the Ozarks (D. A. Murphy, pers. comm.). In the early 1960s, feral hog populations were confined to three small areas in southeastern Missouri. The animals were decreasing at that time and did not exceed 1,000 individuals (McKnight, 1964; Lewis et al., 1965).

On 1 January 1969, the Missouri state legislature put into effect a closed-range law outlawing unconfined livestock. As a result, feral hogs were virtually eliminated in Missouri (D. A. Murphy, pers. comm.). By the mid 1970s all feral hogs had been removed from the state (B. T. Crawford, pers. comm.).

New Mexico

A population of several hundred feral hogs has lived in the San Luis, Animas, and Peloncillo mountain ranges of Hidalgo County (McKnight, 1964; Findley et al., 1975) since the early 1900s (S. J. Dobrott, pers. comm.). These animals were domestic stock that either escaped from pens or were released by local ranchers and farmers during the early part of this century. Later, these people either live-trapped or roped individual animals when they needed meat (B. Donaldson, pers. comm.). Although this population currently occurs primarily on private lands, some individuals had spread into national forest lands by 1981, where they were considered pests (W. Conley, pers. comm.; W. A. Snyder, pers. comm.).

North Carolina

Feral hog populations are scattered throughout North Carolina, particularly on national forest lands in the western part of the state, where they have been introduced several times in the past 40 years by private individuals (R. B. Hamilton, pers. comm.). In the early 1960s, feral hogs were numerous in both the mountainous western part of the state and the swampy regions of the lower coastal plain (Hanson and Karstad, 1959; McKnight, 1964). Since that time, there has been a decrease in the coastal plain animals (Wood and Lynn, 1977), and most populations may no longer be extant (J. M. Collins, pers. comm.).

In the early 1700s, feral hogs were reported to be "more numerous" in North Carolina than in any other English colonial province (Brickell, 1968). Several plantations were said to have populations numbering in the hundreds, and "vast numbers" of feral hogs were encountered in the woods (Brickell, 1968).

In the 1920s, feral hogs were ranging over much of the area around Hooper Bald, Graham County. These animals were largely later displaced by introduced wild boar (Jones, 1959).

Feral hogs on the McKay National Wildlife Refuge, Currituck County, originated from domestic stock released on open range in the 1920s and 1930s. The unclaimed stock remaining in the area after the refuge was formed in the 1930s became the present-day feral population (I. W. Ailes, pers. comm.). The SCWDS (1988) survey, however, no longer indicates a population persisting in this portion of the state (Appendix E-1).

In the summer of 1980, feral hogs were released into the Shining Rock Wilderness Area of the Pisgah National Forest, Haywood County, by an unknown individual. The present status of these animals is unknown. None have been sighted since the close of the fall 1980 hunting season (D. Rhodes, pers. comm.).

Oklahoma

Oklahoma has a small population of feral hogs in the southeastern portion of the state (Caire et al., 1989; G. Bukenhofer, pers. comm.). These animals are the result of local free-range rearing of domestic swine, a practice once common in the forested areas of such counties as Pushmataha, Le Flore, Latimer, McCurtain, Sequoyah, and Adair (Young, 1958). Oklahoma's feral hogs were estimated to number fewer than 1,000 individuals in the early 1960s (McKnight, 1964). Statewide open range existed legally until 1965. At that time, most feral hogs were removed from Oklahoma's forests (G. Bukenhofer, pers. comm.). Some areas in the southeastern part of the state are still frequently used by local people for free-ranging ear-marked domestic swine. These semiwild domestic swine currently share much of this area with the remainder of the previously established feral hog populations (R. E. Thackston, pers. comm.). In 1981, a few feral hogs were known to occur on the Weyerhauser Paper Company lands in this corner of the state (R. E. Rolley, pers. comm.). There have been recent reports of feral hogs in the Arbuckle Mountains in south-central Oklahoma (Caire et al., 1989).

All feral hogs were removed from the Ouachita National Forest in eastern Oklahoma in the late 1960s. Occasionally, domestic swine are illegally released or wander into U.S. Forest Service lands. These animals are either immediately hunted down by the local residents who learn of their existence or killed by hunters during the fall deer season (G. Bukenhofer, pers. comm.).

Oregon

A small population of feral hogs existed in the southwestern part of the state, especially in the Rogue River drainage, during the first half of this century (De Vos et al., 1956; McKnight, 1964). These animals originated from escaped domestic swine (Russell, 1947). A large feral boar (MVZ 107740) was collected from Curry County along the Rogue River early in 1947 (Russell, 1947). In the early 1960s, these animals numbered fewer than 100 individuals (McKnight, 1964). This population has decreased in recent years, and there is no evidence of its continued existence (P. Ebert, pers. comm.).

Scheffer (1941) noted that feral hogs competed with beavers for acorns in southeastern Oregon. No other references in the literature have noted the existence of a wild pig population in this part of the state.

South Carolina

Feral hog populations in South Carolina are found along the Savannah River and in scattered populations throughout the coastal plain. Feral hogs have been established in the state since the time of the first settlements (Golley, 1966). The first permanent European colonists in the state bought pork from local Indians who were already hunting feral hogs in the woodlands (Towne and Wentworth, 1950). Most feral hogs in South Carolina are probably descendants of domestic "range" swine that were turned loose by farmers to forage on their own. Some of these animals were never caught and became feral (Wood and Lynn, 1977).

The first introduction of domestic swine into the state may have occurred as early as 1526, when a group of 500 Spanish colonists organized by the licentiate Lucas Vásquez de Ayllón attempted to establish a colony at the head of Winyah Bay, Georgetown County. The colony, called San Miguel de Gualdape, was located on a site that became known as Hobcaw Barony some 200 years later (Wood and Lynn, 1977; Stephenson, 1979). Although available information lists only horses as the expedition's livestock, it is conceivable that swine were also brought along for meat. Wood and Lynn (1977) have speculated that Hobcaw may have been the site of the first introduction of swine into the continental United States, but this remains unsubstantiated. Although subsequent introductions of domestic swine into this population were likely, there have been no such stockings for the last 60–80 years (Singer and Stoneburner, 1977). A large population of feral hogs now occurs on this tract of land (Wood and Lynn, 1977).

The first definite importation of domestic swine into South Carolina occurred in 1566 with the establishment of the Spanish colony of Santa Elena on the salt marsh edge of the Beaufort River at the present site of the Parris Island Marine Base in Beaufort County (Reitz, 1980). This colony was one of several established by the Spanish admiral Menéndez in an effort to expel the French from the New World (Towne and Wentworth, 1950). The colonists at Santa Elena were supplied with provisions and livestock, including domestic swine, from Menéndez's base at St. Augustine, Florida. Santa Elena was abandoned by the Spanish in 1587 (Reitz, 1980). With the exception of the domestic swine stolen by the Indians, none of the colony's animals survived Santa Elena's demise (Towne and Wentworth, 1950).

Domestic swine and cattle were introduced into the area around the Francis Marion National Forest in the early 1600s by the first settlers and were traditionally turned loose to forage on their own. The U.S. Forest Service discontinued open range in this area in the late 1940s. The free-ranging cattle were easily captured and removed, but the swine evaded capture and are now feral (R. Tyler, pers. comm.).

The region of the Francis Marion National Forest is a source of mule-foot, or syndactylous, feral hogs, although their specific origin is unknown (Figure 3). The mule-foot breed of domestic swine was never popular or widely distributed in the United States. Officially, the breed was started in Ohio in 1908, but this morph has been known since about 350 B.C. (Darwin, 1867; Gray, 1868; Day, 1915; Ensminger, 1961; Mason, 1969). As a domestic breed, it was most numerous shortly after the breed was initially registered (Ensminger, 1961). It is possible that some of these animals were turned loose into the central coastal area of South Carolina after the turn of the century. This population also may have been the source of the mule-foot hogs brought to Santa Catalina Island, California, in the 1950s. The southeastern states were thought to be the source of these animals, and the area around the Francis Marion National Forest is the only locality in this region where this morph has been reported to occur in any numbers. Other mule-footed feral hogs, however, have been observed in the Savannah National Wildlife Refuge (J. W. Reiner, pers. comm.) and in northern Florida (W. B. Frankenberger, pers. comm.).

The syndactylous morph was commonly observed in the Francis Marion feral hog population before the 1970s. Extreme periods of drought in 1979–80, however, have apparently reduced the overall size of this population. The last evidence of the persistence of the mule-foot morph in the free-living state was a set of tracks seen in the spring of 1980 (R. Tyler, pers. comm.). The last wild-collected South Carolina mule-foot morph, which is represented by a specimen known to the authors, was a 180+-kg male shot by C. H. Andrews of Sumter, South Carolina, near the Black River in Williamsburg County, South Carolina, in March 1977. This animal was black with white on the forelegs and feet; all four feet were syndactylous. A trophy head and four feet of this specimen were mounted and are currently in Andrews's possession (C. H. Andrews, pers. comm.). The area bordering the portion of the Black River where this animal was taken continues to be cited by local hunters as the most likely place for free-living mule-foot hogs to persist in South Carolina.

In 1682, Thomas Ashe stated the domestic swine were escaping from plantations and that "great numbers" of them were running wild in the woods of South Carolina (Salley, 1911). During the early 1700s, the English settlers recognized the value of the abundant acorn crops in South Carolina for fattening free-ranging domestic swine. From early local records, it is known that the barrier islands in the state were highly prized for this husbandry method. These islands not only had a greater acorn crop than the adjacent mainland, but as islands, they were essentially "enclosed areas" into which stock could be released and confined without fencing. Numerous domestic swine, cattle, and sheep were turned loose on the islands by early colonists (W. P. Baldwin, pers. comm.).

Bull Island, Charleston County, one of the barrier islands in central coastal South Carolina, was granted a charter of colonization by the English government in about 1690. Domestic swine and cattle were probably introduced there in the early 1700s (W. P. Baldwin, pers. comm.). On 30 December 1700, John Lawson, surveyor-general of North Carolina, stopped at Bull Island during a trip from Charleston to North Carolina. Lawson stated that the island was owned at that time by a Colonel Cory of South Carolina and had "a great number" of both wild cattle and hogs on it. While on the island, his Indian guides "had killed two more deer, two wild hogs, and three raccoons, all very lean except the raccoons" (Lawson, 1937; Penney, 1950). Subsequent owners of the island in the 1800s and early 1900s periodically released new domestic swine to "improve the strain." In the late 1930s, Bull Island became part of the Cape Romain National Wildlife Refuge. The U.S. Fish and Wildlife Service decided to remove all feral livestock shortly after the refuge was established. This program occurred primarily between 1938 and 1940. About 400–500 feral hogs were live-trapped in wood pens. Toward the end of the program, the wariest animals evaded the traps and had to be shot. The last two feral hogs were said to be enormous animals that defied capture or shooting. Constant pursuit finally caused these two animals to attempt to swim the half-mile inlet separating Bull Island from Caper's Island to the south. One apparently made it; the carcass of the other later washed up on shore. The removal program has been successful, and no feral hogs are currently on the island (W. P. Baldwin, pers. comm.).

The feral hogs in the Black Swamp, located between South Carolina Highway 119 and U.S. Highway 321 in Jasper County, South Carolina, have had a varied history. A portion of the swamp is occupied by the Bantam-Harper Hunt Club. This property was a Tory plantation before the American Revolution. When the war broke out, the plantation was destroyed; its buildings were burned and its livestock, including domestic swine, escaped. These swine became feral and inhabited the bottomland areas throughout the property (J. Hubbard, pers. comm.). There is some question as to whether or not this population survived up to the mid 1900s (T. Clary, pers. comm.). When the hunt club was started in the 1970s, the hardwood bottomlands had been uncut since the plantation was destroyed. Large numbers of feral hogs were harvested on the property during the first few years (J. Hubbard, pers. comm.). In the winter of 1973–74, 30 domestic sows and 12 wild pigs were released into the area of the swamp bisected by Sandbar Ferry Road. These animals were mostly all black, all red/brown, and red/brown-and-black spotted (T. Clary, pers. comm.). Since that time, most of the bottomlands have been cleared and drained by the owner of the land, Georgia Pacific, Inc. The feral hog population in this area is a remnant of the former numbers present there.

Salley (1911) reported that feral hogs were occasionally present in the lower part of South Carolina in the early 1900s. The feral hogs on the Santee Coastal Reserve and the rest of the Santee River delta south of South Carolina Highway 41, Georgetown and Charleston counties, are the descendants of free-ranging stock released by the early farmers in the area. Hampshires and Duroc-Jerseys were the two breeds that made up most of the original stock. Many of the more recently established feral animals escaped from a penned impoundment where they were being used in snipe management (Rutledge, 1965; T. Strange, pers. comm.).

The feral hogs in the river swamps below Augusta, Georgia, originated from domestic stock free-ranged in this area by local farmers. This last occurred in the early 1940s, during years when feed prices were high. Some farmers left the area and never reclaimed or tried to catch the stock they had released (J. W. Reiner, pers. comm.). Feral hogs were numerous in this area in the 1960s and early 1970s, but heavy local hunting pressure caused this population to decrease during the early to mid 1980s (J. W. Reiner, pers. comm.).

When the U.S. Atomic Energy Commission (AEC) established a nuclear facility at the Savannah River Plant (SRP), in Aiken, Barnwell, and Allendale counties, large numbers of domestic swine were left behind by the resident farmers when they were moved off the nearly 775 km^2 of land that was closed to the public in 1952. One landowner claims to have lost more than 100 domestic hogs that he was unable to catch (Jenkins and Provost, 1964). Considering the free-range livestock practices common in the South at that time, it is likely that feral hogs were already present in the swamps along this portion of the Savannah River. This, however, has not been documented. Since the land was closed to the general public by the AEC, these animals have thrived and multiplied (Plate 1). In 1960–61, the population density of feral hogs on the SRP land ranged from one hog per 81 hectares up to one hog per 24 hectares. Because of the damage done to pine tree plantations by these animals, an active live-trap-and-removal program was initiated by the U.S. Forest Service after the SRP became operational in the early 1950s. By 1964, 280 feral hogs had been removed (Jenkins and Provost, 1964). In 1970, 300–400 feral hogs were estimated to exist on the SRP site (Sweeney, 1970). At that time, the population was concentrated in the bottomland swamps along the Savannah River and the adjacent pine plantations (Kurz, 1971). Crouch (1983) estimated the 1979 SRP population density to be one hog per 70 hectares. A 1982 estimate of this population's size was between 400 and 800 animals (T. W. Hughes, pers. comm.). In the late 1970s, a small population of feral hogs was discovered along Upper Three Runs Creek and Tinker Creek in the northern part of the plant, south of South Carolina

Highway 278. According to the SRP annual hunt records, the first of these animals was collected on the SRP public deer hunts in 1980 (Figure 4). In 1982, these animals numbered fewer than about 100 individuals and may have represented escaped stock from a farm along the northern boundary of the plant. Field observations in this area suggest that this population is expanding and spreading throughout the bottomland drainage systems of the entire upper half of the plant. Animals in the invading northern population are morphologically distinct from those in the more southerly population in the Savannah River swamp forest system. Northern animals are generally lighter in body weight, have proportionately longer snouts, are proportionately taller at the whithers, shorter in body length, and have longer guard hairs and shorter, more erect ears as compared with the animals in the southern SRP area. This suggests that the northern animals have not been derived from the more southern population.

The feral swine that occupied Kiawah Island, Charleston County, were descendants of domestic stock introduced on this barrier island (Pelton, 1975). The dates of these introductions are unknown. Wild pigs were hunted on the island during the 1940s (T. Clary, pers. comm.; S. Cornelison, pers. comm.). These hogs were described as "small in size, solid black in color, with curly hair and a high ridge down the middle of their backs" (T. Clary, pers. comm.). A 1974 study described feral hogs as abundant on the island and utilizing every habitat type available (Pelton, 1975). Pelton (1975) recommended removal of all feral hogs from the island by shooting. The population has been completely eradicated as part of the recent development of Kiawah (J. W. Gibbons, pers. comm.).

Feral hogs have been present in the northwestern corner of South Carolina, especially in Oconee County, since about 1974 or 1975. Rumors that these animals are of wild boar ancestry are apparently unfounded (W. B. Conrad, pers. comm.).

Tennessee

The origin of the feral hogs in Tennessee is unclear. It is thought that they are descended from the domestic swine allowed to forage in the woodlands until the 1940s. During that decade, the state legislature passed a bill called the Fence Law, which required all farmers to keep livestock in confinement or in fenced areas. It is believed that not all domestic swine were rounded up after the law was passed and that some reverted to the wild. Populations of these animals in the southeastern corner of the state and in the Cumberland Plateau have since hybridized with wild boar. In the early 1970s, non-hybridized feral hogs in Tennessee were found only in Overton, Pickett, Scott, Bledsoe, and Sequatchie counties (R. H. Conley, pers. comm.).

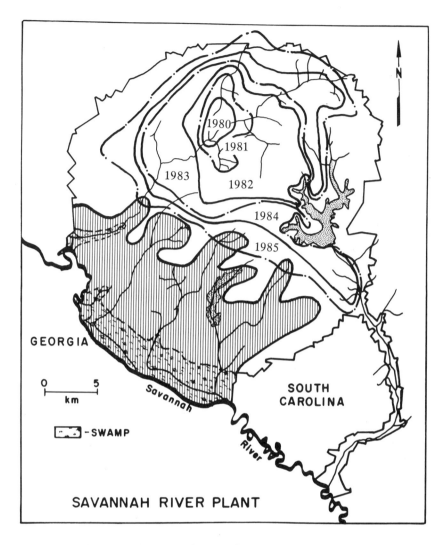

Figure 4. Distribution of the feral hog (*Sus scrofa*) population on the Savannah River Plant (SRP), South Carolina. The 1952 to present distribution of the bottomland swamp-forest feral hog population is indicated by the crosshatched area. The dated, enclosed contours represent the expanding distribution of the upland SRP feral hog population of individuals showing external morphotypes contrasting with those of the bottomland population. This expanding range is depicted through the year in which the two populations became overlapping. These map contours are based on both hunter-kill locations and reported sightings.

Free-ranging domestic swine inhabited the mountains in Monroe County when imported wild boar escaped from Hooper Bald, North Carolina, during the early 1920s (Stegeman, 1938). Interbreeding between the two forms occurred freely after that time (R. H. Conley, pers. comm.).

Feral hogs have been extirpated from several areas in Tennessee (C. J. Whitehead, pers. comm.; J. Murrey, pers. comm.). Before 1960, feral hog populations existed in Humphreys and Hickman counties (Schultz et al., 1954; McKnight, 1964). There are no recent records of persisting populations in this area.

In the early 1970s, a wild pig–trapping program was undertaken in the Tellico Wildlife Management Area. In an effort to establish the wild boar phenotype, live-trapped males with feral hog characteristics were castrated before being released (C. J. Whitehead, pers. comm.).

Texas

Feral hogs are found over most of Texas, although populations are light and scattered in the western deserts and panhandle (C. J. Winkler, pers. comm.). Texas has some of the most extensive wild pig range in the southern United States, with earlier estimates of populations approaching one million animals (Jackson, 1964; Wood and Lynn, 1977). Some of these populations have since hybridized with introduced Eurasian wild boar in the coastal plain, Edwards Plateau, and Rio Grande plains sections of the state.

Feral hogs in Texas date back to the 1500s and 1600s. The earliest origins of these animals are attributed to domestic stock brought into the state in 1542 by an expedition of De Soto's men (Benke, 1973). In 1565, Spanish colonists at Saltillo, Mexico, drove their domestic swine across the Rio Grande into Texas to avoid losing them to local Indian raids. Some of these swine must have come into Indian ownership, since the conquistador Sergeant-Major Vicente de Zaldivar tried to buy hog's lard from the Picuris Indians in Texas in 1598 (Towne and Wentworth, 1950).

In the late 1670s, domestic swine were brought north from Mexico as a food source for the inhabitants of a series of missions that had been established along the Texas coast by Fernando del Bosque in 1675. The early missionaries and colonists in the state usually allowed their domestic swine to range freely around their settlements. The Spanish mission domestic swine multiplied rapidly and soon "exceeded computation." The animals ranged the countryside surrounding the missions for miles around, and many reverted to the wild (Towne and Wentworth, 1950).

In 1685, René-Robert Cavelier, Sieur de La Salle, attempted to establish a French settlement, Fort St. Louis, in the coastal region of Texas on a creek off of Lavaca Bay, Calhoun County. Among the livestock La Salle brought to his colony were domestic swine. The colony failed because of Indian attacks, disease, and snakebite. In late 1688 or early 1689, two Cibolo Indians reported seeing swine running wild near the ruins of La Salle's fort (Benke, 1973; Weddle, 1973). Later, on 22 April 1689, General Alonso de León, leading a Spanish expedition of soldiers looking for the colony, found only many dead pigs around the deteriorating buildings at the site (Weddle, 1973). Although domestic swine from the La Salle expedition may have been the founding stock for the area's present feral hogs, (Benke, 1973), it is unknown exactly how long free-ranging domestic swine from this colony survived in that area.

In the mid 1800s, while traveling through Texas, Frederick Law Olmsted reported that free-ranging hogs stole food from almost every camp that his party made in the state. As quoted by Towne and Wentworth (1950), Olmsted wrote while in a camp near Crocket, Houston County: "At this camp we were annoyed by hogs beyond all description. At almost every camp we were surrounded by them; but here they seemed perfectly frantic and delirious with hunger. These animals proved, indeed, throughout Texas, a disgusting annoyance."

In the late 1800s, the country around Mason, Mason County, had a dense population of feral hogs. These animals sustained themselves by scavenging on the carcasses of cattle that had been killed and skinned for their hides (Towne and Wentworth, 1950).

In the early 1900s, Mearns (1907) noted that feral hogs were numerous in many parts of Texas along the Rio Grande.

Since 1900, farmers and ranchers in Texas have free-ranged domestic swine on large tracts of open range. At the turn of the century, local domestic swine commonly called Texas rooter hogs were released on open range near Miele (or Melle) Dietrich Point on the Blackjack Peninsula of Aransas County by Mr. Smith from Rockport (Halloran, 1943; Springer, 1975). Free-ranging descendants of these animals were present there in 1919, when the area was purchased by the San Antonio Loan and Trust Company. Although it is unknown whether any additional domestic swine were released, those feral hogs present in 1919 were allowed to range on these lands until the late 1930s (Halloran, 1943). This population later hybridized with introduced wild boar (Halloran and Howard, 1956).

From the depression years until the mid 1940s, free-ranging domestic swine was a common practice by ranchers south of San Antonio. These animals multiplied and spread out over the neighboring ranches. They are currently quite common in south Texas and are locally considered serious economic pests because they prey on range calves and goats (Cooney, 1976; R. T. Layton, pers. comm.; W. Walker, pers. comm.; J. B. Wise, Jr., pers. comm.).

Feral hogs around the Rockport area of coastal Texas are also the result of this free-ranging husbandry practice with domestic swine. These animals have been feral for many generations (Bauer, 1970b).

During the 1920s and 1930s, it was a common practice in the wilds of east Texas to systematically round up free-ranging hogs at certain times of the year and move them by barge to rail centers from which they were sent to the Chicago markets. Many of these hogs were noted for having wattles—fleshy protuberances with a gristlelike inner core measuring about 5–10 cm in length and hanging from the neck just posterior to the angle of the jaw (Anon., 1987; Prentice, 1987). Although these wattles occur on feral swine from other populations (e.g., those in Hillsborough County, Florida [I. Convers, pers. comm.; G. Johnson, pers. comm.]), these characteristic appendages so typified many of the east Texas hogs that they were named for them when they were later developed into domestic bloodlines. Some registries have used different spellings of the term, thus the Red Waddle Hog and the Red Wattle Hog, but all originate from the same east Texas feral populations.

As indicated by the breed names, these hogs are most commonly red, although black markings are also often present. The animals in the domestic lines are noted for attaining large body sizes (in excess of 700 kg) and for their gentle nature and exceptionally lean meat (Anon., 1987). The captive Red Waddle Hog bloodline was primarily initiated in the mid 1960s, by Harry Wengler on a small farm in east Texas. Later, the Endow and Timberline bloodlines were developed by breeders in Iowa (Prentice, 1987). As a boy in 1907, Wengler had know the red-wattled hogs when they were found in a bottomland between two rivers called Thousand Hells Acres. Initially the red-wattled hogs were said to have originated on the island of Espíritu Santo in the New Hebrides, where they are now believed to be extinct, but from where they were imported to southern Texas in the late 1800s by a wealthy landowner (Anon., 1987). The current status of animals of this type in the feral state is unknown.

Wild pigs on the Gus Engeling Wildlife Management Area are descendants of local escaped domestic stock (Cox, 1981). The presence of Eurasian wild boar ancestry in this population has been rumored but remains unsubstantiated (Henderson, 1979; Anon., 1980b). In 1977, the feral hog population on this 4,450-hectare wildlife management area was estimated to be as high as 350 animals (Henderson, 1979). Although individuals from this population were harvested during the fall hunting season, their numbers had increased to the point where this species was negatively affecting wildlife habitat and disrupting nesting efforts by game birds within the wildlife management area (Henderson, 1979). In an effort to control the growth of this feral hog population, a special hunt was implemented for these animals in the spring of 1977 by the Texas Parks and Wildlife Department (Henderson,

1979). This annual hunt continues through the present (C. K. Winkler, pers. comm.).

No records describe the origin of the wild pigs on and around the Welder Wildlife Refuge, San Patricio County. It is assumed that these animals are the descendants of escaped or free-ranged domestic swine from local farms and ranches since about 1900. These animals occur only seasonally in any numbers on the refuge. During the spring growing season, local wild pigs move out of the refuge onto adjacent ranches to forage in the grainfields. When present on the refuge, these animals number at most only a few hundred individuals (L. Drawe, pers. comm.).

Virginia

Feral hogs in Virginia date back to the English settlements on the lower James River in the early 1600s, but they are found at present only in the southeastern corner of the state. The colonists began to free-range domestic swine because of the shortage of feed, and some animals became wild and multiplied. In 1627, the wild pigs in that area of Virginia were described as "innumerable" (Towne and Wentworth, 1950).

Free-ranging branded domestic swine was a common practice in the Dismal Swamp in the early 18th century. During a boundary survey in March 1728, William Byrd observed both free-ranging hogs and cattle in the local marshes and swamps (Handley, 1979). During a visit to the area in the early 1920s, Stansbury (1925) found that wild hogs were considered game animals in the Dismal Swamp. There has been no mention of feral hogs in the Dismal Swamp in the recent literature (Handley, 1979).

The area of Princess Anne County around the Back Bay National Wildlife Refuge, Barbours Hill Game Management Area, and False Cape State Park had open range in the early 1920s and 1930s. Most, if not all, the feral hogs now in these localities are descended from this early stock. When the refuge was established in 1938, all range stock were removed by their owners. Unclaimed animals remained to give rise to the present-day feral population (I. W. Ailes, pers. comm.; R. Duncan, pers. comm.). In the early 1960s, this population was estimated to be decreasing and to number fewer than 250 animals (Lewis et al., 1965). More recently, their numbers have increased, and this population was estimated to consist of approximately 1,000 animals in 1981 (I. W. Ailes, pers. comm.) and 400–500 animals in 1986 (Duncan and Schwab, 1986).

Feral hogs have also been reported in the western mountains of Virginia. In the late 1940s, their range in the United States was stated as including "the Blue Ridge Mountains of Virginia" (Anon., 1949). McKnight (1964) mentioned the presence of feral hogs in the interior mountains of the state, but no further specific information was given. There have been no recent reports of any feral hogs in this area (R. H. Cross, Jr., pers. comm.), and no areas in this part of the state are indicated as having any wild pig populations by SCWDS (1988).

West Virginia

The presence of feral hogs in West Virginia is uncertain. Blennerhasset (1907) described an encounter between a female bobcat and several feral hogs near Parkersburg, Wood County, West Virginia. McKnight (1964) reported that feral swine were widespread but scattered in the state. Lewis et al. (1965) stated that West Virginia had no wild hogs. There have been no references to the existence of these animals in West Virginia in the recent literature. R. L. Miles (pers. comm.) similarly found no evidence to support the occurrence of feral hogs in the state at the present time.

Wild Boar Introductions

Eurasian wild boar and wild boar × feral hog hybrids have been introduced into at least eight states in the continental United States. Individuals from two of these populations have since dispersed naturally into four adjacent states. The majority were introduced as exotic big game animals for sport hunting purposes. Most introductions have been successful in establishing wild populations. In all but a few cases, the wild boar were initially introduced into areas that contained previously established feral hog populations, and hybridization between these two forms has occurred. At the present time, however, all these populations have probably been hybridized to some extent.

The Corbin's Park Introduction

The first pure wild boar were brought to the United States in the late 1800s. In 1886, Austin Corbin, the millionaire founder of the Long Island Railroad and Coney Island, purchased approximately 12,141 ha in the mountains of Sullivan County, New Hampshire (Spears, 1893; Martin, 1970). In 1890, Corbin incorporated his estate as the Blue Mountain Forest Association, with himself as the president and a few friends as charter members (Martin, 1970). A 9,300- to 9,700-ha portion of the property, known as Corbin's Park, was enclosed with 58 kilometers of 3-m-high wire-mesh fence (Spears, 1893; Baynes, 1931; Silver, 1957; Manville, 1964; Taylor, 1980). This fence was also buried 60 cm underground to be "wild boar-proof" (Martin, 1970). The fenced-in area encompassed two mountains and included parts of the townships of Grantham, Croydon, Newport, Cornish, and Plainfield (Spears, 1893; Silver, 1957; Taylor, 1980).

In 1889, Corbin purchased 14 juvenile European wild boar from Karl Hagenbeck of Hamburg, Germany. They were from the Black Forest region of Germany and cost Corbin $1,000 (Baynes, 1931; Manville, 1964; Martin, 1970). Their origin in the Black Forest indicates they were of the subspecies *Sus scrofa scrofa* (Appendix B). One died at sea in transit to the United States (Manville, 1964), and in September 1890, the remaining 13 animals were released into Corbin's wild game preserve (Spears, 1893; Baynes, 1931). A darker, larger race of Russian wild boar from the Ural Mountains was supposedly later stocked in the park, but the details of this introduction are unknown, and its authenticity remains uncertain. By 1891, the wild boar had increased to 18 and had divided into several small herds within the park (Spears, 1893). According to Silver (1957), local residents, threatened by Corbin's acquisition and believing they were being forced out by the rich, began to cut openings in the fence surrounding the park. About five years after the wild boar were introduced, several escaped through one of these openings. The animals were not recovered, and shortly afterward one of Corbin's employees, Theophilus Wylie, found evidence of the wild boar's presence around Keene City and Peterboro in southern New Hampshire (Silver, 1957; Figure 5).

In May 1896, Corbin died in a carriage accident, and his son, Austin Corbin, Jr., assumed the management of the park (Martin, 1970). In late 1896, the wild boar there were estimated to number at least 500 animals. Also during that year there was a failure in the beechnut crop. The park wild boar were not artificially fed during the winter of 1896–97, and all but 50 apparently died (Manville, 1964). By 1903, however, the wild boar inside the park had again increased to about 400–500 animals (Silver, 1957). The park's history during the next 30 years is obscure (Silver, 1957). In 1931, Austin Corbin, Jr., stated that there were hundreds of wild boar remaining in the park, which was still owned by the Corbin family (Baynes, 1931); however, an unknown number of wild boar escaped after trees felled by a hurricane in 1938 knocked down sections of the fence (Presnall, 1958; Jones, 1959; Laycock, 1966; Taylor, 1980).

In 1940, the wild boar inside Corbin's Park were estimated to number at least 50 animals. At that time, although the park's wild boar seemed to be in fair condition, it was necessary to artificially supplement their winter diet with tons of corn (Silver, 1957). The "outside" wild boar, as they were called locally, had increased to about 25–30 in number and ranged over the towns of Croydon, Plainfield, Grantham, and Cornish, staying within about 24 km of the park (Silver, 1957; Figure 5). A few spread farther north to Alexandria in Grafton County and west to Concord in Merrimack County (Silver, 1957; Taylor, 1980; Figure 5). In 1941, wild boar on state and private lands in New Hampshire were estimated to number 200 animals (Jackson, 1944). By 1948,

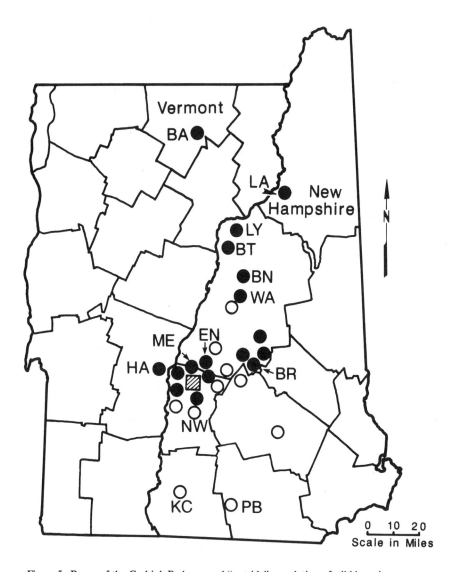

Figure 5. Range of the Corbin's Park escaped "outside" population of wild boar in New Hampshire and Vermont between 1895 and 1989. Black circles indicate localities where animals were killed. Open circles represent reported sightings of wild boar outside the park. The hatched square indicates the locality of Corbin's Park. Locality abbreviations are as follows: BA = Barton, BN = Benton, BR = Bristol, BT = Bath, EN = Enfield, HA = Hartland, KC = Keene City, LA = Lancaster, LY = Lyman, ME = Meridan, NW = Newport, PB = Peterboro, and WA = Warren.

48

the outside wild boar increased to their largest number ever, and one hunter, Palmer Read, of Plainfield, managed to shoot seven of them (Taylor, 1980; H. A. Laramie, pers. comm.). One herd of 13 outside wild boar, the largest seen to date, was observed during that period. Wild boar damage to local corn and potato fields also reached a peak at that time (H. A. Laramie, pers. comm.). In 1949, the citizens of Plainfield raised $200 toward finding a way to prevent or eliminate further damage by these outside wild boar. A lawyer from Claremont was hired, only to inform the town's selectmen that there was little that could be done to solve the problem (Taylor, 1980). Later that year, however, the New Hampshire state legislature was presented with a bill that would require the persons responsible for the introduction of the wild boar to attempt to capture or exterminate them. The bill also stated that the owners and keepers of the wild boar would be held responsible for all damage done by these animals subsequent to April 1950 (Silver, 1957). The bill became law later that year.

In early 1955, the range of the outside wild boar was described as beginning 48 km north of Concord, extending west to Claremont, east to Danbury or beyond, and north to Canaan (Lineaweaver, 1955). Two outside wild boar were found to have dispersed into Vermont during the mid 1950s (Figure 5). One was shot across the Connecticut River in Hartland, Windsor County, Vermont, in 1955 (Angwin, 1955). Another was shot farther north in Barton, Orleans County, Vermont, in 1956 (J. D. Stewart, pers. comm.). The outside wild boar drifted from place to place following the most plentiful supplies of food. The hunter harvest of outside wild boar reached a peak during this period (H. A. Laramie, pers. comm.). This increase lasted until the late 1950s. Heavy hunting was apparently the cause for the subsequent decrease in the size of the outside population (H. A. Laramie, pers. comm.). In 1957, the outside wild boar were estimated to number between 20 and 40 individuals (Presnall, 1958). The reported hunter kills of outside wild boar also decreased after that time, with 25 in 1958, 20 in 1959, 12 in 1960, and 3 in 1961 (Manville, 1964; Godin, 1977). In the early to mid 1960s, some of these animals could still be found around the outside of the park in northern Sullivan County (Waters and Rivard, 1962; H. A. Laramie, pers. comm.). The unfenced herd remained small, and during the late 1960s and early 1970s, no records showed that wild boar existed outside the preserve (H. R. Siegler, pers. comm.). In 1962, an estimated 250 wild boar inhabited Corbin's Park (Manville, 1964).

After 1975, following more damage to the park's fence by local vandals, there was an increase in the number of outside wild boar. In addition, the stream floodgates that now comprised sections of the fence surrounding the park occasionally became jammed open after floodwaters subsided, allowing another potential avenue of escape for the park's wild boar (E. P. Orff, pers. comm.). Most of these animals were killed in the towns around the park.

A few were killed to the east in the Plymouth-Bridgewater area of Grafton County in the late 1970s (H. A. Laramie, pers. comm.; Figure 5). In 1979, three outside wild boar were killed: one each in the towns of Warren, Enfield, and Bristol in Grafton County (Taylor, 1980; H. A. Laramie, pers. comm.; R. F. Dufour, pers. comm.). Early in 1980, a pair of wild boar were sighted in Warren, and signs of others were discovered around Meridan in Sullivan County. Later that year, after doing damage to a farm in Newport, one outside wild boar was recaptured and returned to the park by the preserve's staff (Taylor, 1980).

In the late summer of 1980, two wild boar were seen at a Boy Scout camp on Route 25-A in Grafton County (E. P. Orff, pers. comm.). On 12 November 1980, a subadult male wild boar (Plate 2) was killed on a farm one mile southeast of Partridge Lake in Lyman, Grafton County, about 60 miles north of the park. This animal was part of a small herd of about six wild boar that had been present in the Littleton-Lyman area since the summer of 1980 (R. F. Dufour, pers. comm.; Poliquin, 1982). Small numbers of these animals have continued to exist in that area of Grafton County. In November 1981, an adult male wild boar was killed by a deer hunter near Long Pond in Benton, Grafton County (E. P. Orff, pers. comm.). In January 1982, after being hunted for more than a month, another adult male wild boar was killed in Bath, Grafton County (Poliquin, 1982). A small number of wild boar were reported from the area around the park in the early to mid 1980s. One outside wild boar was killed in the town of Lancaster on 5 November 1986 (E. P. Orff, pers. comm.). There have been no recent reports of wild boar from areas outside of Corbin's Park. In an effort to control and possibly eliminate the future escape of animals from the park, the Blue Mountain Forest Association is currently encompassing the entire park with a chain-link fence (E. P. Orff, pers. comm.). If this endeavor eliminates recruitment of individuals into the outside population, the free-ranging population of wild boar in New Hampshire will probably be eliminated within a few years because of hunter harvest and apparent minimal reproductive success of the outside animals. No estimates exist of the size of the population of outside wild boar in New Hampshire.

After the death of Austin Corbin, Jr., in the mid 1940s, Corbin's Park remained under private ownership but fell into disrepair owing to the lack of finances for its maintenance. In 1946, a group of 30 private sportsmen re-activated the Blue Mountain Forest Association, took over the ownership of the preserve, and have managed it to the present (Silver, 1957; Martin, 1970). From 1948 to 1955, 317 wild boar were killed by hunters within the park. Seventy-nine, the largest single annual kill, were taken in the 1951–52 season (Silver, 1957). An annually harvest population of wild boar still exists within the park (E. P. Orff, pers. comm.). In the mid 1970s a male and female wild pig from the stock being maintained at the Bronx Zoo were obtained through an animal dealer in Massachusetts. Both animals were paid for by one of the association's members (A. Dobles, pers. comm.; H. J. McCarthy, pers. comm.). These animals were from the illegal release in New York State in 1972 (see "The Indian Lake Introduction"). This introduction of new genetic stock into Corbin's Park was made in an effort to correct an inbreeding problem expressed for some time as changes in coat coloration of the park's wild boar (A. Dobles, pers. comm.; E. P. Orff, pers. comm.). This new pair of wild pigs had a black basal coloration and were reportedly purebred animals (H. J. McCarthy, pers. comm.). The pair were kept in the sanctuary area of the park for breeding purposes and eventually died of natural causes (H. J. McCarthy, pers. comm.). The introduction of these two animals may have permanently altered the gene pool of what had been the oldest and purest Eurasian wild boar population still found in the United States. The wild boar population within Corbin's Park was estimated to number 500–600 individuals in the early 1980s (M. S. Garabedian, pers. comm.). With additional artificial feeding year-round, the numbers in the park had increased to between 2,000 and 2,500 animals by the mid 1980s (E. P. Orff, pers. comm.; K. Kammermeyer, pers. comm.). During that same time, the association enlisted the aid of researchers at the U.S. Department of Agriculture's Beltsville Parasitological Laboratory to assist the park staff with containing an outbreak of *Trichinella spiralis* within the park's wild boar population. Treatment in the form of preventive medication was added to the artificial feed supplements in an effort to reduce the spread of this parasite within the herd (K. Kammermeyer, pers. comm.; E. P. Orff, pers. comm.).

Silver (1957) disproved a story that the outside wild boar were descended from animals escaped from a railroad wreck en route to the park. This story had apparently grown out of inaccurate information distorted by the passage of time. Her check of the records of the Boston and Maine Railroad, over whose lines all of Corbin's shipments must have passed after entering the state, turned up no evidence to support this story. Spears (1893) reported that 26 white-tailed deer were killed in a train collision while en route to the park, and this accident perhaps formed the basis for the rumor concerning the wild boar.

The Litchfield Park Introduction

Around 1900, millionaire Edward H. Litchfield released 15–20 European wild boar from Germany on his estate, Litchfield Park, in the Adirondack Mountains of Hamilton County, New York (Bump, 1941). These animals disappeared into the rugged forests of the area and managed to maintain themselves in the wild for several years (Anon., 1972). A. Joachim (pers.

comm.) has described these wild boar as having survived in this area for 3 years; Bump (1941) has estimated a 20-year survival.

The Hooper Bald Introduction

The origin of the well-known wild boar population in Tennessee and North Carolina has been attributed to several sources. Some claim that an adventurer named Barnes imported the wild boar from Germany's Black Forest or that two North Carolina brothers brought them back from Russia after World War I (Martin, 1945); however, the most detailed account of the history of the origin was presented by Jones (1959). Unless otherwise cited, the following historical account was taken from that source.

This introduction began as a business venture. On 24 February 1908, the Whitting Manufacturing Company, an English concern, bought from the Great Smoky Mountain Land and Timber Company a vast tract of land in the mountains of southwestern North Carolina. Included in this tract was a 1,659-meter peak called Hooper Bald in Graham County. A financial advisor, George Gordon Moore of St. Clair, Michigan, was hired to transact the purchase of the land. As payment for his work, Moore was allowed to establish a game preserve on several thousand acres surrounding Hooper Bald. The work on the preserve was begun in 1909 and completed in 1911. Two enclosures—one 600 hectares and the other 200–240 hectares—were constructed along with several buildings. The smaller lot, referred to as the boar lot, was enclosed with a chestnut split-rail fence nine rails high. Moore hired a local man named Garland "Cotton" McGuire to be the caretaker of the preserve (Shaw, 1941). In April 1912, game animals for the preserve began to arrive, and deliveries continued throughout that summer. Fourteen Eurasian wild boar, including 11 females and 3 males, were among the first animals to arrive. These were young animals, weighing only 27–32 kg. One of the wild boar died on the 40-km trip to the preserve. The remaining animals were put into the 200–240 hectare enclosure.

The origin of these animals has been something of a mystery. Stegeman (1938), Towne and Wentworth (1950), and De Vos et al. (1956) attribute the origin to northern Germany, probably the Harz Mountains. Holman (1941) states that the imported wild boar came from the Ural Mountains of Russia. Vinson (1946) places the source as both the Ural Mountains of Russia and the Black Forest of Germany. As quoted by Jones (1959), Moore discussed the origin of the wild boar in a 1949 letter to McGuire: "I got my knowledge of wild boar through Walter Winants, the famous sportsmen [*sic*], who had his own wild boar hunting lodge in Belgium. He told me that the largest and fiercest wild boar were from the Ural Mountains in Russia. I purchased the boar through an agent in Berlin who represented that they were from

Russia." More recently, Conley et al. (1972), Pine and Gerdes (1973), and Igo (1977) have also maintained that the Ural Mountains were the supposed source of these animals. Bratton (1977) argued that this was not possible, since wild boar are not known to occur in the Ural Mountains, and that a more likely source was either Germany or Poland, as the agent probably sold Moore local (German) wild boar rather than importing them from the Urals. Bratton (1977) went on to state that around the turn of the century, owners of German hunting preserves were concerned about inbreeding in their wild boar stock. To avoid potential problems, these individuals purchased animals from dealers for the purpose of outbreeding their wild boar. Bratton stated that these wild boar were frequently sold as "Russians" but were almost invariably Polish or western European in origin, implying that Moore's agent had done the same thing. No references were cited for any of this information.

In addition, Bratton (1977) was not entirely correct about the absence of wild boar in the Ural Mountains. Bobrinskii (1944) stated that wild boar formerly occurred in the pine woods of northern Kazakhstan, in districts around the southern end of the Ural Mountains. Heptner et al. (1966) also stated this and indicated that wild boar were originally found in the southern extremes of the Ural range. Sokolov (1959) also indicated that wild boar formerly occurred in the southern Ural Mountains. It is possible, therefore, that these animals could have at least come from areas surrounding the southern Urals in northwestern Kazakhstan. But the fact remains that this entire argument is academic, since Moore stated only that the animals were supposedly from Russia, with no mention of a specific locality. Unfortunately, this uncertainty of origin does create a taxonomic problem. Because the possible origin of these animals may have been either Germany, Poland, western Europe, or western Russia, these original wild boar could have been any one of six or seven different subspecies (Appendix B).

From the very beginning, the wild boar were able to root out of their enclosure at Hooper Bald and pass in and out at will. It seems likely, however, that the majority chose to remain within the enclosure, where they were allowed to reproduce unmolested for a period of 8–10 years (Stegeman, 1938). The first time Moore and his guests set dogs on the animals, some of the wild boar leaped over low places in the enclosure fence and escaped into the surrounding area. Forest fires also weakened the boar-lot fence during this period, allowing still more wild boar to escape (Vinson, 1946).

About 1921 or 1922, Moore became indifferent to the preserve on Hooper Bald and his visits ceased. He also neglected to continue paying his caretaker, Cotton McGuire. McGuire took legal action against Moore, who then asked McGuire to come to him in New York. When McGuire got there, Moore gave him the title for the preserve and a check for $1,000. After his return to the preserve, McGuire organized a hunt for the wild boar, which were believed to number between 60 and 100 in the enclosure. At the conclusion of the hunt, only two of the animals had been killed. The remainder had broken down the fences shortly after the start of the hunt and escaped. These animals spread into the surrounding area, and many dispersed into Tennessee.

The escaped wild boar extended their range through the uplands of both Tennessee and North Carolina for a considerable distance. Domestic swine and feral hogs were running loose in much of this area at that time, and crossbreeding occurred freely (Stegeman, 1938). All degrees of intergradation between wild boar and feral hogs began to appear in the population (Stegeman, 1938). In spite of frequent hunting by locals, the wild boar became numerous in their new range. In 1932, an epidemic of hog cholera spread from local domestic swine to the wild boar population and reduced their numbers (Stegeman, 1938).

From the mid 1930s to the present, the wild boar have increased their range in North Carolina and Tennessee through natural dispersal and as a result of numerous releases by state fish and game agencies and private individuals. In 1937, Stegeman (1938) estimated that the wild boar numbered about 230 animals in southeastern Tennessee and southwestern North Carolina.

Kellogg (1939) stated that the northern limit of the range of the wild boar was still to the south of the Great Smoky Mountains National Park. Arthur Stupka, the park's naturalist at that time, believed that the Little Tennessee River, which separates the park from the southern half of the Cherokee National Forest, had stopped the northern dispersal of these introduced animals (Kellogg, 1939). By the end of 1939, the U.S. Forest Service's annual big game census estimated the populations of these animals in the Nantahala and Cherokee national forests at 130 and 500 individuals, respectively (Rachford, 1939).

Shaw (1941) reported that the range of the wild boar in Tennessee in 1940 was restricted to the north and south forks of Citico Creek, the Tellico River, the North River, and the Bald River. In 1941, the U.S. Fish and Wildlife Service estimated that there were 662 wild boar in Tennessee and North Carolina (Jackson, 1944).

About 1943, a few wild boar were released in the Green River Cove section of Polk County, Tennessee (Hamnett and Thornton, 1953; Jones, 1959). Around the same time, a female wild boar and four or five young piglets were released in the Mount Mitchell section of Yancey County, North Carolina, by local sportsmen (Jones, 1959). In 1946, there were an estimated 700 wild boar in the Tellico Wildlife Management Area in Tennessee (Vinson, 1946). In 1948, a U.S. Forest Service wildlife census estimated that there were 100 wild boar in the Nantahala National Forest in North Carolina and 1,000 in the Cherokee National Forest in Tennessee (Towne and Wentworth, 1950).

On the basis of information provided by park records and interviews with retired park staff, Bratton (1977) found that the wild boar were first observed near the western and southwestern boundaries of the Great Smoky Mountains National Park in the late 1940s. Pelton (1984) stated that the movement of wild boar into the park was probably not a natural invasion, but rather the result of releases of these animals near the park boundary. As early as 1950, wild boar were seen occupying the grassy balds inside the park (Bratton, 1975). In 1959, the park's wild boar population was concentrated in the area between Cades Cove and Fontana Lake, in the southern end of the park. During August of that year, there were about 500 wild boar in the park (Linzey and Linzey, 1971). Also in 1959, local newspapers and the park staff became concerned about the extreme rooting damage to the western grassy peaks of Parson Bald and Gregory Bald in the park. As a result, control efforts against the wild boar were begun. Between August 1959 and 1975, 626 wild boar were live-trapped and removed from the park, and an additional 301 were shot (Singer, 1976). Despite these measures, the wild boar occupied the southern third of the park by the mid 1960s (Lewis et al., 1965). In 1973, the wild boar extended their range north of U.S. Highway 441, which divides the park approximately in half. Between then and 1976, the wild boar increased their range to the extreme northeastern corner of the park, then occupying all the park except the southeastern corner. This last range expansion was aided by private introductions of wild boar near the town of Cosby, Cocke County, Tennessee. In 1976, the wild boar population inside the park was estimated at 1,920 animals (Singer, 1976). Animals from this last invasion have since expanded their range north of the park into the northern half of the Cherokee National Forest during the mid 1970s (C. J. Whitehead, pers. comm.).

At the end of 1977, a controversy over the shooting and removal of wild boar from the park led to a moratorium on the killing of these animals (Coleman, 1984). Trapping, however, has continued through the present. In April 1978 the moratorium was lifted and direct reduction of this population was again implemented as a control effort (Coleman, 1984). In 1979 densities of the wild boar in the park were estimated to be 7–9 animals per square kilometer in the northern hardwood areas of the high western ridgeline and 2 animals per square kilometer in low-elevation pasture and oak-pine habitats (Singer, 1981). The highest densities in the park have been reported seven years after the initial occupation of an area, with stabilization occurring after 20–27 years (Tate, 1984). Between 1978 and 1983, more than 1,600 wild boar were removed from the park through shooting and live-trapping (Tate, 1984).

The impact of the wild boar on the vegetation of the park and surrounding areas has been well documented and has been largely negative (e.g., Howe et al., 1981; Singer et al., 1984). Lacki and Lancia (1986), however, demonstrated enhanced growth of beech (*Fagus grandifolia*), apparently as a result of enhanced nutrient mobilization in soils disturbed by wild boar rooting.

In the mid 1950s, the size of the wild boar population in Tennessee and North Carolina was increasing, but estimates varied as to the exact number. As reported in a popular hunting magazine, in 1955, an estimated 3,000 wild boar were thought to occur in the mountains of both states (McNally, 1955). In contrast with this previous estimate, between 1956 and 1957 the combined North Carolina and Tennessee wild boar population was estimated to number 500 or more individuals (Presnall, 1958). In 1956, hunters killed 42 wild boar in Tennessee. Compared with the general abundance of the wild boar in this area, this latter number was felt to be low because of the difficulty in hunting them and their limited trophy appeal (Presnall, 1958). As of April 1956, Swift (1957), the director of the U.S.D.A. Division of Wildlife Management, estimated that there were 1,100 wild boar in the two states.

Several wild boar were reported to have crossed into northern Georgia during the 1950s (Jones, 1959) and have since interbred with the remnant populations of feral hogs in this area (Golley, 1964). In the early 1970s, these animals occurred in the northern parts of Towns and Rabun counties in Georgia (J. Scharnagel, pers. comm.), but more recently, they have been restricted to two wildlife management areas in Rabun County. They are fall and winter residents, and move to higher elevations in adjacent North Carolina in the summer. Current fall and winter estimates of the density of this population are one to two animals per square kilometer (R. W. Whittington, pers. comm.).

Since the early 1960s, private individuals have stocked wild boar into northeastern Tennessee in the vicinity of the junction of Washington, Greene, and Unicoi counties, and in Carter County near the Laurel Fork Wildlife Management Area. Since then, this population has apparently become established. Litters are being produced, and six animals had to be removed as "pests" from the Clark's Creek area of Washington County by the Tennessee Game and Fish Commission in the spring of 1971 (Smith et al., 1974).

Private releases also occurred in North Carolina during the early 1960s. Feral hogs and wild boar were reportedly stocked into three areas of western North Carolina. These releases were made discreetly by unknown individuals, and the results were still indefinite by the mid 1960s (Lewis et al., 1965).

In May 1962, 26 pen-raised wild boar were released near Crossville, Cumberland County, Tennessee, by the Tennessee Game and Fish Commission in an attempt to establish another huntable population in the state. The stocking failed when the released animals proved to be too tame to become wild-living and had to be recaptured (Lewis, 1966).

In 1964, a few so-called wild boar were released by private individuals into a commercial shooting preserve in Hyde County, North Carolina (Lewis et al., 1965; R. B. Hamilton, pers. comm.). No recent reports indicate a persisting population in that part of the state, and these animals have almost assuredly been extirpated (R. B. Hamilton, pers. comm.).

In 1965 and 1966, the Tennessee Game and Fish Commission again introduced wild boar into Cumberland County, this time successfully. During this stocking effort, a total of 46 wild boar were live-trapped in the Great Smoky Mountains National Park and released on the Catoosa Wildlife Management Area (Lewis et al., 1965; J. Murrey, pers. comm.). These animals have since dispersed into other counties of the Cumberland Plateau region of Tennessee (R. H. Conley, pers. comm.). Some have reportedly dispersed into Kentucky (Hines, 1988).

In the late 1960s, a small number of wild boar expanded their range into two localities of the Blue Ridge Parkway adjacent to the Nantahala National Forest in western North Carolina. Control efforts by the U.S. Park Service have since eliminated these two local populations (Singer, 1981).

Between 1969 and 1971, the Tennessee Game and Fish Commission released 15 wild boar live-trapped in the Great Smoky Mountains National Park into the Tellico Wildlife Management Area (J. Murrey, pers. comm.). This relocation effort was conducted in conjunction with the control program being operated by the National Park Service.

In the early 1970s, some of the wild boar from North Carolina were reported to have dispersed into the northwestern corner of South Carolina (C. J. Whitehead, pers. comm.), but authorities in South Carolina discredit this report as unfounded (W. B. Conrad, pers. comm.).

Between March 1971 and July 1972, the Tennessee Game and Fish Commission released 115 wild boar into the Ocoee Wildlife Management Area of the Cherokee National Forest and the Hiwassee Ranger District in Polk County as part of a big-game stocking program. The first release consisted of 51 wild boar that were live-trapped in the Great Smoky Mountains National Park and released into the Ocoee Wildlife Management Area (R. H. Conley, pers. comm.; C. J. Whitehead, pers. comm.). The second release involved 64 animals captured in the Catoosa Wildlife Management Area and released into the Ocoee Wildlife Management Area and the Hiwassee Ranger District (J. Murrey, pers. comm.). During this same time period, the Cleveland Bear and Boar Club of Cleveland, Tennessee, released a total of 46 pen-raised wild boar in the Ocoee Wildlife Management Area (J. Murrey, pers. comm.); some have since dispersed south into the Cohutta Mountains of northern Murray and Gilmer counties in Georgia (J. Scharnagel, pers. comm.). By 1981, these animals had established a population that was distributed from eastern Murray County through northern Gilmer County and into western Fannin County.

The estimated population density is one to one and one-half individuals per square kilometer. Few of these animals exhibited characteristics of pure wild boar (R. W. Whittington, pers. comm.).

In 1979, live-trapped wild boar from the Great Smoky Mountains National Park were also used to stock the Mississippi River bottomlands of the 12,200-hectare Anderson-Tully Wildlife Management Area of Lauderdale County, Tennessee (Wood and Barrett, 1979). They have since multiplied and spread into the lands surrounding the wildlife management area. In 1981, 54 wild boar were harvested from this population (Lawrence, 1982). Most (63%) were killed outside the wildlife management area on private agricultural lands. The hunter harvest of this population has declined since then; the total harvest on the management area and surrounding lands ranged between two and eight animals per year between 1982 and 1986 (J. Murrey, pers. comm.). The present population is almost entirely restricted to the management area (J. Murrey, pers. comm.).

The Eurasian wild boar in California also originated from Hooper Bald animals. In 1924, George Gordon Moore wrote to Cotton McGuire and requested that he capture several wild boar from around Hooper Bald and ship them to California, where Moore had purchased some property (Laycock, 1966). With the help of Paul Lovin, a neighbor, McGuire was able to trap 12 animals: 3 males and 9 females (Jones, 1959; Pine and Gerdes, 1973). It should be noted that these animals were trapped by McGuire at least two years after the last of the imported wild boar and their descendants had escaped from the Hooper Bald enclosure. There was thus a low probability that any of these animals were of pure Eurasian wild boar ancestry; more likely they were hybrid stock. After the wild boar arrived in California in late 1925 or 1926, Moore released them on his property, the San Francisquito Ranch, along San Jose Creek between Carmel Valley and the Los Padres National Forest in Monterey County. This property is now called the San Carlos Ranch (Bruce, 1941; Pine and Gerdes, 1969; 1973; M. E. Stanley, pers. comm.). Shortly after that, Stuyvesant Fish, owner of the neighboring Palo Carona Ranch, obtained some wild boar from Moore's herd and released them on his own property (Amato, 1976). From these two ranches the wild boar spread south into the Santa Lucia Mountains (Jones, 1959; Pine and Gerdes, 1969; Seymour, 1970) and bred with the already established feral hog population in the area, producing offspring with varying characteristics (Pine and Gerdes, 1973).

In 1932, 24 yearling wild boar from the San Francisquito Ranch were transplanted to the Carmel watershed of the Los Padres National Forest in Monterey County (Shaw, 1941). From here the animals began to extend their range southward and increase in number. By 1937, wild boar were well established in this area and had reached the headwaters of the Arroyo Seco

River, 42 km from the 1932 release site (von Bloeker, 1938; Pine and Gerdes, 1969). Two years later, there were about 100 wild boar in the Los Padres National Forest (Rachford, 1939). By 1941, the Arroyo Seco was still the southern boundary of their range, and the U.S. Fish and Wildlife Service estimated that 350 wild boar were in existence on state, private, and national forest lands in California (Bruce, 1941; Jackson, 1944). In 1948, according to a U.S. Forest Service wildlife census, there were an estimated 150 wild boar in the Los Padres National Forest alone (Towne and Wentworth, 1950). Before 1952, William Randolph Hearst, Sr., reported in a letter to George Gordon Moore that the wild boar had already reached the Hearst ranch near San Simeon, in San Luis Obispo County (Pine and Gerdes, 1969). By the end of April 1956, wild boars had increased to about 600 in the Los Padres National Forest (Swift, 1957). Shortly after that, private individuals and state agencies began to stock other parts of the state with live-trapped wild pigs from Monterey County.

In 1957, an unknown number of wild boar from the upper Carmel Valley were released into the Lower Mill Creek Flat in southeastern Fresno County. Attempts were made to trap and remove these animals in 1959 at the request of the U.S. Forest Service; they numbered about 40–50 individuals in 1968 (Anon., 1968).

In the spring of 1957, 25 wild boar were trapped by the California Department of Fish and Game at Big Sur State Park because of depredation problems. They were transported and released 23–40 km to the south in the Los Padres National Forest, at Lacimiento Summit, and in other localities in Monterey County (Pine and Gerdes, 1973; T. M. Mansfield, pers. comm.).

Harold Eade, owner of the Hepsadam Ranch in San Benito County, purchased some wild pigs with European wild boar phenotypes from a ranch in Monterey County and brought them to his ranch, where they escaped around 1961 and became established as a wild-living population (Bish, 1967; Anon., 1970).

About 1961, two wild boar were introduced a few miles north of Willits, in Mendocino County (Anon., 1968). In 1965, two additional introductions of an unknown number of wild boar were made into San Benito County: one on the Quien Sabe Ranch, and the other in the Hernandez Valley. In 1968, each of these populations numbered about 75 animals (Anon., 1968).

Since about 1965, hunters have shown an increasing interest in the wild pig as a game animal in California (Barrett, 1977). This has led to unregulated introductions of the wild boar throughout the oak-woodland zone and central coast range of the state by private landowners and other individuals (Pine and Gerdes, 1973; Barrett, 1977). Sometime during the 1960s, three wild boar were released into the Whiskey Creek section of western Shasta County to improve the wild pig hunting in the area (Anon., 1970).

In 1968, William S. Keeler, the landowner of the Dye Creek Ranch in Tehama County, began breeding captured feral sows from his ranch with "European type" wild boar males from Monterey County. By January 1970, over 20 feral sows had been bred and released into the Dye Creek population. The wild boar phenotype is now present in this area (Barrett, 1978; R. H. Barrett, pers. comm.). In the spring of 1974, the ranch was estimated to have 1,200 wild pigs (French, 1974). The ranch has since been purchased by the Nature Conservancy (R. H. Barrett, pers. comm.).

In the spring of 1970, wild boar from Monterey County were released north of the King's River area of Fresno County (Anon., 1970). During the 1976–78 drought in central California, the wild boar in Monterey and San Benito counties seemed better adapted to these adverse conditions than the animals exhibiting hybrid characteristics. This was evident in both the numbers observed in the field and those harvested during this period. Over the last 10–15 years, there has been a decrease of both the wild boar and hybrids in their historic habitat along the central California coast. This area includes all lands between San Simeon in the south and Salinas in the north, west of U.S. Highway 101. In contrast to this, these animals have dramatically increased in number in the areas east of U.S. Highway 101, from Parkfield in the south to Hollister in the north. This area is generally made up of private lands and is considerably more arid than areas immediately to the west along the coast (P. A. Dubsky, pers. comm.).

In Tennessee, especially in the vicinity of Tellico Plains, several people maintain captive herds of "local" (Hooper Bald stock) European wild boar for sale or training dogs (Lewis et al., 1965). Animals from some of these herds have been used successfully to start or upgrade wild pig populations in at least three other states. There have been three such introductions in Florida. In the mid 1950s, a few male wild boar purchased from one of these captive herds were mated to live-trapped feral sows in central and southeastern Florida. Just before parturition, the pregnant sows were released on the Avon Park Air Force Range in Polk and Highlands counties, and on the J. W. Corbett Wildlife Management Area, Palm Beach County. According to David Austin, the Florida state wildlife biologist in the Avon Park area, the hunting pressure on the wild pigs there was so great after these introductions that little to no traces of the hybrid stock reportedly were able to survive (R. C. Belden, pers. comm.).

The second introduction was initiated in December 1960, when 10 male wild boar were purchased from R. S. Heins of Cleveland, Tennessee, by the wildlife management unit at Eglin Air Force Base, Florida. In 1961, these wild boar were bred with live-trapped feral sows that had been captured on the base. After being bred, the sows were released back onto the base (C. Maloy, pers. comm.) to upgrade the wildness of the feral hog population there (Crossman, 1965; Lewis et al., 1965). In March 1962, 6 of the 10 male wild boar themselves were also released on the reservation. Later, the 3 remaining animals were released into the same area (one had died previously in captivity of unknown causes [C. Maloy, pers. comm.; J. R. Knowles, pers. comm.]). Approximately 1,000 wild pigs were harvested from these population during the 1966 public hunting season on the base (Dalrymple, 1970).

Wild boar were introduced on a private ranch near West Palm Beach by the Florida Game and Fresh Water Fish Commission during the late 1970s at the request of one of the state commissioners. These boar were reportedly from stock originating from the Tennessee–North Carolina hybrid population (J. L. Schortemeyer, pers. comm.).

In 1970, the wildlife division of the West Virginia Department of Natural Resources decided to introduce wild boar into the southern mountains of their state, hoping that a population could be established to provide huntable big game species in this area where white-tailed deer, wild turkey, and black bear populations have historically always been few in numbers. In January 1971, 15 pen-reared adult wild boar of both sexes were purchased from D. Cole of Tellico Plains, Tennessee. The animals were brought to the French Creek Game Farm, near Buckhannon, Boone County, West Virginia, on 28 January. The wild boar were held there for quarantine and to allow the pregnant sows to farrow. Five of the sows gave birth in captivity between 29 January and 22 March. The Spruce-Laurel watershed in Boone and Logan counties in the southwestern portion of the state was chosen as the release area for these wild boar. A total of 10 adults and 20 young were successfully released between 25 March and 27 April 1971. On 6 March 1973, an additional 5 adults and 7 shoats were released in the same general area. The animals spread from the release site, and in 1975 occupied a 335-km^2 area of Boone, Logan, and Wyoming counties. Litters have been produced each year subsequent to these releases. Despite an initial tameness exhibited by the released wild boar and some mortalities due to poaching, the animals were successfully established in the area by the late 1970s (Igo, 1977; Igo et al., 1979).

There have been several introductions of hybrid stock from the Tennessee–North Carolina population into South Carolina. In the early 1970s Jarrell Brown, owner of Cowden Plantation near the town of Jackson in Aiken County, South Carolina, purchased a male and female "wild boar" and released the two animals on his plantation to upgrade the resident feral hog population. However, both animals were killed by the plantation's lease hunters within a year of their introduction (J. W. Reiner, pers. comm.; H. W. Arnold, pers. comm.; G. Hogsed, pers. comm.). One or two years later, Brown bought nine "wild boar," including one adult male, two adult females, and six piglets, from an animal dealer in Alliance, South Carolina. These animals were reportedly descended from the Tennessee–North Carolina hybrid population (J. W. Reiner, pers. comm.). The "wild boar" phenotype has since been bred into this population. In an effort to control the population, Brown live-trapped and sold hogs to other plantations and hunt clubs. In 1982, he removed more than 360 hogs from his property (J. W. Reiner, pers. comm.). In 1983, several wild pigs exhibiting wild boar skull characteristics and coat coloration were harvested from the portion of the Savannah River Plant adjacent to Cowden Plantation (Mayer, 1984). Between 1983 and 1988, the wild boar phenotype spread throughout the river swamp population of wild pigs on the installation (P. E. Johns, pers. comm.). In February 1983, a yearling boar was captured on a private farm in Burke County, Georgia, immediately across the Savannah River from Cowden Plantation. This animal exhibited partial wild boar coat coloration and probably indicates that the hybrid phenotype has also spread across the river from the Cowden Plantation population. The Savannah River does not represent a dispersal barrier to wild pigs. On nine different occasions, the senior author has observed wild pigs successfully swimming across the river in this area of the border between Georgia and South Carolina in an effort to elude hunting dogs.

The second introduction into South Carolina occurred in the river swamps of the Congaree River south of Columbia in Richland and Calhoun counties, into an area with an established population of feral hogs. These feral hogs originated from free-ranged domestic stock released by the former landowners in the area (F. T. Ehrlich, pers. comm.). The "wild boar" stock introduced into this area was obtained from the captive wild boar herd maintained at the University of Georgia by Dr. Robert W. Seerley which originated from the captive hybrid population maintained by the Tennessee Game and Fish Commission at Tellico Plains. The animals introduced into the bottomland areas of the Congaree River were released on lands owned by the Cedar Creek Hunt Club. This area was later purchased by the federal government and became the Congaree Swamp National Monument. In the late 1970s and early 1980s, wild pigs were common on the monument property (Singer, 1981), and by the mid to late 1980s they had expanded into a dense population. The damage being done to monument property resulted in measures to control the size of this population (R. S. McDaniels, pers. comm.). These animals were also very common on properties adjacent to the monument lands and were both live-trapped and hunted by the landowners to control property damage (J. W. Reiner, pers. comm.). The hybrid phenotype is prevalent among the animals in this population (J. W. Reiner, pers. comm.). Juvenile striping has also been observed among the piglets from this area (R. E. Mancke, pers. comm.).

A third location in South Carolina where the wild boar phenotype has been introduced into an existing feral hog population is Groton Plantation,

Allendale County. These animals were purchased from an animal dealer whose stock also originated from the Tennessee–North Carolina hybrid population. Stocking of these hybrid animals has been occurring since 1978, and the hybrid phenotype is very common on the plantation (H. L. Stribling, Jr., pers. comm.; H. O. Hillestad, pers. comm.).

Wild pig stock with some wild boar ancestry was released on Bostick Plantation, Hampton County, South Carolina, to improve the quality of the existing feral hog population (J. W. Reiner, pers. comm.).

Two hybrid juvenile males from Cowden Plantation were released on the Bantam Harper Hunt Club property, Jasper County, South Carolina, in August 1983. One was killed later that same year. The fate of the other hybrid animal is unknown.

The Denman Introductions

Two introductions of wild boar have been made into central coastal Texas. In 1919, the St. Charles Ranch on the Blackjack Peninsula of Aransas County, Texas, was acquired by the San Antonio Loan and Trust Company. In about 1923, Leroy G. Denman, Sr., chairman of the board of the bank, assumed management of the ranch. Denman prohibited all hunting on the ranch and built a game-proof fence across the northern part of the peninsula. Several species of exotic birds and mammals were introduced into this area. Between 1930 and 27 June 1933, 11 European wild boar were released (Halloran and Howard, 1956). Denman had obtained these wild boar from the Brackenridge Park Zoo in San Antonio, Texas, which supposedly had "pure" wild specimens (M. D. Springer, pers. comm.; L. G. Denman, Jr., pers. comm.). Unfortunately, the San Antonio Zoological Society did not maintain records prior to 1969, so the origin of these individuals will probably remain a mystery (R. Smith, pers. comm.). The wild boar freely interbred with the feral hog population, which had been present in the area since the early 1900s (Halloran and Howard, 1956). When the area was purchased by the U.S. Department of the Interior to be used as a national wildlife refuge, the former owners were allowed to trap and remove as many of the descendants of the wild boar and feral hogs as possible. Between 1 October 1936 and 30 July 1939, 3,391 hybrid animals were trapped and either shipped out for sale or butchered at the site (Halloran and Howard, 1956). On 31 December 1937, the ranch and much of the surrounding area officially became the Aransas National Wildlife Refuge. The pig population has increased since then, and the wild boar phenotype is very common (Springer, 1977). These animals are now found on the refuge and the adjacent properties (Coombs and Springer, 1974). In the late 1960s and early 1970s, 400 wild pigs were estimated to exist on the refuge lands (Springer, 1975). By the early 1980s, these animals had become very common on the refuge despite persistent year-round control shooting by refuge personnel and hunting by the general public during annual hunts. Approximately 100 pigs were killed each year during this period. Groups of up to 40 or more wild pigs were spotted from the air in the nonpublic areas of the refuge during the annual whooping crane aerial censuses of the early 1980s (K. Butts, pers. comm.).

In 1939, Denman also acquired some additional property of his own, the Powderhorn Ranch, between Port O'Connor and Seadrift, in Calhoun County, Texas. Shortly after the purchase of the Powderhorn, he stocked a number of exotic big game species onto the ranch, including 10–15 European wild boar. These animals were also purchased from a zoo, probably again the Brackenridge Park Zoo (L. G. Denman, Jr., pers. comm.). Unlike the St. Charles Ranch, feral hogs were not present on the Powderhorn Ranch before the wild boar introduction (L. G. Denman, Jr., pers. comm.), and the wild boar phenotype is still the only one observed there. Denman prohibited the hunting of all wildlife on his own ranch, just as he had done on the St. Charles. This was also a condition of his will, and the Powderhorn Ranch remains today a semipreserve for the introduced wild pigs, although some poaching does occur (J. C. Smith, pers. comm.). The Powderhorn wild boar population has expanded into the surrounding ranches. In 1964, 400 unfenced wild boar were reported on three ranches in Calhoun County (Jackson, 1964). Some of the wild boar from the Powderhorn Ranch crossed Espiritu Santo Bay during low tides and are now present on Matagorda Island. The purity of the wild boar phenotype in this population, however, is uncertain, since the northern and western range boundaries of these animals have become contiguous with feral hog and known hybrid populations (J. C. Smith, pers. comm.).

Descendants of one of the Denman wild boar introductions were stocked on San Jose Island, Aransas County, by private individuals in an effort to establish a huntable population there (J. C. Smith, pers. comm.). The details of this introduction are unknown.

In 1973, 17 wild boar obtained from the Powderhorn Ranch were accidentally released along the intracoastal waterway in Chalmette, Louisiana. These animals have since dispersed into the surrounding area (B. L. Goatcher, pers. comm.).

The Edwards Plateau Introduction

Sometime during the early 1940s, Harry Brown, a rancher in the Scenic Loop area west of Boerne, Bexar County, Texas, purchased several wild boar and released them into a large fenced-in section of his ranch (K. Schwarz, pers. comm.; Anon., 1979). The exact origin of these animals is unknown. Tinsley (1968) claimed that they were purchased from the owner of a traveling zoo who sold his wild boar to help pay his debts. Several years after the wild boar were purchased, a flood washed out portions of the fence and the animals escaped into the surrounding hill country of the Edwards Plateau (Maly, 1976; K. Schwarz, pers. comm.; P. W. Payne, pers. comm.). Since their escape, these animals have spread south to Helotes, Bexar County, and west to just north of Hondo, Medina County. As of 1981, the population was still restricted to areas north of U.S. Highway 90 and west of Interstate 10 (K. Schwarz, pers. comm.). Some of the wild boar dispersed as far west as the ranch country near Utopia in Bandera County (Maly, 1976). These wild boar have interbred with the feral hogs in most areas of the Edwards Plateau (K. Schwarz, pers. comm.). In 1964, 175 animals were reported in Bexar County, and hogs were described as occurring in heavy concentrations in Medina County (Jackson, 1964).

Although there has never been a natural expansion of this hybrid population to the east of U.S. Highway 10, in the winter of 1984–85, 22 wild pigs descended from the Edwards Plateau hybrid stock were released on the KWW Ranch, 16 km east of Boerne in Comal County (R. F. Smart, pers. comm.). The success of this introduction is unknown.

During the 1960s and 1970s, several south Texas ranchers purchased wild boar × feral hog hybrids from individuals in the Edwards Plateau and released them on ranches in Webb County in an effort to improve the feral hog populations there. At present, most of the animals in this area are feral hogs; only a few exhibit any wild boar characteristics. These animals are now dispersed over five counties along the Rio Grande in south Texas (Benke, 1973; K. Schwarz, pers. comm.).

Animals from the Edwards Plateau hybrid population were released on north Texas ranches in Throckmorton and Haskell counties during the 1970s and have spread over a wide area in that part of the state since that time (J. R. Hunter, pers. comm.). Descendants of these releases were found in the original two counties and in Cottle, Foard, King, Knox, Young, Shackelford and Stephens counties (SCWDS, 1988; J. R. Hunter, pers. comm.).

In 1979, ranchers in LaSalle County, Texas, introduced three "pure" European wild boar to upgrade the quality of the feral hogs in that area. Although the origin of these animals is unknown, it is thought that they were hybrids purchased from ranches in the Edwards Plateau (K. Schwarz, pers. comm.).

The Indian Lake Introduction

In June 1972, wild boar began appearing in the forests of the Adirondack Mountains around the town of Sabael, near Indian Lake in Hamilton County, New York (Anon., 1972). Six of these animals were either shot by landowners for destruction of private property or hit and killed by automobiles. One adult male, two adults females, two yearling males, and two yearling females were captured and sent to the Bronx Zoo by the New York State Department of Environmental Conservation in late June and early July of that year (A. Joachim, pers. comm.; Stone, 1973; J. G. Doherty, pers. comm.). James G. Doherty, general curator of the Bronx Zoo (pers. comm.), noted that these animals seemed to be very tame when they arrived at the zoo, suggesting that they may have been of captive origin. In addition, their behavior in and around Sabael, New York, was similar to that of pen-reared hybrids that were released near Crossville, Tennessee, in 1962, and in southwestern West Virginia in 1971 and 1973 (Lewis, 1966; Anon., 1972; Igo et al., 1979). The origins of these animals and the name of the person responsible for this illegal release are still unknown. Initial theories about them being remnants from the Litchfield Park population have been discounted (Anon., 1972). For a short time there was some question as to whether or not there were wild pigs remaining in the Adirondacks (A. Joachim, pers. comm.); however, there have been no recent reports of this species from that area.

The Skagit River Introduction

During the winter of 1981, an estimated 40–60 wild boar were illegally released by unknown individuals along the Skagit River in Skagit and Whatcom counties in north-central Washington State. Investigators from the Washington State Game Department have been unable to determine who was responsible for this introduction and the size and racial origin of this population of wild boar (Park, 1982). Some of the wild pigs in the Skagit River release were killed by Game Department personnel; others were killed by deer hunters (R. Johnson, pers. comm.). There have been no recent reports of these animals in this area, and it is assumed that all of them were eliminated (R. Johnson, pers. comm.).

Additional Introductions

Other, lesser-known, introductions of Eurasian wild boar have supposedly been made into the United States. Unfortunately, few details are available, and some of the accounts of these introductions are of questionable authenticity.

There have been several reports of wild boar introduced on the coastal islands of at least two southeastern states. Hanson and Karstad (1959) stated that individuals wishing to increase the sporting value of the feral hogs on the coastal islands of Georgia and North Carolina introduced a few European wild boar into the feral herds there. No other details were given. Diehl (1965) indicated that 300 wild boar were sent by King Humbert of Italy to the Jekyll Island Club, which apparently released them on Jekyll Island, Glynn County,

Georgia. There have been no recent reports of this population. Singer and Stoneburner (1977) stated that European wild boar had previously been introduced into the feral hog population on Cumberland Island to improve the sporting quality of those animals. This statement was apparently based solely on the presence of a juvenile striped coat coloration found in some of the island's current wild pig population. Again, no details of this introduction were given. It is now known, moreover, that certain domestic coat color phenotypes can also produce a striped juvenile pelage (Keller, 1902; Hetzer, 1945; Crandall, 1964; Searle, 1968). Johnson et al. (1974) stated that European wild boar have been released on the Georgia islands of Cumberland, Little Cumberland, Jekyll, St. Simons, Little St. Simons, Ossabaw, and Wassaw. The authors noted further that there were probably no true wild boar persisting on the Georgia coastal islands by the early 1970s, because the wild boar had either been extirpated from those islands or genetically integrated into the feral hog populations already present. There are no records of any wild boar populations existing in recent times in any coastal areas of North Carolina (R. B. Hamilton, pers. comm.).

Foreyt et al. (1975) stated that the "local feral hogs" on the Welder Wildlife Refuge, San Patricio County, Texas, and the P. H. Welder Ranch, Victoria County, Texas, were "descendants of crosses between wild domestic hogs (*Sus scrofa domesticus*) and the European Russian boar (*Sus scrofa cristatus*)." Authorities at the Welder Wildlife Refuge, however, believe that the wild pigs there are strictly feral hogs (L. Drawe, pers. comm.). It is possible, however, that the wild pigs on the Welder Ranch farther to the north could be hybrids descended from animals that dispersed from the sites of the Denman introductions in the neighboring counties along the coast during the 1930s.

Several people have claimed that European wild boar have also been released on Santa Cruz Island, California (Ingles, 1954; Palmer, 1954; Haltenorth and Trense, 1956; Rue, 1969; Wooters, 1973, 1975; Elman, 1974; Whitaker, 1980). No proof has ever been offered, and the owner of the island disagrees, stating that the Santa Cruz wild pigs are strictly feral hogs (C. Stanton, pers. comm.).

In the early 1970s, the owner of the Hatchet Creek Hunting Preserve, Alachua County, Florida, purchased approximately 25 "wild boar" from a defunct hunting preserve in northern Georgia. These animals were put into a 5-acre enclosure in Alachua County on the hunt club site north of Gainesville (W. B. Frankenberger, pers. comm.) and released on the hunt club property about one year later. A couple years later, staff from the Florida Game and Fresh Water Fish Commission were asked to trap wild swine from agricultural fields on a farm near the hunt club. All the hogs trapped in that area were reported to exhibit hybrid characteristics (W. B. Frankenberger, pers. comm.).

Wild pigs were reported to have been deliberately introduced into the central-eastern portion of Cherokee County, Alabama, to start a huntable wild population. The original males in this introduction had some "Russian" wild boar stock in their ancestry, while the introduced females were all feral hogs (J. R. Davis, pers. comm.). The eastern range of this population now includes Chattooga and Walker counties in Georgia (R. Davis, pers. comm.). Hybrid stock from the North Carolina–Tennessee population has also supposedly been stocked into south Alabama (Barrows et al., 1981), but no specific details were given. James R. Davis, Alabama's former district game biologist for that area, claims that these animals are all feral hogs (pers. comm.).

During the late 1960s and early 1970s, the Mississippi Department of Wildlife Conservation released "wild boar" into a number of areas in Mississippi (J. H. Phares, pers. comm.; R. K. Wells, pers. comm.). Most of these introductions were on private hunt club lands, but some animals were released into the Copiah County Game Management Area (R. K. Wells, pers. comm.). In September 1971, four female and two male wild boar × feral hog hybrids were captured in Copiah County and released by the department into the Pearl River Game Management Area, Madison County, Mississippi (R. K. Wells, pers. comm.). Feral hogs had been present on the management area prior to the introduction of this hybrid stock (J. Lipe, pers. comm.). The hybrids were immature animals, each weighing only about 27 kg (R. K. Wells, pers. comm.). Hunting was prohibited until 1974, when these animals had increased in number and spread into private lands and the Natchez Trace federal property adjoining the management area. The animals damaged local soybean fields on private farmland and wildlife food plots on the management area (R. K. Wells, pers. comm.; J. Lipe, pers. comm.). The uprooted areas on the management area and federal lands have caused damage to equipment operating on the food plots and, by serving as fire lanes, have created difficulties in conducting controlled burns in the pine flatwoods of this area (J. Lipe, pers. comm.). Starting in 1974, the state began control shooting of these animals. One of the hogs killed during these efforts was a male that weighed 193 kg (R. K. Wells, pers. comm.). As an additional control measure, some wild pigs were live-trapped on the Pearl River Game Management Area and released on private lands in Rankin County, across the Pearl River from Madison County (R. K. Wells, pers. comm.). These animals in Madison County are harvested during the deer season on the management area (J. Lipe, pers. comm.). Approximately 20–60 wild pigs have been harvested per year during the recent seasons on the Pearl River Game Management Area. The wild boar phenotype now predominates on the management area; almost all the piglets are striped (R. K. Wells, pers. comm.).

In 1973, wild boar were illegally introduced in the Mississippi River bottomlands near Port Gibson, Mississippi. The numbers of animals released and persons responsible are unknown. These animals have since spread as far south as Ferriday, Louisiana, and the swamps west of Fayette, Mississippi (B. L. Goatcher, pers. comm.).

In the early 1970s, Louis Davis of Douglas, Georgia, purchased two adult male "wild boar" from a commercial shooting preserve in Pine Mountain, Georgia. The larger of these animals weighed 205 kg. The history of these animals prior to the purchase is unknown. They were released on a 400-hectare private hunt club near Dublin, Laurens County, Georgia. The two males have since interbred with the feral hog population that was present on the property before their introduction, and the wild boar phenotype is now dominant in this population. In early August 1981, Roscoe Davis, president of the hunt club, and his brother, Jimmy, both of Douglas, Georgia, captured two "trailer loads" of juvenile wild pigs from this population and released one trailer load each at Horse Creek Wildlife Management Area, Telfair County, and Bullard Creek Wildlife Management Area, Jeff Davis County, in an effort to upgrade the feral hog populations in these areas. In addition, they gave six striped piglets to the staff of the Muskhogean Wildlife Management Area, Telfair County. It is not known, however, whether any of these latter individuals were actually released on the wildlife management area's lands (R. Davis, pers. comm.).

In 1975, reportedly "pure" European wild boar were stocked in an unfenced area along the Ocmulgee River in Houston County, Georgia, by Bobby Tuggle of Perry, Georgia (R. W. Whittington, pers. comm.; R. Snow, pers. comm.). These animals were purchased from a source in Ohio (R. W. Whittington, pers. comm.) and are now established in the area between Perry and Hawkinsville along the Ocmulgee River corridor (R. W. Whittington, pers. comm.; R. Snow, pers. comm.). Most are said to exhibit pure wild boar characteristics (R. W. Whittington, pers. comm.). One animal collected from this population and examined by JJM did have a wild-grizzled coat coloration typical of Eurasian wild boar.

During the 1970s, wild pigs of "Russian" wild boar origin were stocked along the Ocmulgee River in Ben Hill County, Georgia. Feral hogs were present in this portion of the river before this introduction, and the effect of this introduction on the morphology of wild pigs along the Ocmulgee River in Ben Hill, Telfair, Wilcox and Dodge counties is unknown (R. Davis, pers. comm.).

The wild boar phenotype has been reportedly introduced in several areas in Louisiana. The animals released were descendants of the stock introduced into Louisiana in the early 1970s. The areas where hybrid stock occurs in the state now include Pearl River; Jean Lafitte National Historic Park and Preserve; Yscloskey; Belle Chasse Naval Air Station; Gust Island (Madisonville); Slidell; Manchac Swamp; New Orleans East (proposed Bayou Savage National Wildlife Refuge); Caldwell Parish; Winn Parish; and near Franklinton, in Washington Parish (B. L. Goatcher, pers. comm.).

There is a population alleged to be European and/or Russian wild boar in the bottomlands of the Sulphur River in extreme northeast Texas and southwestern Arkansas. The hunters in that part of Arkansas supposedly distinguish between these wild boar and the local feral hogs (Wooters, 1973); however, the existence of wild boar in this region remains unconfirmed (Wooters, 1975).

The San Diego Zoo Importation

For more than sixty years there were no new importations of Eurasian wild boar into the United States, and the only pure wild boar in the country were those descended from the importations described in the preceding sections. By the early 1980s, however, most or all of these strains had been hybridized by either deliberately or inadvertently crossing them with animals having some degree of feral hog ancestry. The lack of new importations during this period was the result of a strict governmental ban on the importation of all live suids, both wild and domestic, in an effort to protect the nation's livestock industry from exposure to a number of diseases that may have been carried by such animals.

In the mid 1980s, however, some importation restrictions were relaxed so that, by adhering to strict quarantine regulations, certain suids could be brought into the United States on a limited basis. Because of the expense associated with the lengthy period of quarantine required, most suids imported under these new regulations have been species of unique scientific and/or exhibition value to zoos (e.g., babirusa, *Babyrousa babyrussa*; and wart hogs, *Phacochoerus aethiopicus*).

On 16 June 1986, however, a trio of wild boar successfully completed quarantine and arrived at California's San Diego Zoo (M. L. Jones, pers. comm.). These animals (one male and two females) had been purchased from the Tierpark Berlin in East Berlin, where they had been born in 1986 (M. L. Jones, pers. comm.). It has not yet been possible to determine the degree of genetic interrelatedness of these three imported animals, or to determine the exact geographic location of capture of the original wild stock from which they were produced at the Berlin Zoo. However, they are almost certainly of the subspecies *Sus scrofa scrofa* (M. L. Jones, pers. comm.).

Over the past three years, these three imported boar have been bred at the San Diego Zoo, and their offspring have been sold to individuals throughout the country. The majority of these offspring (approximately 13–16) were sold to Earl Tatum of Holly Springs, Arkansas, and through him to Ross Wilmoth of the Wild Wilderness Safari Park of Gentry, Arkansas (E. Tatum, pers. comm.). A single male offspring from the San Diego Zoo trio was sent to the Detroit Zoo, and a male/female pair to the Catskill Game Farm,

Catskill, New York (M. L. Jones, pers. comm.). At least two other offspring were sold to Jim Fouts, an animal dealer from Wichita, Kansas, from whom they eventually were acquired by an unidentified individual in Oklahoma (J. Fouts, pers. comm.).

Recently, a pair of offspring from the San Diego Zoo imports was sent to Aiken County, South Carolina, where they are now the property of a syndicate that plans to use them in breeding programs designed to continue to propagate this pure strain of wild boar in captivity as well as to upgrade the percentage of wild boar ancestry in various present-day hybrid populations (J. W. Reiner, pers. comm.). There is no record of any of the descendants of the San Diego Zoo wild boar having been released into the wild.

3. Comparative Morphology

The varied origins of the wild-living pigs in the United States have produced populations with varied phenotypic characters. This is especially true among the feral populations (McKnight, 1964). Although represented by only one species, these populations are composed of stock whose origins trace back to at least two, and possibly more, subspecies of the wild ancestor. This compilation is due to the possibly varied origins of the introduced Eurasian wild boar. Bratton (1977) stated that more morphological work was needed on this species in the United States to determine the subspecific makeup of these populations. The purpose of the portion of the study discussed in this chapter was to determine the morphological variation present and thus possibly provide more information concerning the taxonomic origins of these various populations. This involved an analysis of morphometric and discrete characters of the skull, external body dimensions and proportions, coat coloration patterns, and hair morphology.

The specific objectives of the work discussed in this chapter were to (a) determine the characteristic phenotypes or morphotypes of Eurasian wild boar, domestic swine, feral hogs, and wild boar × feral hog hybrids; (b) determine if any morphological variation exists between long-term and short-term feral hog populations; (c) determine to what extent sexual dimorphism in this species is statistically significant; (d) compare the morphotype of feral hogs from the United States with those from other countries; (e) compare the morphotypes of the free-living pigs with those of their captive counterparts; and (f) obtain further insight concerning the taxonomic identity of U.S. wild pig specimens with little or no known history.

At the present time, the stocking and translocation of wild pigs is commonplace in the southeastern United States, Texas, and California (Barrett, 1977; K. Schwarz, pers. comm.; J. W. Reiner, pers. comm.). Along the Savannah River, for example, hunters using dog packs frequently catch feral hogs alive in the river marshes and swamps. These feral hogs are released at other localities to improve hunting along the entire length of the river from Augusta to Savannah, Georgia. In many cases, the hunters themselves are uncertain exactly where they released the animals they have stocked (J. W. Reiner, pers. comm.). Because these new introductions and stockings are so frequent and usually unreported, it has become increasingly more difficult to be completely certain of the history of any wild pig population in the United States today. For this reason, morphological criteria that would separate the several types of wild pigs would be useful. These criteria could then be used to provide further insight into the possible histories of populations with unknown or unclear origins. These same criteria could also prove useful in helping to make legal determinations of the types, and hence ownership (e.g., state versus private), of swine from populations of unknown origins or derivations.

Methods

Known-History Populations

The populations described in chapter 2 were considered to have known histories with regard to the types of wild pigs involved in their foundation and development. Specimens from these populations (Table 1) thus formed the base groups for determining morphotypic diagnostic characters. Because of the lack of definitive information about the wild boar introduced on Cumberland Island, specimens from that population were included in the feral hog group. The wild pigs on Santa Cruz Island were treated in the same manner for the same reason. The animals on the Pearl River Game Management Area in Mississippi are known hybrids and were treated as such in the following analyses. Specimens from the remaining populations mentioned in the subsection on Additional Introductions were treated as unknowns.

Morphotype Determination

Specimens Examined. Specimens represented 474 individuals from 23 populations with known histories in the United States (Table 1); 46 Eurasian wild boar specimens from central and northern Europe, western Asia, and the Middle East; and 149 domestic swine from the United States, Canada, Germany, and Turkey. These specimens were used to determine and define the four target groups.

Specimens were either collected in the field or obtained on loan from public and private collections in the United States and Canada. These specimens consisted of varying combinations of cleaned crania, cleaned mandibles, and preserved skins (either tanned flat skins or dried study skins). Numbers preceding the collection acronym indicate the number of specimens examined from that locality. All collection acronyms are given in the list of abbreviations.

Table 1. Extant wild-living populations of *Sus scrofa* in the United States with known histories as Eurasian wild boar, feral hogs, or wild boar × feral hog hybrids. Specimens examined from these populations were thus considered to be of the types indicated and formed base groups for determining morphotypic diagnostic characteristics in the present study. NWR = national wildlife refuge; GMA = game management area.

Locality	Type of *Sus scrofa*		
	Eurasian Wild Boar	Feral Hog	Hybrid
Alabama			
Cherokee County			
central-eastern portion			X
Arizona			
Mohave County			
Havasu NWR		X	
Arkansas			
Ouachita National Forest		X	
Ozark National Forest		X	
Felsenthal NWR		X	
southwestern counties		X	
central-eastern counties		X	
California			
Humboldt County			
Pilot Ridge		X	
Rainbow Ridge		X	
Showers Pass		X	
Klamath River		X	
Siskiyou County			
Klamath River		X	
Shasta County			
eastern foothills		X	
Whiskey Creek			X
Tehama County			
Dye Creek Ranch[a]			X[b] 1968
Paskenta area		X	
Lake County			
Walker Ridge		X	

Table 1—*Continued*

Locality	Type of *Sus scrofa*		
	Eurasian Wild Boar	Feral Hog	Hybrid
Cow Mountain		X	
Middle Creek		X	
Colusa County			
Colusa NWR		X	
Sonoma County			
northwestern quarter		X	
Elder Creek			X
Santa Clara County			
hills east of Los Altos		X	
Fresno County			
north of King's River			X
Lower Mill Creek Flat			X
Mendocino County			
north of Willits			X
Monterey County[a]			X[b] 1925–26
San Benito County			
Hepsedam Ranch			X
Quien Sabe Ranch[a]			X
Hernandez Valley			X
San Luis Obispo County			
Huasna Creek		X	
Lopez Canyon		X	
Stoney Creek		X	
northern end of county[a]			X[b] 1940s
Santa Barbara County			
Santa Cruz Island[a]		X	
Santa Rosa Island[a]		X	
Los Angeles County			
Santa Catalina Island[a]		X	
San Clemente Island		X	
Florida			
Eglin Air Force Base			X[b] 1960
Franklin County			
St. Vincent Island NWR		X	

Continued on next page

Table 1—*Continued*

	Type of *Sus scrofa*		
	---	---	---
Locality	Eurasian Wild Boar	Feral Hog	Hybrid
Wakulla County			
St. Marks NWR		X	
Osceola National Forest		X	
Lake Woodruff NWR		X	
Chassahowitzka NWR		X	
Manatee and Sarasota counties			
Myakka River State Park[a]		X	
Avon Park Air Force Base			X[b] 1950s
Palm Beach County			
J. W. Corbett GMA			X[b] 1950s
Kissimmee River valley		X	
Georgia			
Towns and Rabun counties			X[b] 1950s
Murray and Gilmer counties			X[b] 1950s
Chatham County			
Ossabaw Island[a]		X	
Liberty County			
St. Catherines Island[a]		X	
Hawaii			
Islands of Niihau, Kauai, Oahu, Molokai, Maui, and Hawaii[a]		X	
Kentucky			
Southeastern corner of state		X	
Louisiana			
Cameron County			
Sabine NWR		X	
Catahoula NWR		X	
Mississippi			
Madison County			
Pearl River GMA			X[b] 1971
Gulf Islands National Seashore			
Horn Island		X	

Table 1—*Continued*

Locality	Type of *Sus scrofa*		
	Eurasian Wild Boar	Feral Hog	Hybrid
New Hampshire			
Sullivan County			
Corbin's Park[a]	X		
rest of county	X		
Grantham and Merrimack counties	X		
New Mexico			
Hidalgo County[a]		X	
North Carolina			
Currituck County			
McKay NWR		X	
Graham, Swain, Cherokee, Clay,			
Macon, and Jackson counties			X[b] 1912
Oklahoma			
Southeastern corner of state		X	
South Carolina			
Georgetown County			
Hobcaw Barony		X	
Francis Marion National Forest		X	
Savannah River Plant[a]		X	
Santee Coastal Reserve		X	
Tennessee			
Polk, Monroe, Blount, and			
Sevier counties[a]			X[b] 1920s
Cumberland Plateau			X
Lauderdale County			
Anderson-Tully GMA			X
Texas			
Aransas NWR			X[b] 1930s
Edwards Plateau[a]			X[b] 1940s
Rio Grande valley[a]			X
Calhoun County[a]	X[c]		

Continued on next page

Table 1—*Continued*

| | Type of *Sus scrofa* | | |
| | Eurasian | Feral | |
Locality	Wild Boar	Hog	Hybrid
Virginia			
Princess Anne County			
Back Bay NWR		X	
Barbours Hill GMA		X	
False Cape State Park		X	
West Virginia			
Boone, Logan, and Wyoming counties			X

a Populations for which specimens were available for this study.
b Hybrid populations that resulted from Eurasian wild boar stock being introduced into an area with a feral hog population. The date indicated is when the wild boar introduction occurred.
c The animals were represented as "wild boar" by the Brackenridge Park Zoo from whom they were purchased in the early 1930s (see text).

Eurasian Wild Boar. UNITED STATES: New Hampshire, Sullivan County: Corbin's Park (1, USNM; 1, YPM; 1, UMASS; 1, MOU; 3, FMNH; 1, JJM). Texas, Calhoun County: Powderhorn Ranch (2, JJM). FRANCE: Ardennes (1, AMNH); Lamargelle (1, USNM; 1, AMNH). GERMANY: Harz Mountains (1, USNM); Sichenbergen (1, AMNH); Sendichirfi (1, AMNH); Sachenwald (1, AMNH); Friedrichsruh, Sachenwald (1, AMNH); Berlin, Grunewald (1, AMNH); 14 km N of Wildflecken (1, FSM); Brandenburg, Potsdam (1, UCONN); no specific locality (1, USNM; 3, AMNH; 3, MCZ). IRAN: Turkman Desert, Dar Kaleh (3, AMNH); vicinity of Hamadan (1, USNM); Fars, 17.5 km SW of Yasoodj (1, FMNH); Fars, 24 km E of Yasoodj (1, FMNH); Isfahan, Kuhrang (1, FMNH); Kerman, 32 km SW of Zabol (2, FMNH); Gorgan, 16 km ESE of Gorgan (2, FMNH); Khuzistan, E of Kermanshah (2, FMNH). IRAQ: Jabel Baradost, Ishkufti, Khurwatan (1, MCZ); 5 km S of Baghdad (1, FMNH); Darband area between Kirkuk and Sulimaniyah Liwa (1, FMNH); Khanagin (2, FMNH); Place Ramlla, 16 km from Khanaquin (3, FMNH); Kikuk Liwa, Wadi Hostocki, 0.8 km S of Towaka Zhari (1, FMNH); Chahala, near Amhara (2, FMNH); Baradost (1, FMNH); 30 km N of Hilla (1, UMMZ). POLAND: Bialowieza National Park (1, FMNH). The Eurasian specimens represent the subspecies *Sus scrofa attila, Sus scrofa falzfeini,* and *Sus scrofa scrofa.*

Domestic Swine. CANADA: Ontario, Bruce County: Kincardine Twp (2, ROM); York County: near Toronto (1, ROM); no specific locality (1, ROM). GERMANY: Brandenburg, Berlin (1, AMNH). TURKEY: no specific locality (1, YPM). UNITED STATES: California, Alameda County: Berkeley Market (1, MCH); Yolo County: Davis (2, MCH; 1, UCLA). Connecticut, Hartford County: Marlboro (2, JJM); New London County: N. Franklin (1, JJM); Tolland County: Mansfield, University of Connecticut (8, JJM; 1, RD); Windham County: Chaplin (1, UCONN; 4, JJM); Sterling (1, JJM); no specific locality (3, JJM). Illinois, Gallatin County, Shawnee Town (1, UIMNH); Du Page County: Napierville (1, FMNH); Cook County: Chicago, Armour and Company stockyards (1, FMNH). Indiana, Tippecanoe County: Lafayette, Purdue University Farm (5, FMNH); no specific locality (2, NUVM; 1, JJM). Iowa, Johnson County (2, IOWA). Kansas, Wallace County (1, KU). Maine, Oxford County: Norway (2, MCZ). Maryland, Montgomery County (1, USNM). Michigan, Ingham County: East Lansing, MSU Campus (9, MSU); Washtenaw County: Ann Arbor (1, UMMZ); no specific locality (1, UMMZ). Missouri, no specific locality (1, MUO). Nevada, Elko County: Walker Canyon (1, AMNH). New Jersey, no specific locality (3, NUVM). New Mexico, Sierra County: 21 km W and 6 km S of Salinas Peak (1, NMSU). New York, Schoharie County: Lawyersville (4, AMNH); Nassau County: Locust Grove (3, USNM); Monroe County (1, MCZ). North Carolina, Wake County: Raleigh (1, NCS). Oklahoma, Payne County (1, OSU); Stillwater, Oklahoma State University (1, OSU). Pennsylvania, Chester County (1, USNM); Cumberland County: Carlisle (3, USNM). Texas, Bee County: Beeville (1, UTAVP); El Paso County: Peyton's Meat Packing Company (7, UTEP). Virginia, Montgomery County: Blacksburg (2, USNM); no specific locality (1, USNM). Washington, Pierce County: Tacoma (1, UPS); Fircrest (1, UPS). Wyoming, Goshen County: Torrington (1, MCZ). No specific locality (2, USNM; 1, LSU; 2, UTAVP; 4, UCONN; 2, YPM; 5, AMNH; 11, MCZ; 3, UPVS; 1, NUVM; 1, ANSP; 3, ROM).

Feral Hogs. UNITED STATES: Arizona, Mohave County: Havasu NWR, Topock Marsh (4, JJM). California, Tehama County: Dye Creek Ranch (29, MVZ); Santa Barbara County: Santa Rosa Island (1, UCLA; 11, SBMNH), Santa Cruz Island (3, UCLA; 3, SBMNH); Los Angeles County: Santa Catalina Island (1, CSULB; 1, MVZ; 11, DWB). Florida, Sarasota County: Myakka River State Park (1, AMNH). Georgia, Chatham County: Ossabaw Island (119, JJM; 2, CM; 3, SREL); Liberty County: St. Catherines Island (40, AMNH; 5, JJM); Camden County: Cumberland Island (2, CM; 2, SREL). Hawaii, Hawaii: Parker Ranch (1, USNM), Kilauea Forest Reserve Keauhou Ranch (1, BPBM), Kipuka Puaulu, 4 km NW of Kilanea Crater (1, BPBM); Kauai: Mt. Waialeale, Waimea side (3, BPBM), Mt. Waialeale, upper Olekele Valley (1, BPBM), no specific locality (1, BPBM); Oahu:

Koolau Mtns., Moanalua Valley (1, BPBM), Waiman Valley, one-third way up north branch (1, BPBM), no specific locality (1, USNM). New Mexico, Hidalgo County: 13 km W and 8.5 km N of Hilo Peak (1, NMSU). Oregon, Curry County: 1.6 km E of Wedderburn (1, MVZ). South Carolina, Aiken and Barnwell counties: Savannah River Plant (67, JJM; 5, SREL; 6, UGSF).

Wild Boar × Feral Hog Hybrids. UNITED STATES: California, Monterey County: 6 km SE of Jamesburg (7, MVZ), Parrott Ranch (1, MVZ), Carmel Valley, near Jamesburg (1, CPSU), Hunter-Liggett Military Reservation (3, MVZ), Big Sur, Grimes Creek (1, CPSU); San Benito County: Quien Sabe Ranch, 9 km E of Hollister (1, CPSU); San Luis Obispo County: Camp Roberts (1, CPSU), Camp Natuna, west of Paso Robles (1, CPSU). Mississippi, Madison County: 32 km N of Jackson, Pearl River GMA (1, MMNS). North Carolina, Graham County: Big Santeela Creek (1, NCS); no specific locality (1, NCS). Tennessee, Monroe County: Tellico Plains (3, CM); Blount County: 3 km E and N of Walland (1, MSU); eastern end of state: no specific locality (1, DMNH; 3, UGSF). Texas, Aransas County: Aransas NWR (2, USNM; 4, OSU; 2, ANWR; 152, JJM); Bexar County: west of Boerne (1, JJM); Kendall County: Smart Ranch (1, RFS; 1, JJM), west of Boerne (1, LR; 1, PWP); Kerr County: near Kerrville (2, MO); Medina County: Yancey-Devine area (6, RN); Webb County: near Encinal (1, LN; 2, MO; 1, JJM), no specific locality (2, LN; 1, MO; 3, JJM).

In addition, the feral hog populations were further separated into long-term feral populations (origin and isolation from significant input of domestic animals before 1800) and short-term feral populations (origin and isolation after 1800). This year was selected because it represents the point when careful domestic breed improvement began in the United States (Ensminger, 1961; Warwick, 1962), and the morphological changes and developments that occurred afterward might have affected the morphology of feral populations that originated after that period.

Age classes were determined by the pattern of tooth eruption (Table 2). Assigned age classes represent relative postnatal developmental and growth classes rather than absolute ages. For example, a hybrid pig acquires its complete dentition at about three years, while a domestic pig reaches the same erupted tooth pattern at about two to two and one-half years (Table 3). Both animals would be at the same relative "adult" growth stage but would be of different chronological ages. The relationships between assigned age classes and chronological ages for the various types of *Sus scrofa* are presented in Table 3. Based on a small amount of data, Barrett (1971) found that the teeth of feral hogs at the Dye Creek Ranch in California erupted slightly earlier than those of hybrid animals and slightly later than those of domestic swine. Tooth wear and suture closure were also evaluated to aid in comparisons of specimens in the adult age class. Only molars were used for rating tooth wear.

Table 2. Age classes in *Sus scrofa* L. based on erupted tooth pattern and used to separate specimens for analysis.

Age Class		Erupted Teeth[a]					
Number	Name						
1	Neonate/Infant	di 3/3	dc 1/1				
			to				
		di 123/123	dc 1/1	dp 234/234			
2	Juvenile	di 123/123	dc 1/1	dp 234/234	P 1/1	M 1/1	
			to				
		di 12/12	I 3/3	C 1/1	dp 234/234	P 1/1	M 1/1
3	Yearling	di 123/123	C 1/1	dp 234/234	P 1/1	M 12/12	
			to				
		I 123/123	C 1/1	P 1234/1234	M 12/12		
4	Subadult	I 123/123	C 1/1	P 1234/1234	M 12/123[a]		
			to				
		I 123/123	C 1/1	P 1234/1234	M 123/123[b]		
5	Adult	I 123/123	C 1/1	P 1234/1234	M 123/123		

Note:

di = deciduous incisor; dc = deciduous canine; dp = deciduous premolar

a Lower third molar starting to erupt.

b Upper third molar starting to erupt.

Table 3. Comparisons of the relative age classes (in months) from Table 2 and known absolute age ranges of the various types of *Sus scrofa* L.

Age Class[a]	Domestic Swine[b]	Domestic Swine[c]	European Wild Boar[d]	European Wild Boar[e]	Wild Boar × Feral Hog Hybrids[f]
1	0–9	0–8	0–4	—	0–8
2	9–12	8–12	4–11	12	8–12
3	12–18	12–18	12–23	15–18	12–22
4	18–20	18–20	23–25	24	21–30
5	24+	20+	36+	36+	36+

a As named and defined in Table 2.   c Sisson and Grossman (1938).   e Cabon (1958).
b Pope (1934).   d Boback (1957).   f Matschke (1967).

Wear rating methodology is detailed in Figure 6. The following 24 sutures were checked for closure: interparietal; parietal-supraoccipital; exoccipital-basioccipital; frontal-parietal; interfrontal; parietal-squamosal; supraoccip-ital-squamosal; supraoccipital-exoccipital; basioccipital-basisphenoid; frontal-lacrimal; exoccipital-squamosal; palatal-maxillary; nasal-frontal; malar-lacrimal; maxillary-malar; nasal-maxillary; maxillary-lacrimal; inter-maxillary; internasals; interpalatals; interpremaxillary; maxillary-premaxil-lary; pre-maxillary-nasal; and squamosal-malar. These sutures were rated as follows: open = 0; partially closed = 1; completely closed = 2. The suture score was the numerical sum of all the suture ratings. Only specimens in the two oldest age classes were analyzed separately by sex. Those without a re-corded sex were sexed using the dimorphism of the permanent canine morphol-ogy (Mayer and Brisbin, 1988). Specimens in the younger age classes either could not be sexed by this method of could not be sexed without damaging the specimen. Therefore, the sexes were combined in the younger age classes.

Skull Morphology. Most of the morphotypic analyses were performed through cranial comparisons similar to investigations of North American canids of uncertain taxonomic identification made by Lawrence and Bossert (1967, 1969) and Elder and Hayden (1977). General and nonmensural skull differences between groups were determined and described verbally.

Skull measurements should adequately measure the skull in all its dimensions and at the same time produce the best discrimination between the taxa under study (Lowe and Gardiner, 1974). The initial choice of measurements for this study was a suite of 51 linear measurements and one angle, all of which were taken on each skull. The linear parameters were measured with 150-mm dial calipers (DC) to the nearest 0.5 mm, or with anthropometers (A) to the nearest 1.0 mm. Bilaterally symmetric measure-ments were taken on the right side of the specimen whenever possible.

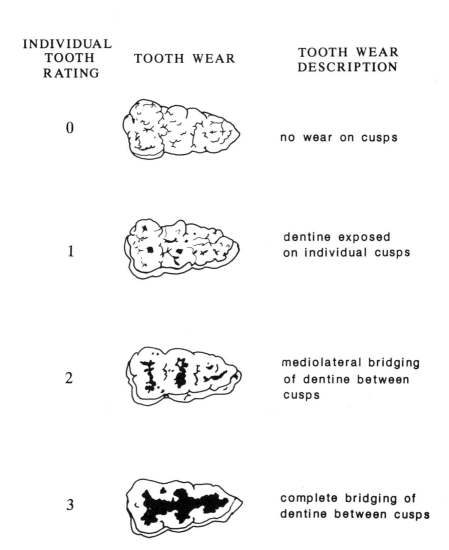

INDIVIDUAL TOOTH RATING	TOOTH WEAR	TOOTH WEAR DESCRIPTION
0		no wear on cusps
1		dentine exposed on individual cusps
2		mediolateral bridging of dentine between cusps
3		complete bridging of dentine between cusps

Figure 6. Methodology used for tooth wear rating. The example illustrated is the lower third molar. The total tooth wear score was the average of the sums of the individual tooth wear ratings of the three molars in the upper and lower tooth rows of the right side. If only the cranium or mandible was available, then only the sum of the ratings of the right molars present in that portion of the specimen were used.

The angle was determined with a metal straightedge and a protractor and recorded to the nearest degree. The profile depth was taken with two metal metric rulers to the nearest millimeter.

Cranial variables, as depicted in Figures 7 and 8, were:

Condylobasal length (A)—from the anteriormost edge of the premaxillae to the posteriormost projection of the occipital condyles.

Condylonasal length (A)—from the anteriormost edge of the nasal bones to the posteriormost edge of the nuchal crest along the midline of the skull.

Occipitonasal length (A)—from the anteriormost edge of the nasal bones to the posteriormost projection of the occipital condyles.

Nasal length (A)—from the anteriormost point of the nasal bones to the posteriormost point at midline.

Length from lacrimal notch to anterior nasal (A)—from the uppermost lacrimal foramen to the anteriormost point of the nasal bones.

Palate length (A)—from the anteriormost ventral edge of the premaxillae to the posterior margin of the palate at midline.

Rostral length—calculated as the square root of the square of difference between the lacrimal notch to nasal length minus the square of one half the least lacrimal notch width.

Zygomatic width (A)—the greatest distance between the outer margins of the zygomatic arches.

Braincase breadth (A)—the greatest breadth of the lateral sides of the parietals, midway between the postorbital processes and the nuchal crest.

Skull height (A)—measured perpendicularly from the dorsalmost edge of the nuchal crest in the midline of the skull to the line formed by the plane of the palate between the first to third molars.

Supraoccipital constriction (DC)—the least distance between the lateral edges of the supraoccipital.

Greatest width of occipital ridge (DC)—the greatest width across the lateral ridges of the nuchal crest.

Parietal constriction (D)—the least distance across the edges of the parietal crest.

Postorbital process width (DC)—the greatest width across the postorbital processes.

Least interorbital breadth (DC)—the least distance between the dorsal margins of the orbits.

Least lacrimal notch width (DC)—the least distance between the dorsal margins of the upper lacrimal foramina.

Premaxillary rostral width (DC)—the greatest breadth across the premaxillae above the upper canine alveolar buttresses.

Maxillary rostral width (DC)—the greatest breadth across the maxillae directly above the infraorbital foramina.

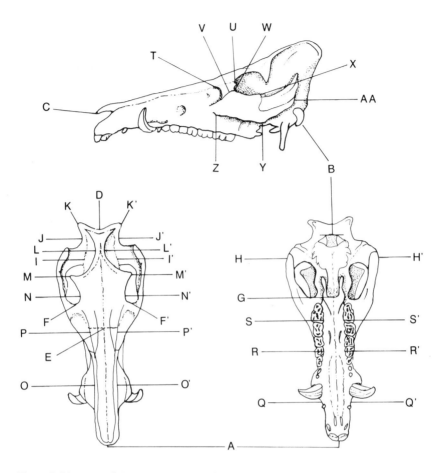

Figure 7. Linear cranial measurements used in the skull analyses: condylobasal length = A to B; condylonasal length = C to B; occipitonasal length = C to D; nasal length = E to C; lacrimal notch to nasal length = F to C; palate length = A to G; zygomatic width = H to H'; braincase breadth = I to I'; supraoccipital constriction = J to J'; greatest width of occipital ridge = K to K'; parietal constriction = L to L'; postorbital process width = M to M'; least interorbital breadth = N to N'; least lacrimal notch width = F to F'; premaxillary rostral width = O to O'; maxillary rostral width = P to P'; palatal premaxillary constriction = Q to Q'; palate width between upper PM4s = R to R'; palate width between upper M3s = S to S'; orbit width = F to M; lacrimo-frontal suture length = T to U; lacrimo-malar suture length = V to W; lacrimal height = U to W; zygomatic arch height = X to Y; zygomatic arch length = Z to AA.

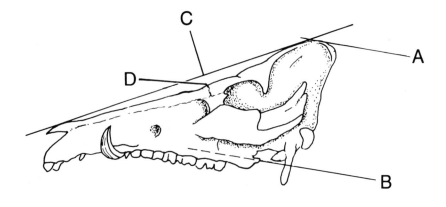

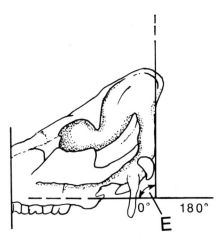

Figure 8. Two additional linear cranial measurements and one cranial angle used in the skull analyses: skull height = A to B; depth of dorsal profile = C to D; angle of the occipital wall = E.

Palatal premaxillary constriction (DC)—the least breadth across the palate at the lateral junction of the premaxillae and maxillae.

Palate width between upper PM4s (DC)—the least distance between the alveoli of the fourth upper premolars.

Palate width between upper M3s (DC)—the least distance between the alveoli of the first cusp row of the third upper molars.

Orbit width (DC)—the least distance between the upper lacrimal foramina and the ventralmost point of the postorbital process.

Lacrimo-frontal suture length (DC)—the greatest length of this suture from the edge of the orbit to the anteriormost point of the suture.

Lacrimo-malar suture length (DC)—the greatest length of this suture from the edge of the orbit below the lacrimal foramina to the anteriormost point of this suture.

Lacrimal height (DC)—the greatest vertical distance between the margins of the lacrimal bones at the edge of the orbit over the lacrimal foramina.

Zygomatic arch height (DC)—the greatest breadth from the ventralmost edge of the malar to the dorsalmost point of the orbital process of the malar in a line perpendicular to the posterior tooth row.

Zygomatic arch length (DC)—the greatest distance from the posterior-most point of the malar to the anterioventral point of the zygomatic process of the maxillae.

Depth of dorsal profile—the depth of the medialfrontal depression of the dorsal profile, perpendicular to the middorsal line formed along the high points of the cranial profile.

Angle of the occipital wall—the angle between the midline of the occipital wall (midline point of the nuchal crest to the midline point of the dorsal edge of the foramen magnum) and the midline of the plane of the palate adjacent and posterior to the third molars.

Mandibular variables, as depicted in Figure 9, were:

Mandibular length (A)—from the anteriormost edge of the mandible (teeth excluded) to the posteriormost plane of both angular processes.

Mandibular height (A)—the perpendicular distance from the plane of the ventralmost point of the angle of the mandible to the dorsalmost point of the condylar process.

Posterior mandibular width (A)—the greatest distance across the angles of the mandibles.

Midmandibular breadth (A)—the greatest distance across the mandible at the lateralmost projections of the middle of the dentaries.

Mandibular constriction (DC)—the least breadth across the anterior dentary in the diastema of the first and second lower premolars.

Mandibular symphysis height (DC)—the least distance from the plane of the dorsal edge of the dentary at the diastema to the ventralmost point of the mental prominence.

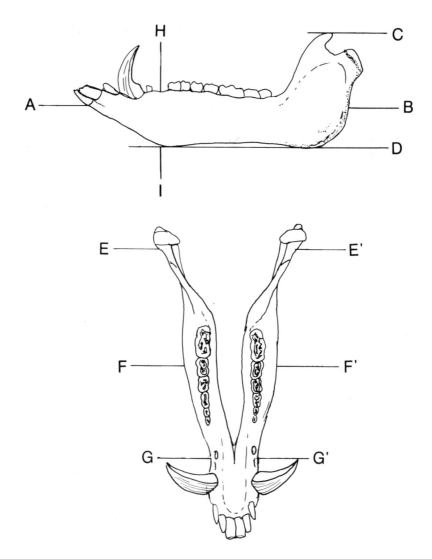

Figure 9. Linear mandibular measurements used in the skull analyses: mandibular length = A to B; mandibular height = C to D; posterior mandibular width = E to E'; midmandibular breadth = F to F'; mandibular constriction = G to G'; mandibular symphysis height = H to I.

Dental variables, as depicted in Figure 10, were:

Lower PM1–PM2 diastema length (DC)—the distance from the posterior margin of the alveolus of the first lower premolar to the anterior margin of the second lower premolar.

Upper mandibular tooth row length (DC)—the greatest length from the anterior edge of the alveolus of the first upper premolar to the posterior edge of the alveolus of the third upper molar.

Lower mandibular tooth row length (DC)—the greatest length from the anterior edge of the alveolus of the second lower premolar to the posterior edge of the alveolus of the third lower molar.

Upper canine alveolar length (DC)—the greatest anterior-posterior length of the upper canine alveolus.

Lower canine alveolar length (DC)—the greatest length of the alveolus of the lower canine.

Lower canine alveolar width (DC)—the greatest width of the alveolus of the lower canine.

Second molar length (DC)—the greatest length of the crown of the second molars, either upper or lower.

Second molar width (DC)—the greatest width of the crown across the posterior cusp row of the second molars, either upper or lower.

Third molar length (DC)—the greatest length of the crown of the third molars, either upper or lower.

Width of first cusp row of third molar (DC)—the greatest width across the first cusp row of the crown of the third molars, either upper or lower.

Width of second cusp row of third molar (DC)—the greatest width across the second cusp row of the crown of the third molars, either upper or lower.

The diagnostic characters of the four general morphotypes of *Sus scrofa* and the two morphotypes of feral hogs were determined using the stepwise discriminant function analysis (BMDP-7M) in the 1979 version of the BMDP Biomedical Computer Programs P-Series packaged programs (Dixon and Brown, 1979). A canonical variate analysis was then used to sharpen the discrimination between groups. This method gives the most weight to those variables that are least variable within and most variable between samples (Lowe and Gardiner, 1974). This analysis was part of the same programed package (BMDP-7M). Only the first two vectors of these analyses were plotted. The four general morphotypes in each age class and of each sex for age classes 4 and 5 were entered first as the *a priori* target groups.

To permit the morphotype determination of unknown specimens that may consist of only a cranium or mandible alone, these two components of the skull were analyzed separately. In contrast with Lawrence and Bossert (1967) and Elder and Hayden (1977), raw data were used as opposed to ratios in these analyses. This was because preliminary investigations indicated that size was important in classifying the four morphotypes of *Sus scrofa*.

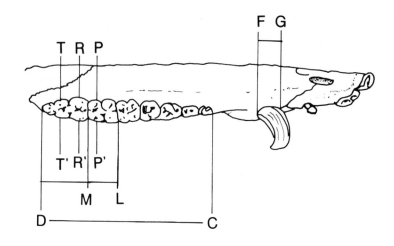

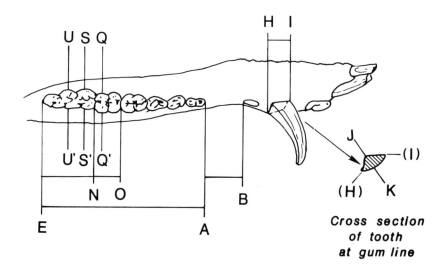

Cross section
of tooth
at gum line

Figure 10. Linear dental measurements used in the skull analyses: lower PM1-PM2 diastema length = A to B; upper mandibular tooth row length = C to D; lower mandibular tooth row length = A to E; upper canine alveolar length = F to G; lower canine alveolar length = H to I; lower canine alveolar width = J to K; upper second molar length = L to M; upper second molar width = P to P'; lower second molar length = N to O; lower second molar width = Q to Q'; upper third molar length = M to D; width of first cusp row of upper third molar = R to R'; width of second cusp row of third upper molar = T to T'; lower third molar length = O to E; width of first cusp row of lower third molar = S to S'; width of second cusp row of lower third molar = U to U'.

90

In addition, because the BMDP program has no method for handling missing variables, each analysis was first run using all the variables to determine those variables most useful for target group resolution. The same analysis was then run again using only those variables identified as aiding in group separation. This permitted many unknown specimens with missing variables to be compared with the four known target groups. Otherwise, these unknown cases would have to have been deleted from the analysis because of missing data.

Sexually dimorphic differences in the skulls from each morphotype were also determined. The skull data (cranial and mandibular) were analyzed both multivariately (BMDP-7M) and univariately (BMDP-3D) to determine if any statistically significant sexually dimorphic mensural cranial characters existed within each morphotype. Both analyses were run on the same data for the following reasons: The stepwise discriminant function analysis (BMDP-7M) will select only those variables necessary for separating the two groups. The univariate Student's *t* test (BMDP-3D), on the other hand, will determine on a variable-by-variable basis all the variables that are significantly different between the sexes. Therefore, to determine the full extent of the sexual dimorphism in this species, both analyses were performed. In addition, the univariate analysis was able to utilize specimens that had been eliminated from the multivariate analyses because of damage or insufficient data. For two reasons, these analyses were restricted to only the two oldest classes. First, many of the animals in the younger classes did not have a known sex. Second, if any such sexual dimorphisms did exist, they would be expected to be most evident in the oldest classes, since mature mammals normally exhibit sexually dimorphic characters more strongly than immature ones.

External Body Dimensions and Proportions. Seven external body measurements (Figure 11) and total body weight also were compared among the four target groups. These data from museum specimens were used when available. For specimens collected in the field, the linear measurements were taken with a steel tape to the nearest millimeter, and the weight was determined to the nearest kilogram.

These dimensions were taken as follows:

Total length—head, body, and tail length in a straight line with the animal on its back.

Head-body length—total length minus tail length; used for comparisons with data from European authors.

Tail length—from the base to the distal end of the last vertebra.

Hind foot length—from the proximal end of calcaneus to the distal end of longest hoof.

Ear length—from the notch to the farthest point of the distal edge.

Shoulder height—the straight-line distance from the dorsal midline of the shoulder to the base of the foot with the forelimb positioned in a normal upright posture.

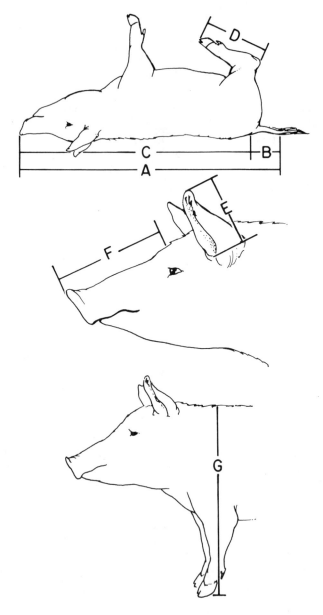

Figure 11. External body measurements used in characterizing *Sus scrofa*: total length = A; tail length = B; head-body length = C; hind foot length = D; ear length = E; snout length = F; shoulder height = G.

Snout length—from the distalmost dorsal point of the rhinarial pad to a point equidistant between the eyes.

Body weight—the total intact weight, including entrails. These data were segregated by sex and age class for comparison of the morphotypes. The data were also compared with published measurements from other populations. Average and maximum body weights for adults of each sex from specific populations were also determined from letters, questionnaires, and personal interviews with private individuals and federal and state personnel.

Coat Coloration Patterns. The coat coloration of the various age classes of each morphotype were determined through (a) specimens examined, (b) field observations, (c) recent literature, (d) letters and questionnaires sent to private individuals and federal and state personnel, and (e) personal interviews with private individuals and federal and state personnel.

General coat coloration patterns were classified into the following categories:

Wild/grizzled—most of the coat light brown to black with white or tan distal tips on the bristles, especially over the lateral portions of the head and dorsal portion of the rostrum; lighter undersides; ears and distal portions of the rostrum, limbs, and tail darker than the rest of the coat, usually dark brown or black with no light tips. This coloration pattern is typical of Eurasian wild boar.

All black—entire coat completely black.

All red/brown—entire coat completely light tan to dark brown.

All white—entire coat completely white.

Black with white shoulder belt—coat black with a white belt across the shoulders and down the forelimbs; also referred to as a Hampshire color pattern.

Red/brown with white shoulder belt—coat red/brown with a white belt across the shoulders and down the forelimbs.

Black-and-white spotted—coat with irregular black and white markings; includes black with white spots and white with black spots.

Red/brown-and-white spotted—coat with irregular red/brown and white markings; includes red/brown with white spots and white with red/brown spots.

Black-and-red/brown spotted—coat with irregular black and red/brown markings; includes black with red/brown spots and red/brown with black spots.

Tricolored spotted—coat with irregular black, red/brown, and white markings.

Black with white points—coat black with white on ears and distal portions of rostrum, limbs, and tail.

Red/brown with white points—coat red/brown with white on ears and distal portions of rostrum, limbs, and tail.

Juvenile striped pattern—coat light gray/tan to brown, with a brown to black middorsal stripe from the back of the head to the medial middorsum and three to four tan to brown irregular stripes with dark margins along the length of the body and the posterior half of the middorsal line; light tan to white undersides and limbs; distal rostrum brown to black; found only in immature individuals (Plate 3).

Other, miscellaneous—includes all other coat coloration patterns that cannot be classified in one of the above patterns; for example, combinations of the above patterns.

The coat coloration patterns defined above were chosen for ease in identification and may not necessarily reflect distinct genotypic series. The frequency of these coat coloration patterns were compared between morphotypes, and between sex and age classes within each morphotype. Variations or combinations of these coat color patterns were also described for each morphotype when found.

Hair Morphology. The type, color, and morphometric variation of each type of hair were determined for each morphotype. The two possible hair types recorded were (1) bristle-type guard hairs, and (2) underhair or underfur (Noback, 1951). Only specimens from the two older age classes were used in this analysis because of the lack of specimens in the younger classes. The length and mid-shaft diameter were determined for a minimum of five bristles removed from the specimen at skin level from the medial middorsal region of the back. Hair length was measured to the nearest 1.0 mm with a steel ruler and the diameter to the nearest 0.01 mm with a calibrated ocular micrometer on a compound microscope. The coloration was noted for each bristle, and the length of each hair color band was measured in millimeters. These bristles were also checked for split or fractured distal tips, the length of any split portion being measured in millimeters. The presence or absence of underfur was also noted, as was the color of the underfur if present.

Comparisons with Non-U.S. Feral Hogs

As a check of the consistency of the feral hog morphotype, specimens from known feral hog populations outside the United States were compared with known feral hogs from within this country. The majority of these specimens consisted of skulls only. These specimens were compared with the four target groups using the stepwise discriminant function and canonical variates analyses. Where available, data on external body dimensions and proportions, coat coloration patterns, and hair morphology were also compared. As before, numbers preceding the collection acronym indicate the number of specimens examined from that locality. Specimens from the main island of New Guinea were excluded from this comparison because of their uncertain taxonomic status.

The non-U.S. feral hog specimens were as follows:

ANDAMAN ISLANDS: Little Andaman, Bimilla Creek (1, USNM); no specific locality (4, MCZ). AUSTRALIA: Northern Territory, 245 km SE of Darwin, near Patonga Hunting Lodge (1, USNM). BELIZE: 84 km W of Belize City near Roaring Creek Park (1, USLBM). COSTA RICA: Corcovada National Park (2, DHJ). ECUADOR: Galápagos Islands, James Island (1, CAS); Indefatigable Island, near Academy Bay (3, MVZ); Charles Island, near Black Beach (1, MVZ); no specific locality (1, AMNH; 1, YPM). GARDENER ISLAND: no specific locality (1, USNM). INDONESIA: Komodo Island (1, AMNH). MARIANA ISLANDS: Guam (2, USNM); Rota Island (1, USNM). MEXICO: Sonora, mouth of the Colorado River (1, USNM); Campeche, Balchaeaj (1, KU); Chiapas, Mapaste-Triumpho, at an altitude of 2,000 m (1, UMMZ). NEW ZEALAND: Auckland County (1, USNM); no specific locality (1, AMNH). NICOBAR ISLANDS: Great Nicobar Island, Ganges Harbor (1, USNM). PAPUA NEW GUINEA: Admiralty Islands, Ponam Islet (2, USNM). SOLOMON ISLANDS: Guadalcanal Island, 0.8 km inland, Nalimbu River (1, MVZ). VIRGIN ISLANDS: St. John Island, Sugar Mill, Reef Bay (1, KU); 0.3 km E of Lameshur (1, KU). ZANZIBAR: Pemba Island, Ngezi Forest (1, MCZ).

Comparisons with Captive Wild Pigs

Specimens of captive wild boar, feral hogs, and wild boar × feral hog hybrids were compared with free-living specimens of the same three morphotypes and also domestic swine to determine the effects of a higher plane of nutrition and reduced activity on the wild pig morphotypes. Again, most specimens consisted of skulls only. This comparison also used the stepwise discriminant function and canonical variates analyses. The specimens of captive wild pigs were as follows:

Wild Boar: Philadelphia Zoological Gardens, Philadelphia, Pa. (1, ANSP); Pittsburgh Zoological Gardens, Pittsburgh, Pa. (1, CM); Bronx Zoological Park, New York (3, AMNH); National Zoological Park, Washington, D.C. (4, USNM).

Feral Hogs: Feral Swine Research Center, University of Georgia, Athens, Ga. (6, JJM).

Wild Boar × Feral Hog Hybrids: Rittimann Ranch, Boerne, Kendall County, Tex. (1, LR).

Identification of Unknown Wild Pigs

Specimens from U.S. wild pig populations with no known history were identified as either wild boar, feral hog, or wild boar × feral hog hybrid by comparison with characteristics of the four morphotypes, as determined from specimens from populations with known histories. This comparison primarily used the cranial analysis with the stepwise discriminant function and canonical variates analyses. These specimens were treated as unknowns and were identified through the calculated Mahalanobis D-square and posterior probability assigned by the discriminant function for each unknown case, as compared with the mean of each target group and through visual relationships of each unknown to the target groups in the canonical plots. Any available data on external body measurements and proportions, coat coloration patterns, and hair morphology from the unknown populations were also compared with those data from the four morphotypes. These comparisons and the results of the cranial analyses were combined to determine the identification of various unknown populations.

The specimens with unknown histories are listed below. Again, numbers preceding the collection acronyms indicate the number of specimens examined from that locality.

ALABAMA, Clark County: Widder Farm (2, JJM). CALIFORNIA, Ventura County: 3.2 km E of Ventura, Hall Canyon (1, CSULB); Mendocino County: near Covelo (1, WFBM); Santa Barbara County: 1.6 km N of Cachuma Lake (2, SDNHM), near Buellton (1, CPSU). FLORIDA, Osceola County: Lake Kissimmee (3, USNM); Highlands County: Kissimmee Prairie, Kicco (1, USNM); Volusia County: New Smyrna Beach (1, MCZ); Indian River County: Felsmere (1, MCZ); Dixie County: Old Town, Suwannee River (1, LMS); Putnam County: Palatka (1, AMNH); Gilchrist County: 11.5 km WSW of Newberry (1, FSM); Taylor County: 53 km N and 5 km W of Steinbetcher (1, FSM); Levy County: Cedar Key (1, FSM); Alachua County: 7.6 km WSW of Gainesville (1, FSM), near Gainesville (4, FSM); no specific locality (1, FMNH). GEORGIA, Baldwin County: Oconee River (14, JJM); McIntosh County: Blackbeard Island (1, USNM; 1, CM); Wheeler County: 3.2 km SE of Glennwood, W of the Oconee River (1, DLA); Thomas County: Thomasville, Chinquapin Plantation (1, AMNH); Columbia County: Martinez (1, JJM); Chatham County: BASF site on the Savannah River (2, SREL); no specific locality (1, FMNH). LOUISIANA, St. Martin Parish: 11.2 km S of St. Martinville (1, USLBM); Lafayette Parish: 8 km S of Lafayette (2, USLBM). MISSISSIPPI, Madison County: 4.8 km N of Madison (1, MMNS). NORTH CAROLINA, Mitchell County: Roan Mountain (1, USNM); Hyde County: Ocracoke Island (1, JJM). SOUTH CAROLINA, Beaufort County: Hilton Head Island (4, USNM); Allendale County: no specific locality (1, USNM). TEXAS, La Salle County: Webb Ranch (1, KS); San Patricio County: Welder Wildlife Refuge (22, WWR; 12, JJM); McMullen County: 32 km S of Tilden (1, TCWC); Hardin County: 6.4 km SW of Saratoga (1, UIMNH); Liberty County: 24 km SE of Cleveland (1, USLBM); Kenedy County: King Ranch, Norris Division (3, OSU); Chambers County: 2.4 km E of Wallesville (1,

UTEP); Travis County: 11.2 km N of Austin (1, UTAVP); no specific locality (1, UTAVP); Frio County: Pearsall (1, KS); Cameron County: Laguna Atascosa National Wildlife Refuge (1, MSB).

Results

Skull Morphology

Significant differences were demonstrated between the four morphotypes for the cranial and mandibular analyses for each sex or combined sex and age class except for the neonate/infant mandibles. No variables were selected by the discriminant function during this latter analysis, indicating that there were no statistical differences between the mandibles of neonates from the four target morphotypes. The percentage of cases correctly classified (PCC) exceeded 66% in all remaining analyses, with the majority of PCCs exceeding 80% (Table 4). Among the 13 canonical analyses, either two or three canonical vectors were significant. The first vector was most useful in all the analyses, accounting for 62%–85% of the total variation (Table 5). In general, with the exception of groups composed of three specimens or less, target group resolution was better in males than in females, and better in older than in younger age classes. Overall, mandibles were less distinct between the morphotypes than the crania and showed less between-group resolution. The cranial and mandibular data were combined during preliminary computer runs comparing the four morphotypes. In every stepwise selection of variables, only the cranial variables were selected to be entered in the function. The mandibular variables were thus less useful in classifying an unknown specimen into one of the four morphotypes.

Overall target group resolution was highest for the adult male crania (Table 4; Figure 12). This group also showed the largest number of variables selected for the discriminant function for group separation. According to the individual case posterior probabilities (PP), group overlap occurred in seven specimens for the first two canonical vectors. Of these, six involved the hybrid target group and its overlap with either the wild boar or feral hog target group. Five of these specimens overlapped with the feral hog target group. Only one hybrid specimen from Kendall County in Texas (PWP 1) had an overlap with the wild boar group (PP = 0.99). Overlap with representatives of parental types should be expected for such a hybrid group. However, 92% of the hybrids were discernible as such in the analysis. Of the four target groups in this analysis, domestic swine had the lowest group PCC (Table 4).

Table 4. Percentage of specimens correctly classified into the target groups they were assigned to originally. The sample size of each group is in parentheses.

Age Class[a]	Sex	Specimen	Percentage Correctly Classified				
			Wild Boar	Feral Hogs	Hybrids	Domestic Swine	Total
5	M	crania	96.4 (28)	93.2 (44)	92.0 (33)	90.0 (14)	93.5
5	M	mandibles	88.5 (26)	88.9 (45)	81.5 (27)	77.8 (9)	86.0
5	F	crania	93.9 (15)	87.9 (33)	92.3 (26)	94.4 (18)	91.3
5	F	mandibles	92.9 (14)	86.1 (36)	93.9 (23)	87.5 (16)	84.3
4	M	crania	100.0 (2)	88.9 (23)	95.2 (21)	80.0 (10)	90.5
4	M	mandibles	100.0 (2)	90.0 (23)	94.4 (18)	66.7 (9)	87.2
4	F	crania	100.0 (2)	80.0 (15)	92.3 (25)	77.8 (9)	86.5
4	F	mandibles	100.0 (2)	75.0 (15)	95.0 (20)	100.0 (8)	90.5
3	M&F	crania	100.0 (2)	87.3 (61)	85.7 (41)	86.7 (30)	86.9
3	M&F	mandibles	100.0 (2)	83.9 (56)	89.2 (37)	80.0 (25)	85.0
2	M&F	crania	77.8 (9)	86.7 (90)	94.5 (9)	89.1 (55)	87.7
2	M&F	mandibles	77.8 (9)	80.0 (90)	85.7 (14)	83.7 (49)	81.5
1	M&F	crania	100.0 (1)	89.5 (20)	84.0 (3)	80.0 (10)	87.5

a Age classes are defined in Table 2.

Table 5. Percentage of total variation between groups accounted for by the first, second, and third vectors of the canonical analyses of the *Sus scrofa* target groups.

Age Class[a]	Sex	Specimen	Percentage of Total Variation Accounted for by the:		
			1st vector	2d vector	3d vector
5	M	crania	74	17	9
5	M	mandibles	69	29	2
5	F	crania	78	21	1
5	F	mandibles	57	43	0
4	M	crania	85	15	0
4	M	mandibles	76	24	0
4	F	crania	82	18	0
4	F	mandibles	81	18	1
3	M&F	crania	78	18	4
3	M&F	mandibles	71	29	0
2	M&F	crania	68	28	4
2	M&F	mandibles	62	35	3
1	M&F	crania	80	17	3

a   Age classes are defined in Table 2.

This was primarily due to one specimen (USNM 35118) that had a PP of 1 as belonging to the feral hog target group. This specimen was collected in Virginia in April 1891 by W. Palmer. Although labeled as a domestic hog, it may have been reared under the free-range conditions commonly used at that time. The low PCC of the domestic swine group was also partly due to its higher variability. Overall length followed by measurements of greatest width were the most important cranial variables in separating the groups, followed primarily by rostral measurements (Table 6).

Adult male mandibles (Figure 12) had a lower total PCC than the crania (Table 4). Overlap was seen in 15 specimens. The graphic resolution on the canonical plot for mandibles was poor compared with that for the adult male crania. The hybrid target group overlapped all three of the other groups,

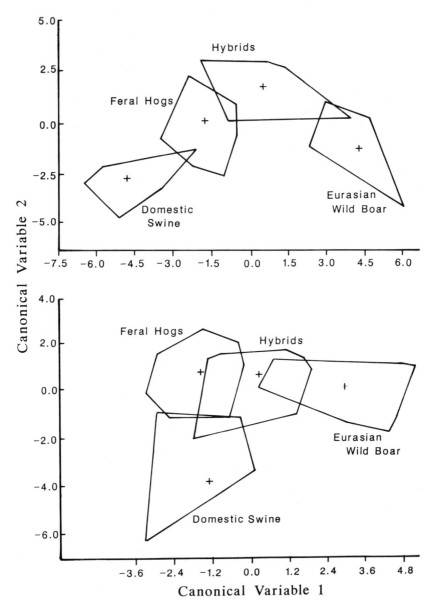

Figure 12. First two canonical variables for adult male crania (above) and mandibles (below) comparing specimens from known populations of the four morphotypes. Target group boundaries were drawn by connecting all outermost points to enclose the maximum convex polygon for all points plotted for that group. The plus signs (+) indicate the locations of target group means.

making the identification of unknowns difficult or impossible. The two hybrid specimens (MVZ 116962 and CPS 1082) that overlapped with the domestic swine target group were from Monterey and San Luis Obispo counties in California, respectively. This was an unexpected occurrence because domestic swine are not a parental stock for this group. Again, the domestic swine sample had the lowest group PCC. As with the crania, overall length was the most useful variable for group resolution. This was followed by measurements of height and then width (Table 6).

The adult female cranial analysis (Figure 13) had a total PCC about 2% lower than that of the adult male crania (Table 4). This was caused by an overlap of eight specimens. There was an increased overlap of the feral target group with the domestic swine and hybrid target groups, as compared with the case of the adult male crania. The extreme overlap of the wild boar and hybrid groups was caused by USNM 241584, a female wild boar from Germany, and a hybrid sow (CPS 2050) from Camp Roberts, San Luis Obispo County, California. The PPs of these were 0.94 for the female wild boar as a hybrid animal, and 0.98 for the hybrid female as a wild boar. The three variables used in the linear function for maximum group separation were a subset of the variables selected for the adult male crania, and these variables were also selected in the same order of importance as for males.

As in the males, the adult female mandibles (Figure 13) were less separable than were the crania. Again, the female mandibles had an overall PCC about 2% lower than the male mandibles, and the feral sample had the lowest group PCC (Table 4). The variables used for separation were also a subset of those of the male analysis, and again, in the same order of importance. Again, the greatest overlap between the wild boar and hybrid target groups was hybrid sow CPS 2050, with a PP of 0.98 as a wild boar. One domestic sow (MCZ 52303) had a PP of 0.52 as a wild boar, making the identification of this particular specimen about as accurate as simple guessing.

Compared with the adult male crania, the subadult male crania (Figure 14) showed about a 3% decrease in the total PCC for all groups (Table 4). This figure was probably a little high because of the small size of the wild boar sample, which had a group PCC of 100%. The domestic sample was highly variable and had the lowest group PCC. One domestic specimen even approached the hybrid target group in the canonical plot. Because of the overlap extremes in this analysis, many unknown specimens that fell into the bounds of the feral hog target group in the canonical analysis would not be discernible as feral, domestic, or hybrid. Rostral length and condylobasal length were the first and second most important variables for target group resolution (Table 6).

Table 6. Measurements, in order of importance, selected by the stepwise discriminant function analysis for separation of the wild boar, feral hog, wild boar × feral hog hybrid, and domestic swine target groups, and the coefficients and constants from the first and second canonical variables of each analysis.

Age Class[a]	Sex	Specimen	Measurement	First Canonical Variable Coefficient	Second Canonical Variable Coefficient
5	M	crania	occipitonasal length	0.10438	-0.07436
			zygomatic width	-0.03494	0.04571
			rostral length	-0.04312	0.20678
			palatal length	-0.02899	-0.14397
			premaxillary rostral width	-0.01726	-0.12071
			palatal premaxillary constriction	0.00719	0.04681
			supraoccipital constriction	-0.03911	-0.02446
			Constants =	-10.63739	9.55382
5	M	mandibles	mandibular length	0.10438	-0.07436
			mandibular symphysis height	-0.11827	-0.11982
			mandibular height	-0.03082	-0.03437
			mandibular constriction	-0.03472	-0.11817
			midmandibular breadth	-0.01308	0.07634
			Constants =	-10.86554	7.50072

Table 6—*Continued*

5	F	crania	occipitonasal length		0.05769	0.02618
			zygomatic width		-0.06474	0.03918
			supraoccipital constriction		-0.04942	0.01721
			Constants	=	-4.60695	-14.78038
5	F	mandibles	mandibular length		0.09051	0.06570
			mandibular symphysis height		-0.16005	-0.21228
			mandibular constriction		0.04915	-0.16835
			Constants	=	-16.48405	1.11264
4	M	crania	rostral length		0.14969	-0.01074
			condylobasal length		-0.08671	0.06230
			Constants	=	-2.27952	-16.57923
4	M	mandibles	mandibular constriction		-0.29793	0.05551
			mandibular length		0.06990	0.04168
			Constants	=	-3.37357	-12.18502
4	F	crania	zygomatic width		-0.12774	0.04538
			occipitonasal length		0.03758	0.04844
			Constants	=	6.86774	-19.44923

Continued on next page

Table 6—Continued

Age Class[a]	Sex	Specimen	Measurement	First Canonical Variable Coefficient	Second Canonical Variable Coefficient
4	F	mandibles	mandibular constriction	-0.36027	-0.08330
			mandibular length	0.06667	0.12012
			mandibular symphysis height	-0.22453	-0.13338
			Constants =	9.27309	-17.07485
3	M&F	crania	zygomatic width	-0.08735	0.08186
			lacrimal notch to nasal length	0.08750	0.02393
			supraoccipital constriction	-0.04422	-0.07941
			Constants =	-0.18452	-9.80509
3	M&F	mandibles	mandibular constriction	-0.12862	0.16443
			mandibular length	0.11698	0.04913
			mandibular height	-0.09166	-0.04659
			mandibular symphysis height	-0.21610	-0.03906
			Constants =	-1.35848	-9.81389
2	M&F	crania	palatal premaxillary constriction	-0.11022	0.13138
			angle of the occipital wall	-0.02453	0.08424
			occipitonasal length	0.07712	0.06255
			zygomatic width	-0.08323	-0.04958
			nasal length	-0.03248	-0.06315
			greatest width of occipital ridge	-0.03604	0.00969
			Constants =	4.35163	-11.89546

Table 6—Continued

2	M&F	mandibles	mandibular constriction	-0.17389	0.22969
			mandibular length	0.08153	0.10192
			mandibular symphysis height	-0.17626	-0.17312
			posterior mandibular width	-0.01520	-0.05692
			mandibular height	-0.03744	-0.04647
			Constants =	3.04780	-9.39529
1	M&F	crania	least interorbital breadth	-0.30668	-0.18038
			braincase breadth	0.62131	0.13305
			parietal constriction	-0.32868	0.39081
			Constants =	-9.85633	-10.74188

a Age classes are defined in Table 2.

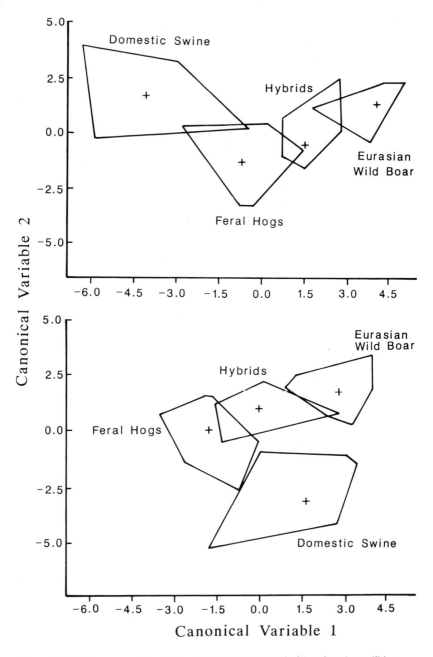

Figure 13. First two canonical variables for adult female crania (above) and mandibles (below) comparing specimens from known populations of the four morphotypes. Target group boundaries and means are drawn as indicated in Figure 12.

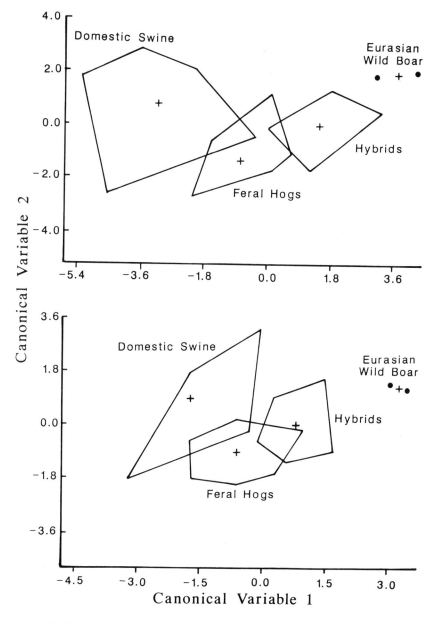

Figure 14. First two canonical variables for subadult male crania (above) and mandibles (below) comparing specimens from known populations of the four morphotypes. Target group boundaries and means are drawn as indicated in Figure 12.

The same generalizations for the subadult male crania also applied to their mandibles (Figure 14). The mandibles again had a lower total PCC compared with the crania for this sex and age class (Table 4). Although the degree of overlap between the feral hog and hybrid target groups appeared to be the same, the hybrid mandibles had a slightly higher group PCC than the crania (Table 4). The mandibles in the domestic sample were again highly variable and had the lowest group PCC of any target group in this analysis (Table 4). The mandibular constriction was the most useful separation variable, followed by mandibular length (Table 6).

Target group separation was better for the subadult female crania (Figure 15) than for the subadult male crania, even though the total PCC was lower than that of any of the other cranial analyses (Table 4). Again the wild boar target group sample was too small to be certain of its adequacy in representing this morphotype. The domestic sample was again highly dispersed on the canonical plot, with a group PCC of 77.8%. In contrast with the subadult male sample, zygomatic width was more important in group separation than any length measurement. This was followed by occipitonasal length (Table 6).

The subadult female mandibles (Figure 15) had the highest total PCC of any of the mandibular analyses (Table 4). This was due to the complete separation of the domestic swine and wild boar from the other two target groups. The feral group had the greatest dispersal on the canonical plot, with a group PCC of 75.0%. The important variables for group resolution were the same as for the subadult male mandibles, with the addition of the mandibular symphysis height (Table 6).

The yearling crania (Figure 16) had the second lowest total PCC of any of the cranial analyses (Table 4). As in the subadult samples, the wild boar group was very small, and it was the only target group to be completely separated from the other groups in the canonical plot. Zygomatic width and lacrimal notch to nasal length were the first and second most important variables for group resolution (Table 6).

The same generalizations hold for the mandibular analysis of the yearlings. The domestic swine sample was the most dispersed and had the lowest group PCC (Figure 16; Table 4). Overlap between the domestic and hybrid target groups occurred within the bounds of the feral target group, making any identification of unknowns that fell into this area of the plot very difficult. The mandibular constriction and mandibular length were the first and second most useful separation variables (Table 6).

The juvenile cranial and mandibular analyses (Figure 17) had the largest number of sample sizes of any of the analyses. Target group resolution of the crania was good, with overlaps occurring only between the expected groups; that is, domestic and feral, feral and hybrid, and hybrid and wild boar.

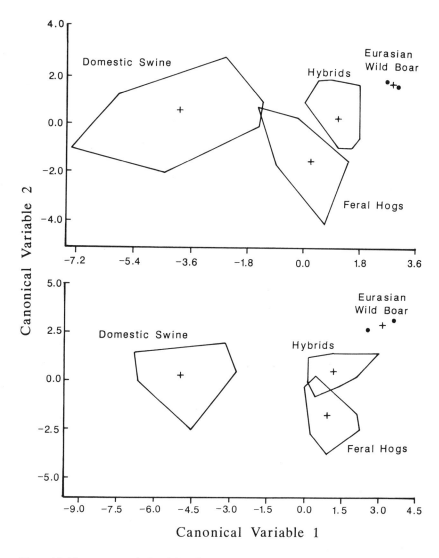

Figure 15. First two canonical variables for subadult female crania (above) and mandibles (below) comparing specimens from known populations of the four morphotypes. Target group boundaries and means are drawn as indicated in Figure 12.

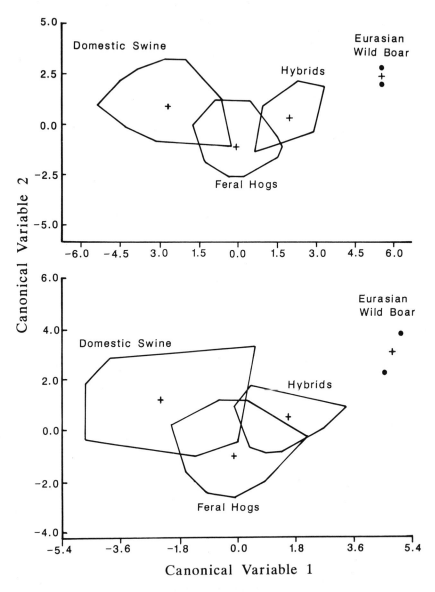

Figure 16. First two canonical variables for yearling (sexes combined) crania (above) and mandibles (below) comparing specimens from known populations of the four morphotypes. Target group boundaries and means are drawn as indicated in Figure 12.

110

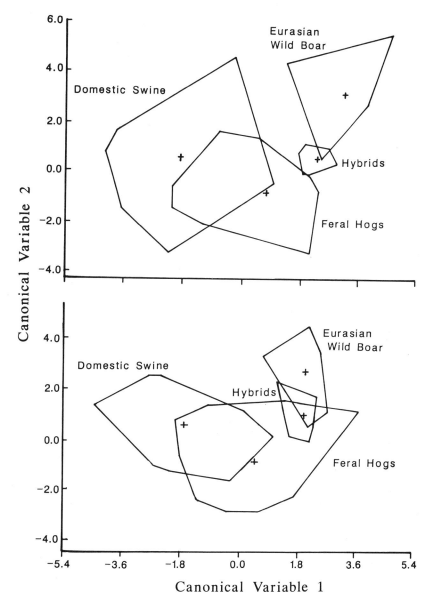

Figure 17. First two canonical variables for juvenile (sexes combined) crania (above) and mandibles (below) comparing specimens from known populations of the four morphotypes. Target group boundaries and means are drawn as indicated in Figure 12.

111

The first and second most useful discriminating variables selected in this analysis were the palatal premaxillary constriction and the angle of the occipital wall, respectively (Table 6). The wild boar sample had the lowest group PCC (Table 4).

The neonate/infant cranial analysis had the smallest total and individual target group sample sizes (Table 4), making the PCCs unrealistically larger than they would have been if larger samples had been used. It is not clear, however, whether the lack of separation was real or merely a result of the small sample sizes (Figure 18). Unlike the other cranial analyses, no rostral or overall length measurements were selected by the discriminant function analysis. Only cranial widths associated with the braincase or splanchno-cranium were useful in group resolution (Table 6). As noted earlier, no differences were found in the morphology of the mandibles between target groups in this age class.

As in the comparative analyses of the four general morphotypes, resolution of the long-term and short-term feral hog target groups was best in the older age classes and better for crania than for mandibles. No variables differed

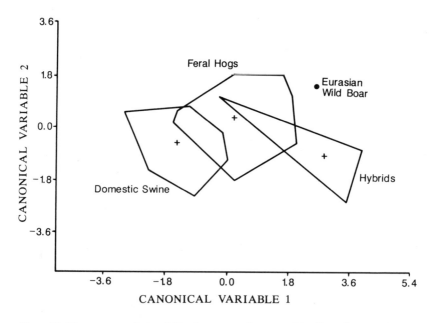

Figure 18. First two canonical variables for neonate (sexes combined) crania comparing specimens from known populations of the four morphotypes. Target group boundaries and means are drawn as indicated in Figure 12.

Table 7. Percentage of specimens correctly classified into the short-term versus long-term feral target groups to which they had been assigned on the basis of their known population histories.

			Percentage Correctly Classified		
Age Class[a]	Sex	Specimen	Short-term Feral Hogs	Long-term Feral Hogs	Total
5	M	crania	100.0 (19)	100.0 (26)	100.0
5	M	mandibles	76.5 (17)	76.7 (30)	76.6
5	F	crania	100.0 (17)	100.0 (17)	100.0
5	F	mandibles	100.0 (18)	94.7 (19)	97.3
4	M	crania	100.0 (12)	100.0 (11)	100.0
4	M	mandibles	60.0 (12)	80.0 (11)	70.0
4	F	crania	83.3 (6)	100.0 (9)	93.3
4	F	mandibles	75.0 (6)	88.9 (9)	84.6
3	M&F	crania	95.5 (22)	97.4 (39)	96.7
3	M&F	mandibles	85.0 (20)	86.1 (36)	85.7
2	M&F	crania	90.6 (53)	87.5 (40)	89.2
2	M&F	mandibles	89.8 (49)	85.7 (42)	87.9

a Age classes are defined in Table 2.

statistically between the two types of feral hogs in the cranial and mandibular analyses of the neonate/infant age class. The differences between the sexes were also not as clear as they were in separating the four general morphotypes. In the adult age class, for example, the female analyses resulted in better resolution of the two groups. In the subadults, however, males had a higher PCC in the cranial analysis, and females had a higher PCC in the mandibular analysis (Table 7).

The adult male crania had complete separation of the long-term and short-term target groups with a PCC of 100% (Table 7; Figure 19). The short-term feral specimens averaged larger than the long-term specimens in all cranial parameters except angle of the occipital wall. (This latter variable will be discussed later.) The variables most useful for group resolution were zygomatic width, zygomatic arch height, and condylobasal length (Table 8). The remaining variables used were three rostral characters and the angle of the occipital wall.

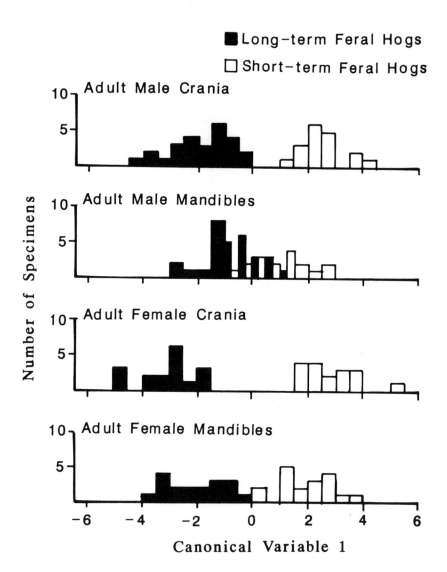

Figure 19. Canonical histograms of the cranial and mandibular analyses of short-term versus long-term adult male and female feral hogs.

114

Table 8. Measurements, in order of importance, selected by a stepwise discriminant function analysis for separation of the short-term versus long-term feral hog target groups, and the coefficients and constants from the first canonical variables of each analysis.

Age Class[a]	Sex	Specimen	Measurement	Canonical Variable Coefficient
5	M	crania	condylobasal length	0.13817
			nasal length	0.11194
			palatal length	-0.28996
			zygomatic width	0.15763
			maxillary rostral width	-0.09954
			zygomatic arch height	-0.23760
			angle of the occipital wall	-0.09974
			Constant =	-2.74628
5	M	mandibles	mandibular height	0.05558
			posterior mandibular width	0.05769
			Constant =	-13.75682
5	F	crania	nasal length	0.16103
			palatal length	-0.26448
			zygomatic width	0.26663
			postorbital process width	-0.22408
			least interorbital breadth	0.35282
			orbital width	0.28340
			zygomatic arch height	-0.17082
			Constant =	-18.96278
5	F	mandibles	mandibular length	-0.05335
			posterior mandibular width	0.19799
			Constant =	-12.80189
4	M	crania	parietal constriction	-0.33003
			rostral length	0.20501
			orbital width	2.74993
			Constant =	-54.24396

Continued on next page

Table 8—*Continued*

Age Class[a]	Sex	Specimen	Measurement		Canonical Variable Coefficient
4	M	mandibles	mandibular constriction		0.38859
				Constant =	-15.86602
4	F	crania	premaxillary rostral width		0.36859
			angle of the occipital wall		-0.12313
				Constant =	-2.38119
4	F	mandibles	posterior mandibular width		0.12378
				Constant =	-14.10087
3	M&F	crania	palatal length		-0.04359
			zygomatic width		0.24033
			zygomatic arch height		-0.24391
				Constant =	-15.35280
3	M&F	mandibles	mandibular height		0.12319
				Constant =	-10.40556
2	M&F	crania	braincase breadth		0.13022
			rostral length		-0.12639
			palatal premaxillary constriction		0.27675
			orbital width		0.53672
				Constant =	-18.70006
2	M&F	mandibles	mandibular length		-0.10611
			posterior mandibular width		0.08680
			mandibular symphysis height		0.28594
				Constant =	-0.53804

a Age classes are defined in Table 2.

In contrast to the crania, the canonical histograms for the adult male mandibles overlapped between the two types of feral hogs (Figure 19). The individual group and total PCCs were the lowest of any of the adult analyses (Table 7). Only two variables were used for group differentiation (Table 8), and overlap occurred in 11 specimens. Short-term feral hogs that were classified as long-term animals were from St. Catherines Island, Georgia (3), and Dye Creek Ranch, California (1). Long-term feral hogs that were classified as short-term were from Ossabaw Island, Georgia (2), Kauai, Hawaii (3), and Santa Rosa Island, California (2).

Adult female crania had slightly better separation in the canonical histogram than did adult male crania (Figure 19). Again, the short-term feral hogs were larger in all measurements except angle of the occipital wall. Zygomatic width was again the most useful variable (Table 8), followed by two rostral variables and then by four variables associated with the orbital/zygomatic region of the cranium.

In contrast to the adult male mandibles, adult female mandibles had complete separation, with PCCs of 100% (Figure 19; Table 7). Again, only two variables were needed for group differentiation (Table 8).

Subadult male crania showed complete separation between the two feral hog morphotypes (Figure 20). Short-term feral hogs were larger in all variables, including angle of the occipital wall. Orbital width was the most useful variable for group separation (Table 8).

The subadult male mandibles had an overlap of three specimens with an overall PCC of 70.0% (Figure 20; Table 7). The overlapping specimens were from St. Catherines Island, Georgia (1), Dye Creek Ranch, California (1), and Oahu, Hawaii (1). The mandibular constriction was selected as the only variable useful for group discrimination (Table 8).

The subadult female crania (Figure 20) had an overlap of one specimen, again a short-term feral hog from St. Catherines Island, Georgia. The long-term sample had a PCC of 100% (Table 7). Premaxillary rostral width and angle of the occipital wall were the first and second most useful variables for group discrimination (Table 8). The short-term sample was again larger in all variables except occipitobasal length, postorbital process width, and angle of the occipital wall.

As in the adult sample, the subadult female mandibles (Figure 20) had a higher PCC than the corresponding male mandibles (Table 7). Posterior mandibular width was the only variable entered into the discriminant function (Table 8). One specimen from St. Catherines Island, Georgia, and another from Hawaii were incorrectly classified.

In the yearling cranial analysis (Figure 21), only two specimens were

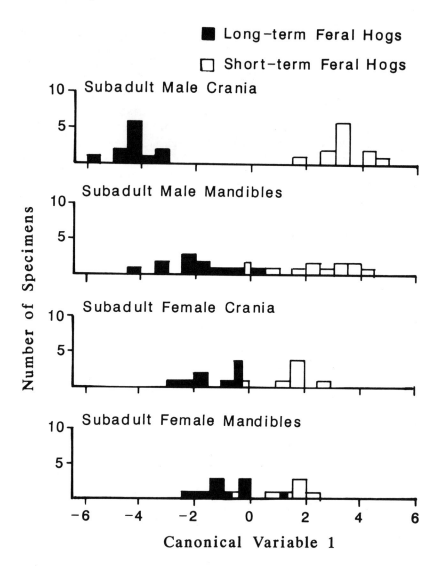

Figure 20. Canonical histograms of the cranial and mandibular analyses of short-term versus long-term subadult male and female feral hogs.

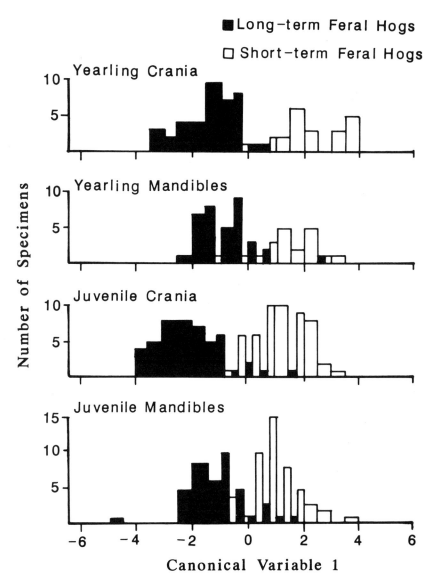

Figure 21. Canonical histograms of the cranial and mandibular analyses of short-term versus long-term yearling and juvenile feral hogs. Sexes were combined in each sample.

incorrectly classified according to the PPs. The overlapping specimens were from St. Catherines Island, Georgia, and Santa Cruz Island, California. The most useful variable was zygomatic width (Table 8). The short-term feral hogs were again larger in all variables except angle of the occipital wall.

The yearling mandibles had a higher PCC than the mandibles of the subadult males and a slightly higher PCC than the subadult females, although overlap was seen in eight specimens (Figure 21; Table 7). The higher PCC, despite the increase in overlap, was a result of an increase in sample size. Overlapping specimens were from St. Catherines Island, Georgia (2), Dye Creek Ranch, California (3), Ossabaw Island, Georgia (1), and Santa Cruz Island, California (2). Mandibular height was the only variable useful for group resolution (Table 8).

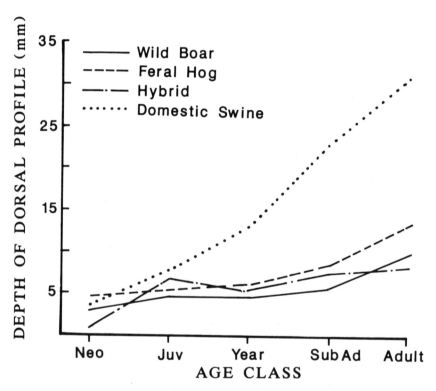

Figure 22. Comparison of the means of the dorsal profile depths (in mm) for wild boar, feral hogs, wild boar x feral hog hybrids, and domestic swine, separated by age class. Sexes were combined in each age class. Age classes are abbreviated as follows: Neo = neonate; Juv = juvenile; Year = yearling; SubAd = subadult.

The PCC of the juvenile feral hog crania (Figure 21) was less than that of the other cranial analyses (Table 7). The overlap also increased to 10 specimens. Braincase breadth was the first variable selected for the discriminant function. (This was the only analysis in which that variable was selected.) This was followed by rostral length and width variables (Table 8). The short-term feral hogs were again larger in all cranial measurements except angle of the occipital wall.

The juvenile mandibles (Figure 21) actually had a higher overall PCC than the mandibles of the two preceding age classes (Table 7). Eleven specimens were incorrectly classified. Three variables, mandibular length being first, were chosen for group separation (Table 8).

The depth of the dorsal profile (Figure 8; Plate 4) is a cranial character often used to compare wild and domestic swine (Epstein, 1971; Bokonyi, 1974); however, this variable was not selected by any of the cranial discriminant functions for group differentiation. Figure 22 shows the variation in this character by graphing of the means of each morphotype. These data were segregated by age class, but not by sex. In the three oldest age classes, domestic swine showed a significantly deeper dorsal profile depth ($P < 0.05$) than wild-living swine. This difference increased with age. There were no significant differences among the three wild-living morphotypes for any age class. Feral hogs, however, did have a higher average than either wild boar or hybrids. The maximum depths for each morphotype in the adult age class with the sexes combined were: wild boar, 19 mm; feral hogs, 32 mm; hybrids, 18 mm; and domestic swine, 50 mm. No significant differences were found by comparing the dorsal profile depth in short-term versus long-term feral hogs (Figure 23). In all but the subadult age class, however, short-term feral hogs had deeper profiles. The difference between these two morphotypes was again greatest in the older age classes. The maximum depths for each type of feral hog in the adult age class, with the sexes combined, were: short-term feral hog, 32 mm; long-term feral hog, 24 mm.

The angle of the occipital wall showed a gradient from wild boar to domestic swine. Wild boar tended to have occipitals that were vertical or posteriorly inclined. This was especially true for the three oldest age classes. The three remaining morphotypes had occipitals that were anteriorly inclined (Figure 24). While wild boar ranged as low as 85° in the older age classes, only one specimen in the other morphotypes had an angle that was greater than 90°; this was a juvenile domestic sow (JJM 76). The slope of the occipital wall was more anteriorly directed in short-term feral hogs than in long-term animals (Figure 25).

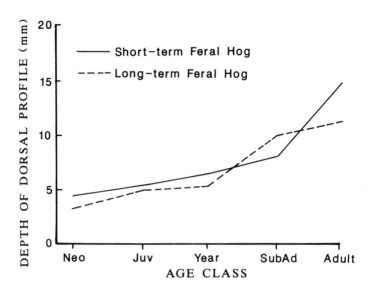

Figure 23. Comparison of the means of the dorsal profile depths (in mm) for short-term and long-term feral hogs, separated by age class. Sexes were combined in each age class. Age classes are abbreviated as in Figure 22.

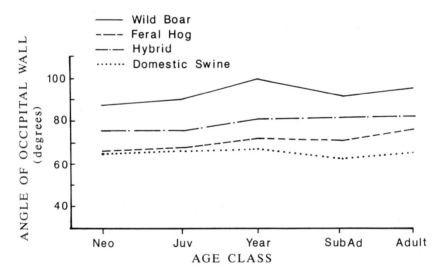

Figure 24. Comparison of the means of the angle of the occipital wall (in degrees) for wild boar, feral hogs, wild boar x feral hog hybrids, and domestic swine, separated by age class. Sexes were combined in each age class. Age classes are abbreviated as in Figure 22.

122

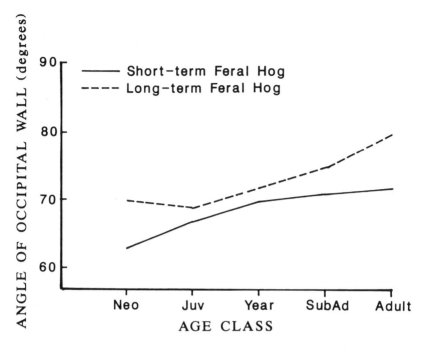

Figure 25. Comparison of the means of the angle of the occipital wall (in degrees) for short-term versus long-term feral hogs, separated by age class. Sexes were combined in each age class. Age classes are abbreviated as in Figure 22.

In terms of general external appearance, the skull differences among the four morphotypes of swine were best shown in the adult male class (Figure 26). This was also demonstrated in the multivariate morphometric analyses just discussed. In overall size, for example, condylobasal length, wild boar are the largest of the four morphotypes. The largest proportion of these lengths are made up by the rostrum (62%–74%). Hybrid crania resemble those of wild boar in proportions, but they are smaller and the occipital wall is sloped more anteriorly. The domestic swine skull is almost as large as that of the wild boar, but it has a proportionally shorter rostrum. The domestic skull is on the average higher and broader than those of any of the wild-living forms. Feral hogs generally have the smallest skulls of the four morphotypes. The average length of the rostrum of a feral hog lies between that of a domestic animal and a wild boar or hybrid.

In addition to the mensural differences among the four morphotypes presented above, the general appearance of the cranial bones of domestic swine differed from those of the three target groups of wild-living *Sus scrofa*.

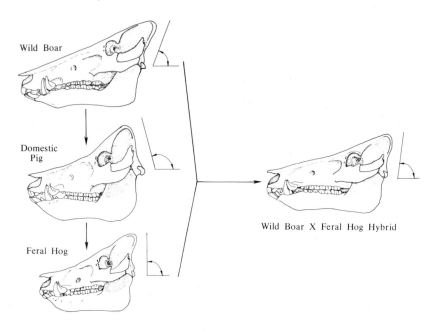

Figure 26. General appearances of the skulls, showing differences among the four morphotypes of adult male *Sus scrofa*. Angled lines behind each skull show the slope of the occipital wall of that specimen. The measured angle of the occipital wall, as used in this study and presented in Figure 24, was equal to 180° minus the angle subtended by the double-headed arrows in this figure.

Bones in the domestic cranium had a convex or inflated appearance between sutures compared with the other morphotypes. In addition, the external surfaces of the compact layer of these bones were more porous and the individual bones appeared more robust in the domestic animals as compared with the three wild-living forms. This latter difference was reflected by the increased skull width measurements of the domestic specimens.

Significant differences were found between the sexes within the wild boar, feral hog, and wild boar × feral hog hybrid morphotypes in the stepwise discriminant function analyses of the skulls comparing the sexes in the two older age classes. More than 80% of the specimens used in these analyses were correctly classified by sex (Table 9; Figure 27). In all measurements selected as useful for separation of the sexes, males were larger than females. These measurements included occipitonasal length, nasal length, braincase breadth, parietal constriction, least interorbital breadth, premaxillary rostral width, palatal premaxillary constriction, angle of the occipital wall, posterior mandibular width, midmandibular breadth, mandibular constriction, and

Table 9. Percentage of correct classifications of male versus female samples of wild boar, feral hogs, wild boar × feral hog hybrids, and domestic swine from stepwise discriminant function analyses using skull measurements comparing the sexes in the adult and subadult age classes.

		Percentage Correctly Classified			
Age Class[a]	Sex	Wild Boar	Feral Hogs	Hybrids	Domestic Swine
5	M	84.0	86.5	89.5	[b]
5	F	85.7	87.1	100.0	[b]
4	M	[c]	88.9	83.3	[b]
4	F	[c]	91.7	88.9	[b]

a Age classes are defined in Table 2.
b No significant differences between the sexes.
c Sample size too small to analyze.

mandibular symphysis height. The subadult age class of the wild boar sample was too small to compare the sexes statistically, but the three male specimens did average and range larger than the two females in all measurements. No significant differences were found between the sexes in the domestic swine sample; however, the males were larger than the females in all but eight of the cranial measurements analyzed.

Univariate analyses of the same data using the Student's *t*-test gave similar results; that is, in wild-living *Sus scrofa*, males are significantly larger than females in most cranial measurements. These measurements included condylobasal length, occipitonasal length, nasal length, length from lacrimal notch to anterior nasal, palatal length, zygomatic width, skull height, rostral length, palatal width between upper PM4s, mandibular length, mandibular constriction, mandibular symphysis height, and lower PM1–PM2 diastema length. In all but the subadult feral analysis, most of the measurements were significantly larger in the males at the 0.01 level of significance (Table 10). Wild-living female *Sus scrofa* were significantly larger than males in only two measurements: adult feral females had a larger upper M2 width, and adult hybrid females had a larger parietal constriction. Both differences were significant at the 0.05 level. In the analysis of the domestic samples, significant sexual differences were found only in the parietal constriction and braincase breadth, with females being larger than males in both measurements.

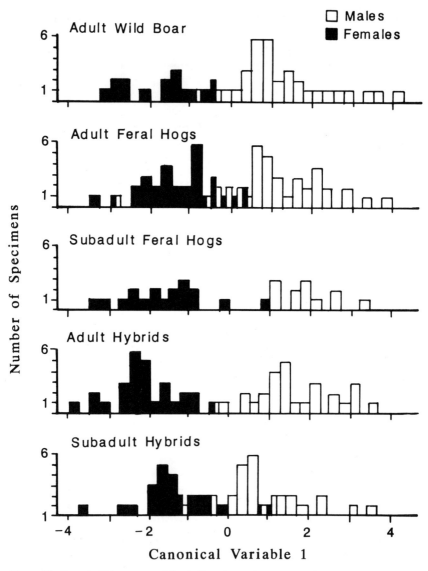

Figure 27. Canonical histograms of the skull analyses of male versus female *Sus scrofa*.

Table 10. Number of cranial measurements significantly larger (at $P \leq 0.01$ and 0.05 levels of significance) in male versus female *Sus scrofa* as determined using Student's *t*-test.

Age Class[a]	Sex	P	Number of Measurements Significantly Larger in Males than in Females			
			Wild Boar	Feral Hogs	Hybrids	Domestic Swine
5	M	0.01	16	21	24	0
		0.05	2	7	4	0
5	F	0.01	0	0	0	1
		0.05	0	1	1	1
4	M	0.01	b	9	25	0
		0.05	b	10	6	0
4	F	0.01	b	0	0	0
		0.05	b	0	0	0

a Age classes are defined in Table 2.
b Sample size too small to analyze.

External Body Dimensions and Proportions

In general, little data were available for external body measurements of animals from the four morphotypes. Measurements and weights were available for 214 specimens from known populations (Table 11). No intergroup statistical comparisons could be performed because of the small sample sizes. However, domestic swine had the largest overall measurements and weights, except for snout length, of any group. This was especially evident in the three oldest age classes.

In total length, the domestic swine were the largest, followed by either the wild boar or hybrid groups (Table 11). Although the feral hogs tended to have the lowest averages for total length and head-body length measurements, they also had the highest maxima for the same variables. The ranges determined for these measurements showed feral hogs were also the most variable of the wild pigs in these two body dimensions. Wild boar and hybrids had proportionally much shorter bodies and proportionally longer heads than either domestic swine or feral hogs. In general, the head-body length is probably a better variable to examine for differences among the morphotypes for the reasons noted below concerning variability in tail length.

Table 11. External body measurements (mm) and weights (kg) of wild boar (W), feral hogs (F), wild boar × feral hog hybrids (H), and domestic swine (D), separated by age class and sex.

Measurement	Age Class[a]	Sex	Type	n	Mean	Range	SD
Total Length	5	M	W	5	1,536.0	1,295–1,715	179.8
			F	4	1,428.5	1,077–1,905	—
			H	2	1,611.5	1,610–1,613	—
	5	F	W	9	1,575.3	1,490–1,695	59.9
			F	3	1,472.0	1,253–1,702	—
			H	2	1,537.0	1,537	—
			D	2	2,197.0	1,994–2,400	—
	4	M	W	1	1,600	1,600	—
			F	1	1,410	1,410	—
			H	2	1,500.0	1,500	—
	4	F	F	1	1,573	1,573	—
			D	2	1,817.5	1,740–1,895	—
	3	M&F	F	18	999.1	821–1,495	169.5
			H	1	1,607	1,607	—
			D	1	1,810	1,810	—
	2	M&F	W	2	885.5	843–928	—
			F	6	847.3	660–1,210	189.4
			H	3	983.0	965–995	—
			D	6	1,360.9	930–1,601	254.2
	1	M&F	W	1	436	436	—
			F	4	404.9	330–538	—
			D	3	698.0	575–870	—
Head-Body Length	5	M	W	4	1,377.5	1,210–1,465	—
			F	4	1,187.8	925–1,626	—
			H	2	1,354.5	1,334–1,375	—
	5	F	W	8	1,360.4	1,234–1,480	74.9
			F	3	1,472.0	1,253–1,702	—
			H	1	1,277	1,277	—
			D	2	1,823.0	1,615–2,032	—

Table 11—*Continued*

Measurement	Age Class[a]	Sex	Type	n	Mean	Range	SD
	4	M	W	1	1,370	1,370	—
			F	1	1,150	1,150	—
			H	2	1,290.0	1,270–1,310	—
	4	F	H	1	1,290	1,290	—
			D	2	1,499.5	1,426–1,573	—
	3	M&F	F	18	796.6	636–1,198	149.1
			H	1	1,308	1,308	—
			D	1	1,480	1,480	—
	2	M&F	W	2	753.0	723–783	—
			F	6	676.9	531–1,000	163.4
			H	3	848.3	835–865	—
			D	7	1,147.1	770–1,320	190.9
	1	M&F	W	1	385	385	—
			F	4	323.7	255–448	—
			D	3	562.7	471–718	—
Tail Length	5	M	W	4	220.0	190–250	—
			F	3	270.3	232–300	—
			H	3	241.3	210–279	—
	5	F	W	8	221.4	190–256	23.2
			F	3	271.7	153–356	—
			H	2	225.0	190–260	—
			D	3	363.7	343–380	—
	4	M	W	1	230	230	—
			F	1	260	260	—
			H	2	210.0	190–230	—
			D	1	400	400	—
	4	F	H	1	283	283	—
			D	2	318.0	314–322	—
	3	M&F	F	18	202.4	162–297	31.7
			H	1	299	299	—
			D	2	325.0	230–330	—

Continued on next page

Table 11—*Continued*

Measurement	Age Class[a]	Sex	Type	*n*	Mean	Range	*SD*
	2	M&F	W	2	132.5	120–145	—
			F	6	170.3	129–212	40.2
			H	3	134.7	100–160	—
			D	7	228.0	160–281	49.6
	1	M&F	W	1	51	51	—
			F	4	81.3	73–90	—
			D	3	135.3	104–152	—
Hind Foot Length	5	M	W	5	284.8	229–320	37.1
			F	2	276	247–305	—
			H	4	271.3	260–280	—
	5	F	W	10	310.8	280–333	16.0
			F	1	248.0	248	—
			H	3	258.0	240–279	—
			D	4	311.2	270–350	—
	4	M	W	1	290	290	—
			H	3	251.7	240–270	—
			D	1	335	335	—
	4	F	D	1	300	300	—
	3	M&F	F	18	205.3	180–245	18.2
			H	3	256.7	230–305	—
			D	3	295.0	280–305	—
	2	M&F	W	2	202.5	191–214	—
			F	6	176.0	130–233	36.3
			H	4	207.8	200–218	—
			D	7	232.1	143–290	54.6
	1	M&F	W	1	107	107	—
			F	4	82.8	65–109	—
			D	3	133.3	110–160	—
Ear Length	5	M	W	4	126.5	116–135	—
			F	11	146.3	114–180	19.4
			H	6	132.5	120–145	9.4

Table 11—*Continued*

Measurement	Age Class[a]	Sex	Type	n	Mean	Range	SD
	5	F	W	7	127.3	103–160	18.3
			F	5	148.2	120–180	29.7
			H	3	129.0	122–140	—
			D	5	195.2	145–241	35.7
	4	M	W	1	120	120	—
			F	1	111	111	—
			H	5	122.0	100–135	14.8
			D	1	194	194	—
	4	F	H	3	130.7	115–142	—
			D	2	197.5	175–220	—
	3	M&F	F	36	119.9	85–175	22.7
			H	2	125.0	117–133	—
			D	5	157.4	135–187	21.5
	2	M&F	W	2	82.5	80–85	—
			F	54	110.7	66–150	19.9
			H	7	112.6	93–135	16.5
			D	12	150.1	105–170	21.5
	1	M&F	F	9	73.2	31–113	29.9
			D	3	83.0	68–106	—
Snout Length	5	M	F	9	220.3	200–250	17.3
			H	6	262.3	240–280	14.2
	5	F	F	5	201.0	195–210	5.5
			H	2	225.0	220–230	—
			D	5	213.4	190–230	17.0
	4	M	W	1	273	273	—
			H	5	225.0	180–250	—
			D	1	230	230	—
	4	F	F	1	195	195	—
			H	3	231.0	223–240	—
			D	2	182.5	165–200	—

Continued on next page

Table 11—*Continued*

Measurement	Age Class[a]	Sex	Type	n	Mean	Range	SD
	3	M&F	F	34	165.4	127–205	22.1
			H	3	221.0	210–228	—
			D	5	186.8	170–205	12.6
	2	M&F	F	53	138.4	98–175	19.0
			H	7	163.0	127–190	21.2
			D	13	166.0	115–195	26.3
	1	M&F	F	8	82.8	45–105	26.0
			D	2	87.5	85–90	—
Shoulder Height	5	M	F	1	686	686	—
			H	1	710	710	0
	5	F	F	1	673	673	—
			D	2	785.0	710–860	—
	4	M	H	1	660	660	—
		F	D	1	680	680	—
	3	M&F	F	15	531.6	459–651	97.1
	2	M&F	F	5	472.0	409–545	54.3
			H	3	439.3	438–440	—
			D	4	514.9	420–610	—
	1	M&F	F	2	151.0	150–152	—
			D	1	360	360	—
Body Weight	5	M	F	7	108.9	31–225	70.1
			H	3	113.0	91–148	—
	5	F	F	3	52.3	39–68	—
			H	2	52.5	46–59	—
			D	5	204.4	159–272	45.1
	4	M	W	1	73	73	—
			H	3	69.7	68–73	—

Table 11—*Continued*

Measurement	Age Class[a]	Sex	Type	*n*	Mean	Range	*SD*
	4	F	D	1	159	159	—
	3	M&F	F	17	24.6	12–78	19.0
			H	1	64	64	—
			D	1	106	106	—
	2	M&F	F	7	24.1	8–52	18.6
			H	5	26.8	23–34	4.3
			D	7	82.3	33–139	39.6

a Age classes are defined in Table 2.

Domestic swine showed the longest tail lengths, in both mean and range, followed by feral hogs (Table 11). These two morphotypes were followed by wild boar and hybrids. Comparisons of tail length were complicated by the fact that tail biting has been observed in both wild-living and domestic swine and could result in shortening the tail. One feral hog on the Savannah River Plant in South Carolina had a tail stump less than 40 mm long. Whether the tail was curled or not was a useful qualitative character only in contrasting Eurasian wild boar with the other three morphotypes. Wild boar tails were always straight, while domestics, ferals, and hybrids might have either straight or curly tails. It should be noted, however, that wild boar can temporarily curl their tails through muscular action.

Wild boar had shorter hind foot lengths than domestic swine but were larger than the two remaining morphotypes in this dimension (Table 11). The feral hog and hybrid data varied in this measurement, although the hybrid group was slightly larger in most cases. In a proportional comparison, wild boar had the largest hind foot lengths of any of the four types of swine.

Ear length decreased from domestic swine to feral hogs to a mixture of results for the wild boar and hybrid morphotypes (Table 11). This measurement may be useful in differentiating between Eurasian wild boar and domestic swine as two extremes but was otherwise of limited value. The normal posture of the ear in swine may also aid in a similar general comparison for morphotype determination. Eurasian wild boar had only erect or upright ears, while the other morphotypes could have either erect or dropped ears.

Wild boar had the longest snout lengths in the one age class where these data were available for comparison (Table 11). Otherwise, hybrids had the longest snouts except in the juvenile age class, where those of domestic swine were longer. Juvenile domestic swine were larger in overall size but did not have proportionally longer snouts. This comparison could not be made in the younger age classes because no hybrid specimens were available. Judging by the average and maximal cranial rostral lengths as determined by this study, the proportional lengths of the snouts in the four morphotypes should vary from wild boar (the longest) to hybrids to feral hogs to domestic swine (the shortest).

Only minimal data were obtained for shoulder height and body weight (Table 11). Domestic swine tended to be taller at the shoulder than the other morphotypes. The relationship between ferals and hybrids was unclear for this variable. Proportionally, the wild boar and hybrid morphotypes tended to have larger shoulder heights than feral hogs as a result of their relatively shorter body lengths.

Domestic swine had the largest body weights. Comparable data from wild boar, feral hogs, and hybrids overlapped, and no comparisons could be made because of the small sample sizes (Table 11). Reported average and maximum body weights of wild pigs were generally largest for wild boar (Table 12). Reported weights of male wild boar average around 120 kg, with the record being 318 kg for an individual killed in the Caucasus Mountains of the Georgian SSR in the Soviet Union (Banoglu, 1952). Average adult males range from about 73 to 166 kg. Female wild boar were reported to be smaller, with a maximum reported weight of 163 kg and an average of between 59 and 68 kg for adult specimens. Feral hogs had the most variable body weights, showing both the smallest average for males and females (about 60 kg and 50 kg, respectively), and also the largest overall weight of any of the three types of wild pigs. The largest feral hog on record was a male from the Santee River delta in South Carolina reported by Rutledge (1965) to weigh 406 kg. Despite rumors that some male feral hogs from the southeastern United States and Texas may approach 450 kg, no reliable data or documentation could be found for these claims. Among the smallest wild-living pigs in the world were feral hogs that averaged 22–24 kg reported by Brisbin et al. (1977a) for Ossabaw Island, Georgia. Hybrids in our samples averaged approximately 92 kg for males and 71 kg for females. The weights of hybrid wild pigs as determined by this study appear to contradict reports that hybrids do not attain the body weights of feral hogs. The maximum reported hybrid weight was 251 kg for a male from the Los Padres National Forest in California. Average body weights for show-quality domestic swine of various breeds ranged between 273 and 500 kg for males and between 205 and 409 kg for females (Ensminger, 1961). The maximum recorded weight for domestic swine is 865 kg (Britt, 1978).

Table 12. Reported average and maximum total body weights for adult male and female wild boar, feral hogs, and wild boar × feral hog hybrids. Data were obtained either from the literature or from personal correspondence. Weights are given in kilograms. Weights that were reported in pounds were converted to the nearest kilogram for comparative purposes.

Location	Average Weights			Maximum Weights			Source
	Male	Female	Both	Male	Female	Both	
Wild Boar							
New Hampshire	—	—	36–41	—	—	124	Silver, 1957
Corbin's Park, N.H.	—	—	68	182	—	—	A. Dobles, pers. comm.
New Hampshire	—	—	—	138	—	—	Siegler, 1962
Powderhorn Ranch, Tex.	73[a]	59[a]	—	102[a]	80[a]	—	J. C. Smith, pers. comm.
Yugoslavia	—	—	—	221	163	—	Romic, 1975
Lettland, USSR	—	—	—	—	—	236	Heptner et al., 1966
White Russian Region	—	—	150	—	—	256	Heptner et al., 1966
Carpathian Mtns., USSR	—	—	120–180	—	—	250	Heptner et al., 1966
Caucasus Mtns., USSR	166	168	—	178	108	—	Heptner et al., 1966
Caucasus Mtns., USSR	—	—	—	255	145	—	Heptner et al., 1966
Volga Delta, USSR	—	—	—	260	—	—	Heptner et al., 1966
Middle Europe	—	—	—	230	—	—	Heptner et al., 1966
Poland	—	—	—	286	—	—	Amato, 1976
France	—	—	—	176	—	—	Amato, 1976
Ardennes, France	—	—	—	82	—	—	Snethlage, 1950
Iraq	—	—	—	227	—	—	Page, 1954
Germany	—	—	—	—	—	301	Shaw, 1941
Syria	—	—	—	136	—	—	Harrison, 1968
Elburz Mtns., Iran	—	—	—	182[a]	—	—	Bauer, 1970a
Turkey	—	—	—	—	—	200	Banoglu, 1952

Continued on next page

Table 12—Continued

Location	Average Weights			Maximum Weights			Source
	Male	Female	Both	Male	Female	Both	
Caucasus Mtns., USSR	—	—	—	318	—	—	Banoglu, 1952
Europe	—	—	—	230	100	—	Gaffrey, 1961
Kazakhstan SSR, USSR	—	—	—	270	—	—	Pfeffer, 1960
Kazakhstan SSR, USSR	—	—	—	183	80	—	Sloudsky, 1956
Iraq	—	—	—	136[a]	—	—	Hatt, 1959
Iran	—	—	—	250	—	—	Harrington, 1977
Feral Hogs							
Ossabaw, Isl. Ga.	21.9	24.0	—	—	—	—	Brisbin et al., 1977a
SRP, S.C.	67.1	55.3	—	—	—	—	Brisbin et al., 1977a
SRP, S.C.	—	—	—	96	66	—	Kurz, 1971
SRP, S.C.	—	—	—	133	122	—	Sweeney, 1970
Hawaii	—	—	91–136	—	—	291[a]	Nichols, 1962a
Santee River Delta, S.C.	—	—	—	406	—	—	Rutledge, 1965
Hobcaw Barony, S.C.	50–60	40–45	—	90+	97	—	Wood and Brenneman, 1977
Burke Co., Ga.	—	—	—	246	—	—	Anon., 1967
Savannah NWR, S.C.	—	—	—	227[a]	—	—	Robertson, 1978
Savannah NWR, S.C.	—	—	86	—	—	198	D. O'Neal, pers. comm.
Levy Co., Fla.	60.4	49.0	—	78.1	72.6	—	Belden and Frankenberger, 1979
Chassahowitzka NWR, Fla.	—	—	73	—	—	182	E. Collinsworth, pers. comm.
St. Vincent NWR, Fla.	—	—	27	—	—	86	M. D. Perry, pers. comm.
St. Marks NWR, Fla.	—	—	47	—	—	92	C. S. Gidden, pers. comm.

Table 12—Continued

Location						Source
Catahoula NWR, La.	—	—	114	—	159[a]	S. K. Joyner, pers. comm.
Sabine NWR, La.	—	—	55[a]	—	114[a]	J. R. Walther, pers. comm.
Colusa NWR, Calif.	—	—	136[a]	—	182[a]	L. Hill, pers. comm.
Back Bay NWR, Va.	—	—	50	—	90	Duncan and Schwab, 1986
Felsenthal NWR, Ark.	—	—	80–82	—	114	L. L. King, pers. comm.
Havasu NWR, Ariz.	—	—	91–114	—	182	R. A. Gilbert, pers. comm.
Wheeler NWR, Ala.	—	—	68	—	136	T. Z. Atkeson, pers. comm.
Merritt Isl. NWR, Ala.	—	—	46	—	105	J. G. Morris, pers. comm.
Southwestern Alabama	—	—	46–57	—	136	J. R. Davis, pers. comm.
Rob Boykin GMA, Ala.	—	36–55	—	157	—	McKnight, 1964
Santa Catalina Isl., Calif.	34	—	59	105	—	D. W. Baber, pers. comm.
Osceola NF, Fla.	—	—	36–46	—	91	D. Hamilton, pers. comm.
Myakka River SP, Fla.	—	—	34–46	—	125	R. Dye, pers. comm.
Hillsborough River SP, Fla.	—	—	36	—	114	R. Danser, pers. comm.
Louisiana	114	91	—	—	193	J. D. Newsom, pers. comm.
Hidalgo Co., N.M.	—	—	46	—	114	S. J. Dobrott, pers. comm.
Southeastern Oklahoma	—	—	91	—	159	R. Thackston, pers. comm.
South-central Texas	—	—	57	337	—	J. B. Wise, pers. comm.
Francis Marion NF, S.C.	—	—	—	114	—	Anon., 1976
Near Rockport, Tex.	—	—	—	—	136[a]	Bauer, 1970b
Hybrids						
Tellico GMA, Tenn.	82	71	103	92	—	Henry, 1970
Tellico GMA, Tenn.	—	—	131	117	—	Conley et al., 1972
Aransas NWR, Tex.	—	—	182[a]	—	—	Gillelan, 1967

Continued on next page

Table 12—*Continued*

Location	Average Weights			Maximum Weights			Source
	Male	Female	Both	Male	Female	Both	
Aransas NWR, Tex.	—	—	57–59	—	—	123[a]	F. Johnson, pers. comm.
North Carolina	102	—	—	205	—	—	Jones, 1959
Nantahala NF, N.C.	—	—	68	—	—	125	R. Stye, pers. comm.
Great Smoky Mountains NP, N.C./Tenn.	—	—	—	—	—	138	Linzey and Linzey, 1971
Monterey Co., Calif.	—	—	—	—	—	221	Pine and Gerdes, 1973
Monterey Co., Calif.	—	—	—	114	—	136	Shaw, 1941
Ft. Hunter-Liggett, Calif.	—	—	58	—	—	131	P. A. Dubsky, pers. comm.
Los Padres NF, Calif.	—	—	57	—	—	251	P. Thomas, pers. comm.
Avon Park AFR, Fla.	—	—	57	—	—	251	P. Thomas, pers. comm.
Pearl River GMA, Miss.	—	—	58	—	—	193	R. K. Wells, pers. comm.
Edwards Plateau, Tex.	—	—	66	—	—	114	R. F. Smart, pers. comm.
Boerne, Tex.	—	—	—	182[a]	—	—	Tinsley, 1968

a Weight estimate.

Coat Coloration Patterns

The juvenile striped coat coloration pattern was found only in museum specimens or live animals belonging to the neonate age class. By about four to six months of age, the coat attains its permanent coloration pattern (Heptner et al., 1966; Conley et al., 1972). Individuals that are not striped as neonates remain the same color for their entire lives. Thus, coat coloration patterns for the four morphotypes can be combined for all individuals from age classes other than the neonate, with the latter presented separately.

The wild boar was the only morphotype that showed only a single adult coat color pattern, namely, wild type or grizzled (Table 13). This included Eurasian specimens and also those from New Hampshire and Calhoun County, Texas. Within this pattern, coloration varied from animals that showed light-tipped bristles only in a saddle pattern on the lateral and middorsal rostral/facial regions of the head to animals that had light-tipped bristles covering most of the body except for the distal limbs, ears, tail, and snout. The animals that had the most complete covering of light-tipped bristles also had the lightest or longest light-tipped bristles in the facial saddle pattern. No spotted, belted, or solid-colored individuals were noted, and no coat color differences were noted between the sexes.

The feral hog coat coloration patterns resemble the variety seen in domestic swine (Table 13). The only domestic pattern not found in the sample was red/brown with a white shoulder belt. None of the feral museum specimens or live feral animals examined exhibited the wild/grizzled pattern of the wild boar. Of the coat coloration patterns observed, all black was the most common (27.9%). This was followed by black-and-red/brown spotted (21.7%) and black-and-white spotted (17.7%). Overall, 42.9% were solid colored, 48.6% were spotted, and 3.1% were belted. This distribution somewhat agrees with the percentage of occurrence of each coat coloration pattern reported for 45 known feral hog populations in the United States (Table 14). All black was the most widespread, occurring in 82.2% of the populations; however, the next most common was all red/brown, in 68.9%. Eighty-six percent of the respondents stated that all black was the most common coat coloration pattern, with the remainder reporting all red/brown as the most common. The least commonly reported were tricolored spotted and red/brown with white points, each occurring in only 20% of the populations. Red/brown with a white shoulder belt also had a low frequency, 26.7%. The apparent rarity of this color pattern and of red/brown with white points agrees with the absence of these patterns among the museum specimens and animals observed in the field. No obvious sexual dimorphism was observed in the coat coloration of this morphotype.

Table 13. Coat coloration patterns of specimens examined and animals observed in the field in nonneonate age classes. The numbers in parentheses are the percentages of the total sample of that morphotype that exhibited the coloration pattern.

Coloration Pattern	Morphotype		
	Wild Boar Total = 59	Feral Hogs Total = 480	Hybrids Total = 116
Wild/grizzled	59 (100.0)	0 (0.0)	53 (45.7)
Solid black	0 (0.0)	134 (27.9)	27 (23.3)
Solid red/brown	0 (0.0)	36 (7.5)	5 (4.3)
Solid white	0 (0.0)	36 (7.5)	2 (1.7)
Black with white shoulder belt	0 (0.0)	15 (3.1)	5 (4.3)
Red/brown with white shoulder belt	0 (0.0)	0 (0.0)	0 (0.0)
Black-and-white spotted	0 (0.0)	85 (17.7)	7 (6.0)
Red/brown-and-white spotted	0 (0.0)	7 (1.5)	0 (0.0)
Black-and-red/brown spotted	0 (0.0)	104 (21.7)	10 (8.6)
Tricolored spotted	0 (0.0)	37 (7.7)	0 (0.0)
Black with white points	0 (0.0)	24 (5.0)	2 (1.7)
Red/brown with white points	0 (0.0)	2 (0.4)	0 (0.0)
Wild/grizzled with white shoulder belt	0 (0.0)	0 (0.0)	2 (1.7)
Wild/grizzled with white points	0 (0.0)	0 (0.0)	2 (1.7)
Wild/Grizzled and tricolored spotted	0 (0.0)	0 (0.0)	1 (0.9)

Table 14. Frequencies of various coat coloration patterns reported for known populations of wild boar, feral hogs, and wild boar × feral hog hybrids in the United States.

Coat Coloration Pattern	Wild Boar Populations ($N_T{}^a$ = 2)		Feral Hogs Populations ($N_T{}^a$ = 45)		Hybrids Populations ($N_T{}^a$ = 31)	
	$N_S{}^b$	% of N_T	N_S	% of N_T	N_S	% of N_T
Wild/grizzled	2	100.0	0	0.0	22	71.0
All black	0	0.0	37	82.2	15	48.4
All red/brown	0	0.0	31	68.9	10	32.3
All white	0	0.0	15	33.3	3	9.7
Black with white shoulder belt	0	0.0	24	53.3	5	16.1
Red/brown with white shoulder belt	0	0.0	12	26.7	1	3.2
Black-and-white spotted	0	0.0	26	57.8	10	32.2
Red/brown-and-white spotted	0	0.0	17	37.8	4	12.9
Black-and-red/brown spotted	0	0.0	21	46.7	8	25.8
Tricolored spotted	0	0.0	9	20.0	2	6.5
Black with white points	0	0.0	17	37.8	4	12.9
Red/brown with white points	0	0.0	9	20.0	1	3.2
Juvenile striped pattern	2	100.0	5	11.1	14	45.2

a N_T = total number of populations in the sample of each type of wild pig.
b N_S = number of populations reported to have a specific coat coloration pattern.

141

As in the pure wild boar, the wild/grizzled coat coloration pattern predominated in the wild boar × feral hog hybrid morphotype (Table 13; Plate 5), and again, this pattern varied from animals that were almost completely covered by grizzled bristles to those that had only a few light-tipped bristles in the facial saddle. The second most common color pattern of the hybrids was all black (23.3%). Some of these animals were observed in the field at a distance and could have had at least a few light-tipped bristles around the facial region that would not have been easily seen under these circumstances. If so, then these should have been included in the wild/grizzled percentage of the sample. The remainder of the sample was scattered among several other coloration patterns. Notably lacking were red/brown patterns with white belts, spotting, or white points. Only five hybrid individuals examined during this study showed combinations of coat coloration patterns. A juvenile sow collected at the Aransas National Wildlife Refuge, Texas, had complete wild boar coloration with a white shoulder belt only on the left side of its body. Although this was only one of two instances where this combination coat coloration pattern was seen during this study, the Aransas staff reported several other hybrid individuals on the refuge with similar patterns. Another animal, a subadult male from the same locality, was all black with some light-tipped bristles in the facial saddle pattern and white front feet. Again, this was only one of two instances in which this coloration was observed during this study. Of the 31 U.S. hybrid populations surveyed during this study, 71% were reported to have animals exhibiting the wild/grizzled coat coloration pattern, with black being the second most frequent coat coloration (Table 14). All the coat coloration patterns designated in this study were reported for hybrid populations. As with the feral populations, the least common patterns were red/brown with a white shoulder belt and red/brown with white points. Again, no differences were noted between the sexes in hybrids.

A sample of 393 domestic swine observed on farms from several localities in Connecticut, Georgia, and South Carolina had the following frequencies of coat coloration: 8% all black, 7% all red/brown, 65% all white, 4% black with white shoulder belt, 1% red/brown with white shoulder belt, 4% black-and-white spotted, 2% red/brown-and-white spotted, and 9% black-and-red/brown spotted. Color frequencies of domestic populations could vary widely however, depending on the popularity of various breeds in the particular herds or geographic regions sampled.

With one exception, the juvenile striped color pattern was observed (by the authors) only in animals from pure wild boar or hybrid populations (Table 15). With the few exceptions noted below, no museum specimens or live animals examined from strictly feral hog populations exhibited this pattern. In the wild boar sample, all juvenile individuals were striped. About one-half of the neonate hybrid animals examined were striped.

Table 15. Specimens examined and animals observed in the field in the neonate age class that exhibited striped juvenile coat coloration patterns in the wild boar, feral hog, and hybrid morphotypes.

Sample	Wild Boar		Feral Hog		Hybrid	
	n	% of Total	n	% of Total	n	% of Total
Total	5	100	23	100	25	100
Striped juvenile pattern	5	100	0	0	13	52

Reports from correspondents and the literature concerning known feral hog populations, on the other hand, show that there are at least eight populations in the United States in which neonate feral animals may also exhibit the juvenile striped pattern (Table 16). Only one of these populations, Cumberland Island, Georgia, has had a suspected but unconfirmed introduction of wild boar and/or hybrids. Barrett (1971) found only two piglets out of several hundred at Dye Creek Ranch, California, that had faint juvenile striping before wild boar were introduced into that population. In the winter of 1988, a striped neonate female was observed on Ossabaw Island. Unlike the striping of true wild boar neonates, however (Plate 3), the color pattern of this animal was predominantly black with only faint suggestions of several light-colored horizontal stripes on each flank. Three littermates of this piglet were completely black, and the mother of the litter was black-and-white spotted. Juvenile striping is also known in litters born to wild-living sows of various adult body colors, ranging from black through slaty gray, from Hillsborough County, Florida (R. M. Wright, pers. comm.). The juveniles from this area show striping over basic body colors that range from pale cream to tan and dark brown. In some of the paler individuals, the striping can be extremely faint and difficult to see unless the animals is examined at close quarters. In two litters born to wild-bred black sows from this area, 3 of 10 and 2 of 6 were striped (R. M. Wright, pers. comm.). Although there have been no records of hybrids released in this part of Florida, color photographs of some of the adult pigs show some suggestion of a wild grizzled coat color pattern with a paler gray facial area. However, none of these animals was available for direct examination by the authors. Striped juveniles associated with a slaty-gray color have also been reported from several counties from southwest Florida, particularly those in the vicinity of the Peace and Caloosahatchee rivers (R. M. Wright, pers. comm.). Of the other four feral populations, juvenile striping was found in less than 50% of the neonate animals.

Table 16. Locations of wild pigs (*Sus scrofa*) populations in the United States with juveniles exhibiting striped coat coloration patterns.

State	Location	Source
Wild Boar Populations		
New Hampshire	Sullivan and Grafton counties	Silver, 1957
Texas	Calhoun County	J. C. Smith, pers. comm.
Feral Hog Populations		
California	Dye Creek Ranch, Tehama Co.[a]	Barrett, 1971
	Santa Catalina Island, Los Angeles Co.	D. W. Baber, pers. comm.
Florida	Hillsborough County[a]	R. M. Wright, pers. comm.
	Southwestern counties in the vicinity of the Peace and Caloosahatchee rivers	R. M. Wright, pers. comm.
Georgia	Cumberland Island, Camden Co.	Z. T. Kirkland, pers. comm.
	Ossabaw Island, Chatham Co.	ILB, pers. observ.
Hawaii	Island of Hawaii	Nichols, 1962a
Hybrid Populations		
California	Monterey Co.	Seymour, 1970
	San Luis Obispo Co.	A. I. Roest, pers. comm.
Florida	Eglin Air Force Base	J. Knowles, pers. comm.
Georgia	Oconee River, Laurens Co.	R. Davis, pers. comm.
	Ocmulgee River, Houston Co.	R. Davis, pers. comm.
Mississippi	Pearl River GMA, Madison Co.	MMNH 6052
North Carolina	Southwestern corner of state	Jones, 1959
Tennessee	Southeastern corner of state	Conley et al., 1972
Texas	Edwards Plateau	JMM, pers. observ.
	Aransas NWR	JJM 348, JJM 350, JJM 351
West Virginia	Boone, Logan, and Wyoming counties	Igo, 1977

a Faint striping.

Table 17. Length and midshaft diameter (in mm) of middorsal bristle samples of wild boar, feral hogs, wild boar × feral hog hybrids, and domestic swine.

Type	*n*	Length			Diameter		
		Mean	Range	*SD*	Mean	Range	*SD*
Wild Boar	191	103.4	54–149	20.6	0.320	0.10–0.52	0.102
Feral Hogs	151	60.2	25–102	17.7	0.256	0.0–0.42	0.063
Hybrids	351	78.6	32–121	13.1	0.241	0.12–0.40	0.042
Domestic Swine	230	61.1	29–91	11.3	0.233	0.12–0.35	0.051

Hair Morphology

Hair shafts of Eurasian wild boar had the longest and thickest middorsal bristles (Table 17), followed by the hybrids and then domestic swine. Feral hogs had a larger range of hair shaft lengths than domestic swine but had about the same average. In hair shaft width, ferals had a larger range and averaged larger than the domestic sample.

Wild boar bristles had the longest split tips of any of the four morphotypes (Table 18). This group was again followed by the hybrid sample. In this variable, the domestic swine both averaged and ranged larger than the feral hogs.

Light-tipped bristles were found only in the wild boar, hybrid, and feral hog samples; no domestic swine exhibited this character (Table 18). The longest light tips in both mean and maxima were in the wild boar sample. This was again followed by the hybrid specimens. One feral specimen, discussed in the next paragraph, had a single light-tipped bristle. In many of the light-tipped wild boar and hybrid bristles, the light tip was either longer or shorter than the split tip.

Hair shaft coloration differed between wild boar and hybrids versus feral hogs and domestic swine, the first two groups having primarily banded bristles, and the latter two having solid bristle coloration (Table 19). In bristles of wild boar and hybrids, most shafts were black with a white to dark tan distal tip. Both morphotypes showed smaller percentages of solid-colored bristles: all black in the wild boar and all black or red/brown in the hybrids. A few (0.9%–1.6%) of these two morphotypes had black bristles with a light-colored band near the distal end of the shaft and a black distal tip.

Table 18. Lengths (in mm) of the split and light-colored tips of middorsal bristle samples of wild boar, feral hogs, wild boar × feral hog hybrids, and domestic swine.

Type	n^a	Split Tip Length			Light Tip Length		
		Mean[b]	Range	SD^b	Mean[c]	Range	SD^c
Wild Boar	191	23.9	0–58	13.0	17.9	0–51	14.5
Feral Hogs	151	7.9	0–25	6.3	0.2	0–35	—
Hybrids	351	18.1	0–40	9.2	8.1	0–53	12.0
Domestic Swine	230	9.1	0–36	7.7	0	0	—

a Number of hairs examined.
b For all hairs, including those with no split tip, which were entered as zeros.
c For all hairs, including those with no light tips, which were entered as zeros.

Table 19. Frequencies of hair shaft colors in hair samples from wild boar, feral hogs, wild boar × feral hog hybrids, and domestic swine.

Color	Wild Boar (N = 191)[a]	Feral Hogs (N = 151)[a]	Hybrids (N = 351)[a]	Domestic Swine (N = 230)[a]
Black with white to dark tan distal tip	169 (88.5)	1 (0.7)	227 (64.7)	0 (0.0)
Black with white band and black distal tip	3 (1.6)	0 (0.0)	3 (0.9)	0 (0.0)
All black	19 (9.9)	125 (82.8)	95 (27.1)	21 (9.1)
All red/brown	0 (0.0)	14 (9.3)	26 (7.4)	16 (7.0)
All white	0 (0.0)	11 (7.3)	0 (0.0)	193 (83.9)

Note: The numbers in parentheses are the percentages of the total sample of that color.
a Total number of hairs examined.

All the domestic sample and 99.4% of the feral sample exhibited solid hair shaft coloration. One hair from a feral hog from Santa Rosa Island, California (SBMNH 1250), had a white-tipped black bristle. No other similarly colored bristles were noted on the skin of this animal. In the rest of the feral sample, most of the bristles were all black. The others were divided about evenly between all red/brown and all white. In the domestic sample, the majority of the bristles were all white. The remainder were almost evenly divided between all black and all red/brown. The preponderance of white bristles in the domestic sample was a consequence of the predominance of the Yorkshire breed and should not be considered as representative of the domestic morphotype in general.

Curly, woollike underfur was found in known populations of wild boar, wild boar × feral hog hybrids, and feral hogs. None of the domestic swine examined had this hair type. All the wild boar examined had dense underfur, as did about 79.3% of the hybrid specimens. Although present in hybrid specimens from Texas, curly underfur was usually sparse on these animals, even in winter. Only one specimen, an adult male from Kerr County, Texas, had a dense coat of underfur. Only 30.4% of the feral hogs had curly, woollike underfur, and all were from long-term feral populations. On the basis of personal observations, personal communications, and the literature, curly, woollike underfur presently occurs in two known wild boar populations, six known hybrid populations, and nine known feral hog populations in the United States (Table 20).

In the wild boar, the underfur ranged in color from smoke gray to dark brown. Sixty-four percent were dull brownish to yellowish gray. The underfur in the hybrid sample was smoke gray/white to black. The most common color in this sample (40%) was black. Feral hog underfur ranged from white to black, with 57% of the sample being black. One of the feral specimens that had underfur was black-and-white spotted. Under the bristles of the black spots the underfur was black, and under the white spots the underfur was white. This was the only spotted individual examined for underfur during this study, so the extent of this consistency of color association between the bristles and the underfur is unknown.

Comparison with Feral Hogs from Outside the United States

Hogs from known feral populations outside the United States compared favorably with the morphotype described in this study for known feral hogs from the United States. Most of these non-U.S. specimens fell within or adjacent to the feral hog target group in the cranial and mandibular analyses comparing the four target groups.

Table 20. Locations of wild pigs (*Sus scrofa*) populations in the United States with individuals in nonneonate age classes exhibiting curly, woollike underfur.

State	Location	Source
Wild Boar Populations		
New Hampshire	Sullivan and Grafton counties	This study[a]
Feral Hog Populations		
California	Santa Rosa Island, Santa Barbara Co.	SBMNH 1250
	Santa Catalina Island, Los Angeles Co.	D. W. Baber, pers. comm.
Florida	Osceola National Forest	D. Hamilton, pers. comm.
Georgia	Ossabaw Island, Chatham Co.	This study[a]
	Ocmulgee River, Coffee Co.	This study[a]
	Cumberland Island, Camden Co.	Z. T. Kirkland, pers. comm.
Hawaii	Island of Hawaii	Nichols, 1962a
	Parker Ranch, Hawaii	USNM 321950
South Carolina	Francis Marion National Forest	R. Tyler, pers. comm.
Hybrid Populations		
California	Monterey Co.	MVZ 89506; MVZ 123594; MVZ 123594; CPSU M-343
Georgia	Ocmulgee River, Houston Co.	Personal observation
Mississippi	Pearl River GMA, Madison Co.	R. K. Wells, pers. comm.
North Carolina	Southwestern corner of state	Jones, 1959
Tennessee	Tellico WMA	Conley et al., 1972
Texas	Edwards Plateau	This study[a]
	Aransas NWR	JJM 222

a   Underfur samples collected.

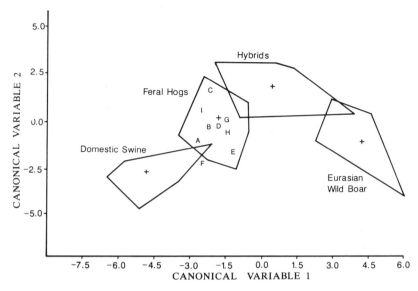

Figure 28. Canonical plot of adult male crania from the four target groups versus the following non-U.S. feral hogs: A = Sonora, Mexico; B-E = Galápagos Islands, Ecuador; F = Northern Territory, Australia; G = Komodo Island, Indonesia; H = Great Nicobar Island, Nicobar Islands; and I = Little Andaman, Andaman Islands. Target group boundaries and means are taken from Figure 12.

In the adult male cranial analysis, all non-U.S. specimens were classified as feral hogs with PPs ranging from 0.51 to 1. Only one specimen (from Australia [USNM 396882]) fell outside the boundaries of the feral group (Figure 28). It plotted adjacent to that group and the domestic group and had a PP of 0.51 as a feral hog and 0.49 as a domestic. This animal was black-and-white spotted and had the following body measurements: total length = 1,905 mm; tail length = 279 mm; hind foot length = 305 mm; and ear length = 165 mm.

The results of the adult male mandibular analysis were not as clear-cut as those of the cranial comparison. Feral boars from Mexico (USNM 60356) and Australia (USNM 396882) were classified as domestics with PPs as domestics of 0.88 and 0.95, respectively, and were plotted in an area of overlap between the domestic and hybrid groups (Figure 29). Two specimens from the Galápagos Islands (MVZ 125511 and 125512) were classified as hybrids with PPs as hybrids of 0.50 and 0.61. As domestics, they had PPs of 0.48 and 0.38, respectively. On the canonical graph, they were plotted in an area of overlap between the feral and hybrid groups. The remainder of the male mandibles were classified within the U.S. feral hog target group.

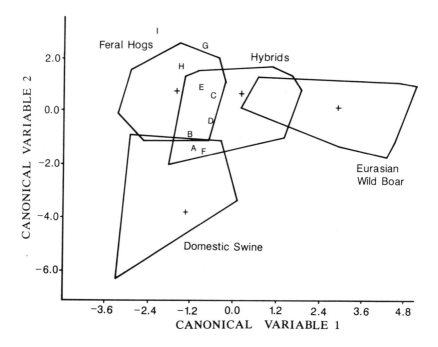

Figure 29. Canonical plot of adult male mandibles from the four target groups versus the following non-U.S. feral hogs: A = Sonora, Mexico; B = Corcovada National Park, Costa Rica; C-E = Galápagos Islands, Ecuador; F = Northern Territory, Australia; G = Komodo Island, Indonesia; H = Great Nicobar Island, Nicobar Islands; and I = Little Andaman, Andaman Islands. Boundaries and means are taken from Figure 12.

Adult female skulls from non-U.S. feral hogs were more variable than those of adult males. In both crania and mandibular analyses (Figures 30 and 31), one sow from the Virgin Islands (KU 74115) classified consistently as a hybrid with PPs of 0.96 and 0.58, respectively. This sow was black and red/brown with the following body measurements: total length = 1,702 mm; tail length = 356 mm; and ear length = 178 mm. Her weight was approximately 68 kg. The sow was from a population that was definitely feral (Bee, 1957) even though it fell in the hybrid target groups in both canonical graphs. Another sow, from Mexico (UMMZ 90802), was also classified by both analyses as a hybrid (PP = 0.57 and 0.72, respectively). On the canonical plot, the cranium fell in an overlap zone between the feral and hybrid target groups (Figure 30). No other data were available for this animal. The remainder of the adult non-U.S. feral sows were classified as feral hogs in both analyses, with PPs of 0.87–1 for the crania and 1 for the mandibles.

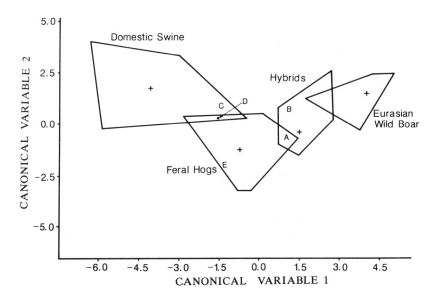

Figure 30. Canonical plot of adult female crania from the four target groups versus the following non-U.S. feral hogs: A = Chiapas, Mexico; B = St. John Island, Virgin Islands; C = Gardener Island, Mariana Islands; and D and E = Guam, Mariana Islands. Target group boundaries and means are taken from Figure 13.

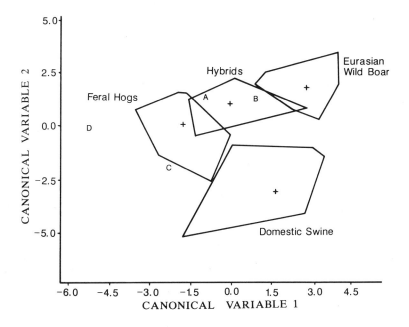

Figure 31. Canonical plot of adult female mandibles from the four target groups versus the following non-U.S. feral hogs: A = Chiapas, Mexico; B = St. John Island, Virgin Islands; C = Guam, Mariana Islands; and D = Andaman Islands. Target group boundaries and means are taken from Figure 13.

151

Plate 1. Adult boar in winter pelage (*top*) and adult sow in summer pelage (*bottom*) from the short-term feral hog bottomland swamp population on the southern portion of the U.S. DOE Savannah River Plant, South Carolina. Photographs by J. J. Mayer.

Plate 2. Subadult male wild boar (*top*) killed near Lyman, Grafton County, New Hampshire, 12 November 1980; rooting damage (*bottom*) done by "outside" wild boar in a field near Partridge Lake, Grafton County, New Hampshire, November 1980. Photographs by J. J. Mayer.

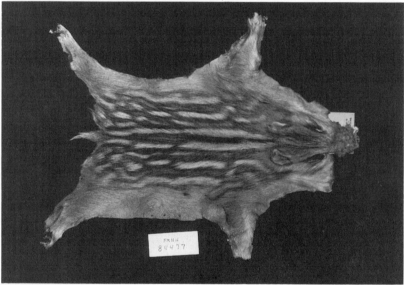

Plate 3. General appearance of the juvenile striped pattern as exemplified by a litter of wild boar × feral hog hybrid piglets from the North Carolina-Tennessee population (*top*) and the flat skin of a Eurasian wild boar piglet (FMNH 84477) from the Middle East (*bottom*). Photographs by J. J. Mayer.

Plate 4. Lateral profiles of the heads of an adult male Eurasian wild boar from Morocco (*top*) and an adult male feral hog from the northern portion of the U.S. DOE Savannah River Plant, South Carolina (*bottom*), illustrating the difference in the depth of the dorsal profile. Photographs by L. W. Robbins (top) and J. J. Mayer.

155

Plate 5. General appearance of the wild/grizzled pelage pattern observed in wild boar ×
feral hog hybrids as exemplified by subadult male hybrids (*top and bottom*) from the hybrid
population in the Edwards Plateau, west of Boerne, Texas. Photographs by J. J. Mayer.

Plate 6. Lateral views of animals showing long-term (*top*) and short-term (*bottom*) feral hog morphotypes. Both animals were collected from the U.S. DOE Savannah River Plant, South Carolina. The short-term animal is typical of those that have inhabited the site since it was closed to the public in the early 1950s. The long-term animal represents a morphotype that first appeared on the site in the late 1970s. It has continued to expand its range there (Figure 4). Photographs by J. J. Mayer.

Subadult male crania plotted either within or immediately adjacent to the feral target group. However, one specimen from Mexico (KU 11991) and another from Zanzibar (MCZ 40952) were classified as hybrids (PP = 0.51 and 0.61, respectively). Both were plotted immediately adjacent to the area of overlap between the feral and hybrid groups (Figure 32). No other data were available for the first animal. The second was all black with the following body measurements: total length = 1,410 mm; tail length = 260 mm; ear length = 111 mm; weight = 46 kg. The remainder of the specimens were classified as feral hogs, all with PPs of 1.

In contrast to this, the available mandibles for these subadult male specimens were all classified as feral hogs with PPs of 0.58–1. This included the two subadult boars from Mexico and Zanzibar (Figure 33).

The one non-U.S. subadult female (DHJ 2) was classified as a hybrid in both the cranial and mandibular analyses. In both canonical plots, however, it fell with the feral target group, although adjacent to the hybrid target group (Figure 34). The PPs resulting from these analyses were 0.68 and 0.62, respectively. This animal was from a feral population descended from domestic stock that escaped from "campesinos," or squatters, in the Corcovada National Park, Costa Rica (D. H. Janzen, pers. comm.). No other data were available for this animal.

Yearling crania ranged from one specimen from the Galápagos Islands (MVZ 125510) that had a PP of 0.99 as a hybrid to another from Rota Island, Mariana Islands (USNM 279015), that had a PP of 1 as domestic. Both were plotted in those target groups, respectively, although the latter specimen fell in a zone of overlap with the feral group (Figure 35). The remainder of the crania were classified as feral hogs.

In the yearling mandibles, the Galápagos specimen was again classified as a hybrid (PP = 0.96) and plotted within that group, while the remainder were classified as feral hogs (Figure 35).

In the juvenile cranial and mandibular analyses, all specimens of non-U.S. feral hogs were classified as feral hogs (Figures 36 and 37). The cranium of one specimen from Guadalcanal Island (FMNH 101683) was plotted in an overlap zone between the domestic and feral target groups (Figure 36). No mandible was available for this animal. This cranium was collected along the Nalimbu River, about 0.8 km inland. It was suspected to be feral, but this was not certain.

Comparisons with Captive Wild Pigs

In general, canonical plots of cranial and mandibular measurements of captive specimens fell within or adjacent to the target group of their wild-living counterparts. Seven of the eight adult male wild boar from zoos had PPs as wild boar ranging from 0.68 to 1 for the cranial analysis.

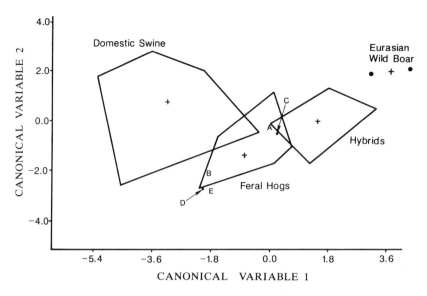

Figure 32. Canonical plot of subadult male crania from the four target groups versus the following non-U.S. feral hogs: A = Campeche, Mexico; B = Belize; C = Pemba Island, Zanzibar; and D and E = Andaman Islands. Target group boundaries and means are taken from Figure 14.

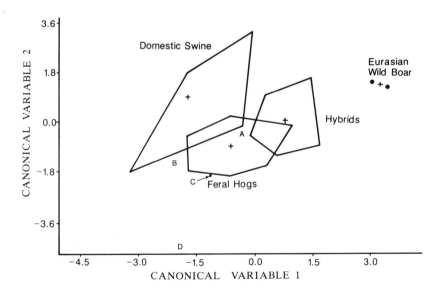

Figure 33. Canonical plot of subadult male mandibles from the four target groups versus the following non-U.S. feral hogs: A = Campeche, Mexico; B = Belize; C = Pemba Island, Zanzibar; and D = Andaman Islands. Target group boundaries and means are taken from Figure 14.

159

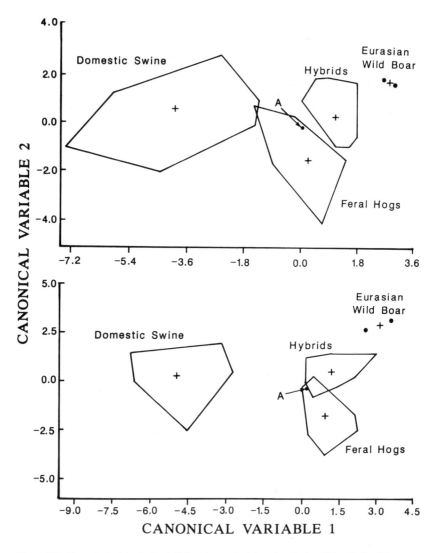

Figure 34. Canonical plot of subadult female crania (above) and mandibles (below) from the four target groups versus the following non-U.S. feral hogs: A = Corcovada National Park, Costa Rica. Target group boundaries and means are taken from Figure 15.

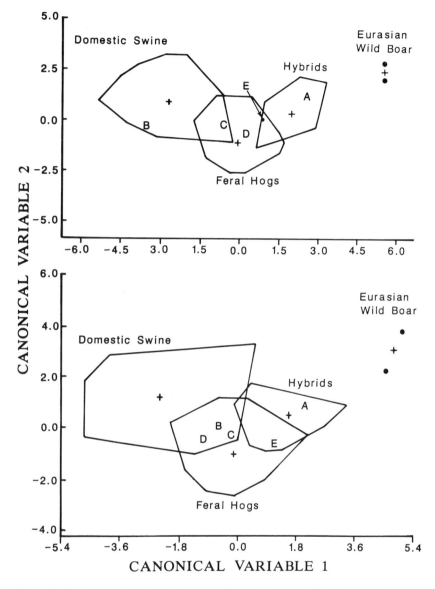

Figure 35. Canonical plot of yearling (sexes combined) crania (above) and mandibles (below) from the four target groups versus the following non-U.S. feral hogs: A = Galápagos Islands, Ecuador; B = Rota Island, Mariana Islands; C and D = Admiralty Island, Papua New Guinea; and E = New Zealand. Target group boundaries and means are taken from Figure 16.

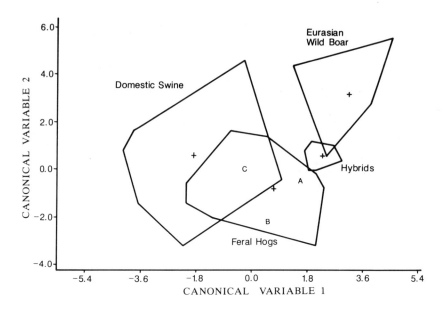

Figure 36. Canonical plot of juvenile (sexes combined) crania from the four target groups versus the following non-U.S. feral hogs: A = Andaman Islands; B = New Zealand; and C = Guadalcanal Island, Solomon Islands. Target group boundaries and means are taken from Figure 17.

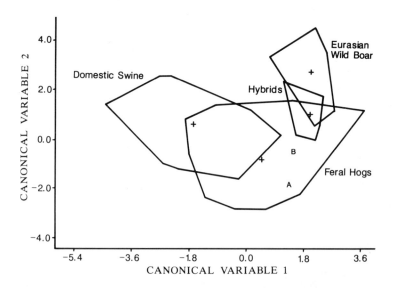

Figure 37. Canonical plot of juvenile (sexes combined) mandibles from the four target groups versus the following non-U.S. feral hogs: A = Andaman Islands; and B = New Zealand. Target group boundaries and means are taken from Figure 17.

162

The remaining specimen was classified as a hybrid (PP = 0.59) on the basis of its cranium (Figure 38). On the basis of the mandibular measurements, five captives were classified as wild boar and three as hybrids (Figure 38). Zoo wild boar average 5.8% shorter and 5.4% broader in the cranial measurements. Six of the eight zoo males had dorsal profile depths within the observed range for Eurasian wild boar. Two zoo males had dorsal profile depths that were even less than the mean for free-living animals. The two zoo animals that had dorsal profile depths that exceeded the maximum profile depth of free-living animals did so by less than 3 mm. Coefficients of variation were greater for the zoo wild boar, although this could have been caused by a small sample size. The zoo specimens were generally older than the free-living wild boar samples, with tooth wear ratings and suture scores of 27–48 versus 6–9 for free-living animals. The cranial bones of the captive wild boar resembled those of domestic swine in being more porous and convex between sutures than those of free-living wild boar.

The two adult female zoo wild boar were classified as wild boar in the cranial analysis (PP = 0.98 and 1) but as hybrids in the mandibular analysis (PP = 0.53 and 0.66; Figure 39). However, in the mandibular analysis, both specimens fell within the bounds of the wild boar target group. Female zoo wild boar were 3.8% shorter in the cranial measurements compared with wild-living females. In contrast with the zoo males, however, the two female zoo animals were 7.2% narrower in the cranial width measurements than the wild boar sample. In addition, both were below the observed mean of the dorsal profile depths for the wild-living females. Again, the zoo specimens were generally older than the wild-living sample.

In the captive feral sample, a four-and-one-half-year-old male that had been captured on Ossabaw Island as a neonate and reared in captivity was classified as a feral hog (PP = 1) in the cranial analysis and as a hybrid (PP = 0.66) in the mandibular analysis (Figure 38). In both analyses, however, it fell within the bounds of the feral hog target group. This animal was within the range of the cranial length, zygomatic width, and dorsal profile depth measurements for adult males from Ossabaw Island. In most of the remainder of the cranial width measurements, however, this specimen was larger than any other adult from this population. All mandibular measurements fell within the observed ranges of the wild-living sample from this locality.

Of two yearling males born in captivity to captive-reared parents from Ossabaw Island, one was classified as feral (PP = 0.91) and the other as domestic (PP = 0.57) in the cranial analysis. One of the specimens, however, fell inside the area of overlap between the domestic and feral target groups in the canonical plot (Figure 40). In the mandibular analysis, both captive animals were incorrectly classified as hybrids (PP = 0.74 and 0.98, respectively; Figure 39). One specimen fell within the overlap zone

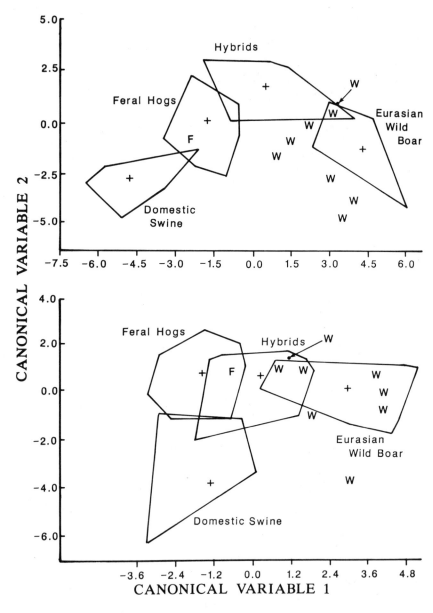

Figure 38. Canonical plot of adult male crania (above) and mandibles (below) from the four target groups versus zoo European wild boar (W) and one captive-reared Ossabaw Island feral hog (F). Target group boundaries and means are taken from Figure 12.

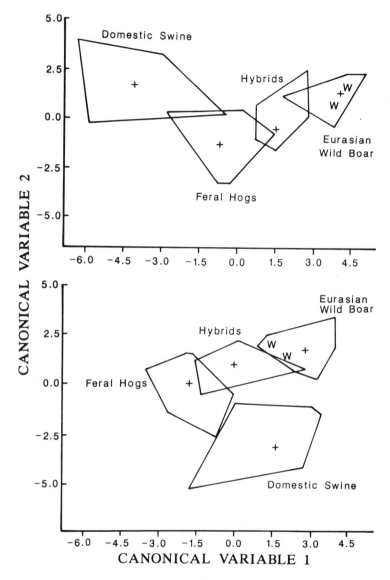

Figure 39. Canonical plot of adult female crania (above) and mandibles (below) from the four target groups versus zoo European wild boar (W). Target group boundaries and means are taken from Figure 13.

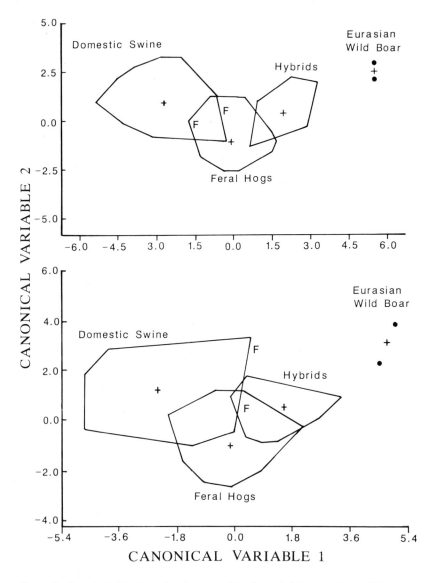

Figure 40. Canonical plot of yearling (sexes combined) crania (above) and mandibles (below) from the four target groups versus two captive-bred and captive-reared feral hogs with Ossabaw Island ancestry (F). Target group boundaries and means are taken from Figure 16.

166

between the hybrid and feral target groups, and the other fell outside all the target groups on the canonical plot but was statistically closest to the hybrid group mean. Both animals had cranial dimensions larger than those of wild-living Ossabaw Island feral hogs of the same age and sex. One specimen equaled and the other exceeded the maximum depths of the dorsal profile for the wild specimens.

One juvenile male, also from captive-reared Ossabaw Island parents, had mixed results. The cranium was classified as feral hog (PP = 0.65), but the mandible was classified as wild boar (PP = 0.88; Figure 41). As with the two yearlings, this animal's cranium was larger than wild-collected specimens from the same population and age group. In fact, it was even larger than wild specimens from the next oldest age class. In addition, this animal had a dorsal profile depth 6 mm deeper than the maximum for the wild specimens.

The one captive hybrid specimen in this analysis, a captive-reared 19-month-old female caught as a neonate in the Edwards Plateau in Texas, was classified as hybrid in both the cranial and mandibular analyses (PP = 0.79 and 0.95, respectively). In the canonical plot, however, this specimen fell outside the hybrid target group in both plots (Figure 42). This animal was larger than the average for wild-living hybrid specimens of the same sex and age class in all skull measurements, but was within the observed range for hybrid specimens within this group. The depth of the dorsal profile was above the mean but below the maximum for wild-living hybrids. The cranial bones of this animal were more robust than those of its wild-living counterparts, again resembling those of domestic swine, especially in the frontal region.

With the exception of externally apparent increased fat deposition, no captive specimens of any morphotype showed external body measurements, coat coloration patterns, or hair morphology outside the range observed for their wild-living counterparts. However, only a limited number (n = 17) of captives were measured, all of them feral hogs.

Identification of Unknown Wild Pigs

We compared 108 wild pigs from U.S. populations whose morphotypic derivation could not be determined from historical records with the four known target groups. The majority of these specimens consisted of cranial material only. Two specimens also had prepared skins. Limited coat coloration data and external body dimensions were available for several other specimens. Specimens were collected after 1960 unless otherwise noted. The results of the comparisons of these specimens with the canonical plots of the four target morphotypic groups are presented below, arranged according to the states from which they were collected.

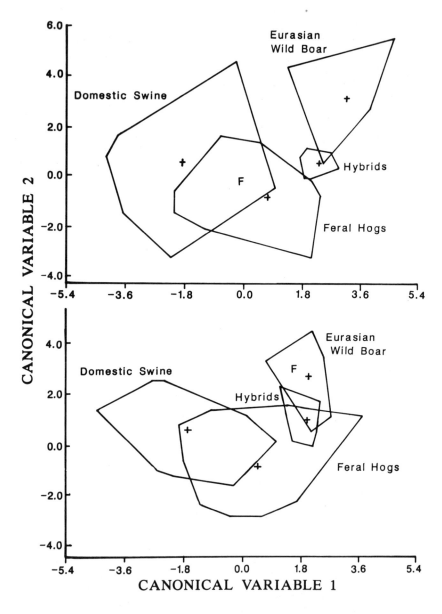

Figure 41. Canonical plot of juvenile (sexes combined) crania (above) and mandibles (below) from the four target groups versus one captive-bred and -reared feral hog with Ossabaw Island ancestry (F). Target group boundaries and means are taken from Figure 17.

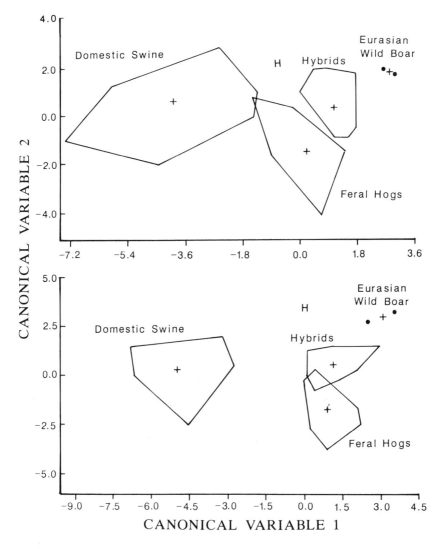

Figure 42. Canonical plot of subadult female crania (above) and mandibles (below) from the four target groups versus one captive-reared wild boar × feral hog hybrid from the Edwards Plateau (H). Target group boundaries and means are taken from Figure 15.

Alabama

The skulls of two adult males from Clark County (JJM 360 and 361) were identified as feral hogs by the cranial analysis and as hybrids by the mandibular analysis. The PPs in the cranial analyses of the two specimens were 1 and 0.97, respectively, as belonging to the feral hog target group; values for the mandibular analyses were 0.47 and 0.97 as belonging to the hybrid group (Figure 43). JJM 360 was also within the bounds of the feral target group in the mandibular analysis. Because of the group overlap in the canonical plot, however, this specimen fell closer to the hybrid than to the feral group mean. No other data were available for these specimens. Typical feral hog coat coloration patterns have been reported for this population in southwestern Alabama. All black and black-and-white spotted are the most common, followed by all red/brown and black-and-red/brown spotted. No individuals with light-tipped bristles or curly underfur have been reported from this area. The piglets do not exhibit the juvenile striped pattern. The wild pigs in this population average 46–59 kg as adults, with a reported maximum of 157 kg (McKnight, 1964; J. R. Davis, pers. comm.). All these characteristics support the identification of the animals from this population as strictly feral hogs.

California

One adult male (CSULB 6467) from Hall Canyon in Ventura County was identified as a feral hog in both the cranial and mandibular analyses (PP = 0.99 and 0.80, respectively; Figure 43).

A juvenile wild pig (WFBM 1343) from near Covelo in Mendocino County was also classified as a feral hog in both the cranial and mandibular analyses (PP = 0.82 and 0.94, respectively; Figures 44 and 45).

A yearling male (SDNHM 21263) and a juvenile female (SDNHM 21262) from a locality near Cachuma Lake in Santa Barbara County could not be clearly assigned to a single morphotype. The male cranium had a PP of 0.91 as a feral animal but was overlapped by the domestic target group (Figure 46). This animal's mandible had a PP of 0.93 as a hybrid but also fell close to the bounds of the feral target group (Figure 47). The female had a PP of 0.47 as a domestic and 0.45 as a hybrid for the cranium, and was also adjacent to feral group in the canonical plot (Figure 44). The mandible for this female had a PP of 0.47 as a feral animal and a PP of 0.34 as a hybrid (Figure 45). Overall, these animals fell within or adjacent to the feral target group more than any other target group in these analyses. The male weighed 46 kg and the female 38 kg. No data on coat coloration or hair morphology were available. Since these two specimens were collected near the point of the southern range boundary of the hybrid population along coastal California, it is possible that they were from mixed wild boar and feral hog ancestry.

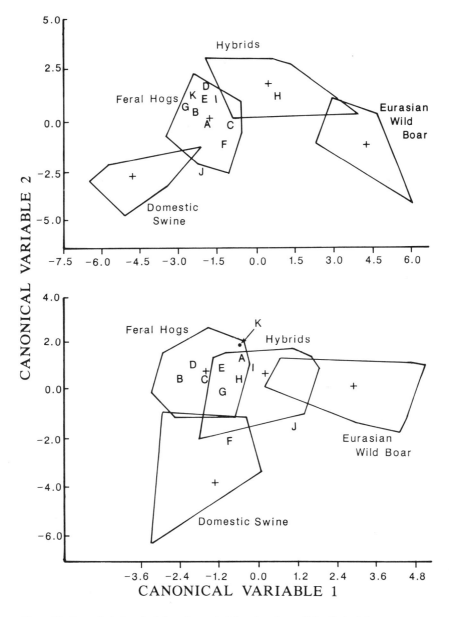

Figure 43. Canonical plot of adult male crania (above) and mandibles (below) from
the four target groups versus the following U.S. unknowns: A and B = Clark Co., Ala.;
C = Ventura Co., Calif.; D = Alachua Co., Fla.; E = Highlands Co., Fla.; F = Baldwin
Co., Ga.; G = Columbia, Co., Ga.; H = Frio Co., Tex.; I = McMullen Co., Tex.; J = San
Patricio Co., Tex.; and K = Kenedy Co., Tex. Target group boundaries and means are taken
from Figure 12.

171

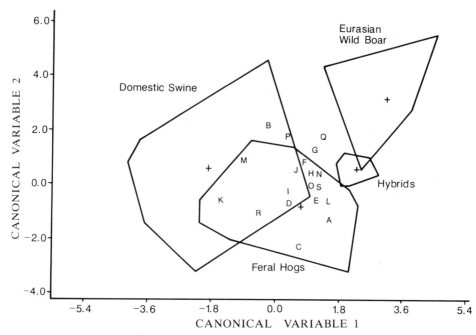

Figure 44. Canonical plot of juvenile (sexes combined) crania from the four target groups versus the following U.S. unknowns: A = Mendocino Co., Calif.; B = Santa Barbara Co., Calif.; C = Thomas Co., Ga.; D-J = Baldwin Co., Ga.; K = Dare Co., N.C.; L = Beaufort Co., S.C.; M and N = Kenedy Co., Tex.; and O-S = San Patricio Co., Tex. Target group boundaries and means are taken from Figure 17.

Both were identified as such by the curatorial staff at the San Diego Natural History Museum where they were deposited. Because of the lack of more descriptive data on these animals and the inconsistencies in the skull analyses, however, more positive identifications of these animals are impossible.

A subadult male (CPSU 1103) from near Buellton in Santa Barbara County also could not be clearly assigned to a single morphotype but again fell most consistently within the feral target group (Figures 48 and 49). The cranium had PPs of 0.42 and 0.58 for the feral and hybrid group means, respectively. The mandible had PPs of 0.59 and 0.25 for the feral and domestic morphotypes, respectively. No other data were available. This specimen was also from near the southern boundary of the hybrid range. The results of the cranial analysis suggest that this animal was a feral hog or at least a hybrid with a high proportion of feral characteristics.

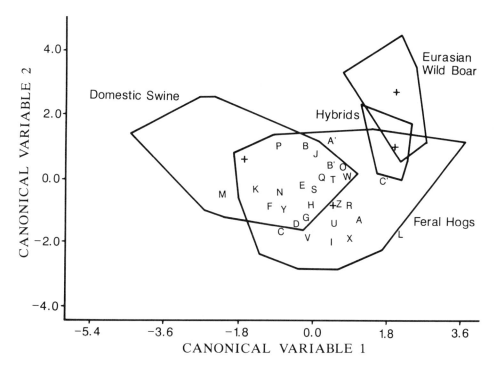

Figure 45. Canonical plot of juvenile (sexes combined) mandibles from the four target groups versus the following U.S. unknowns: A = Mendocino Co., Calif.; B = Santa Barbara Co., Calif.; C = Thomas Co., Ga.; D-J = Baldwin Co., Ga.; K = Dare Co., N.C.; L = Beaufort Co., S.C.; M and N = Kenedy Co., Tex.; and O-C′ = San Patricio Co., Tex. Target group boundaries and means are taken from Figure 17.

Florida

Three specimens—two subadult males (USNM 111366 and 111367) and one subadult female (USNM 111365)—collected from near Lake Kissimmee, Osceola County, in 1901, were classified consistently as feral hogs by the analyses. The PPs for the feral target group were all greater than 0.65 (Figures 48, 49, and 50). Another specimen, an adult male (USNM 254425) collected in 1929 from Kicco, Highlands County, also was identified as a feral hog (PP = 0.96 and 0.71 for the cranium and mandible, respectively; Figure 43).

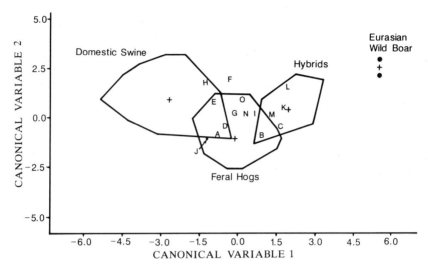

Figure 46. Canonical plot of yearling (sexes combined) crania from the four target groups versus the following U.S. unknowns: A = Santa Barbara Co., Calif.; B = Dixie Co., Fla.; C = Alachua Co., Fla.; D-H = Baldwin Co., Ga.; I = Chatham Co., Ga.; J = Lafayette Par., La.; and K-O = San Patricio Co., Tex. Target group boundaries and means are taken from Figure 16.

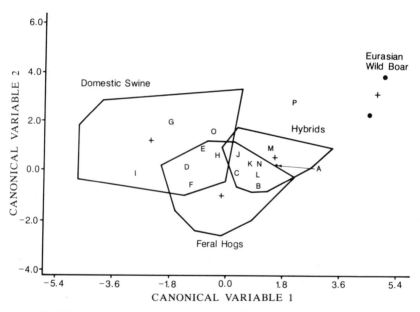

Figure 47. Canonical plot of yearling (sexes combined) mandibles from the four target groups versus the following U.S. unknowns: A = Santa Barbara Co., Calif.; B = Alachua Co., Fla.; C-G = Baldwin Co., Ga.; H = Chatham Co., Ga.; I = Lafayette Parish, La.; and J-P = San Patricio Co., Tex. Target group boundaries and means are taken from Figure 16.

174

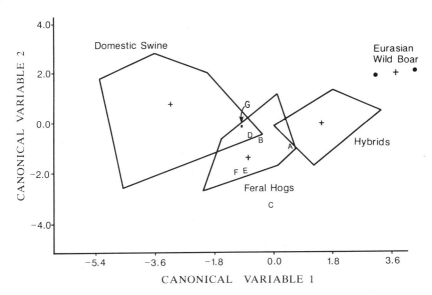

Figure 48. Canonical plot of subadult male crania from the four target groups versus the following U.S. unknowns: A = Santa Barbara Co., Calif.; B and C = Osceola Co., Fla.; D = Taylor Co., Fla.; E and F = Alachua Co., Fla.; and G = no specific locality, Ga. Target group boundaries and means are taken from Figure 14.

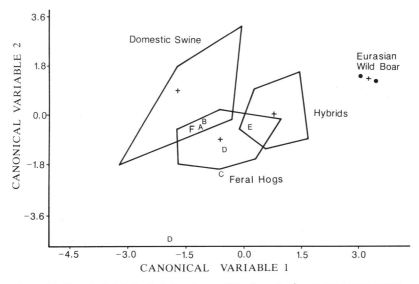

Figure 49. Canonical plot of subadult male mandibles from the four target groups versus the following U.S. unknowns: A = Santa Barbara Co., Calif.; B and C = Osceola Co., Fla.; D = Alachua Co., Fla.; E = LaSalle Co., Tex.; and F = no specific locality, Ga. Target group boundaries and means are taken from Figure 14.

175

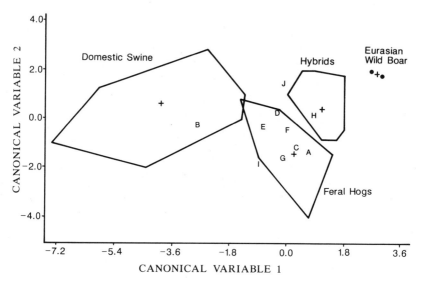

Figure 50. Canonical plot of subadult female crania from the four target groups versus the following U.S. unknowns: A = Osceola Co., Fla.; B = Gilchrist Co., Fla.; C = no specific locality, Fla.; D = Madison Co., Miss.; E and F = Beaufort Co., S.C.; G = Allendale Co., S.C.; H = Harden Co., Tex.; I = Liberty Co., Tex.; and J = Cameron Co., Tex. Target group boundaries and means are taken from Figure 15.

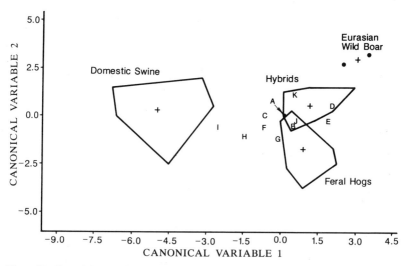

Figure 51. Canonical plot of subadult female mandibles from the four target groups versus the following U.S. unknowns: A = Gilchrist Co., Fla.; B = no specific locality, Fla.; C = Madison Co., Miss.; D and E = Beaufort Co., S.C.; F = Allendale Co., S.C.; G-I = San Patricio Co., Tex.; J = Harden Co., Tex.; and K = Cameron Co., Tex. Target group boundaries and means are taken from Figure 15.

176

The cranium of a subadult female (FSM 2048) collected in 1957 from near Newberry, Gilchrist County, had a PP of 0.99 as a domestic animal, while the mandible was classified between the feral and hybrid target groups (PP = 0.43 and 0.57, respectively; Figures 50 and 51). Because of the better resolution of the mandibular analysis for this sex and age class, this would seem to be the preferred analysis. The mandibular plot of this specimen fell close to both the feral and hybrid target groups but was slightly closer to the hybrid mean as indicated by the PPs. No other data were available for this animal. Since wild boar have never been reported as introduced into this area of Florida, this animal was probably a feral hog. Why the cranium plotted in the domestic group and fell so far outside the wild pig target groups is unknown. One adult female specimen (MCZ 17761) from New Smyrna Beach in Volusia County on the east coast of Florida consisted of a cranium and mandible. It was classified as a hybrid animal with a PP of 0.88 in the cranial analysis (Figure 52). In the mandibular analysis, it had a PP of 0.50 as hybrid and 0.49 as feral (Figure 53). This animal was collected in March 1919, approximately 36 years before any hybrids were reported to have been introduced into the state. No other data were available for this animal.

An adult female (MCZ 39430) collected in 1940 from Felsmere in Indian Beach County was classified as a feral hog by both cranial and mandibular analyses (PP = 0.77 and 0.60, respectively; Figures 52 and 53).

The cranium of a yearling male (LMS 1) from the Suwannee River near Old Town, Dixie County, was classified as a feral hog (PP = 0.70; Figure 46).

A neonate/infant cranium (AMNH 146566) from Palatka, Putnam County, was classified as domestic with a PP of 1. This specimen fell to the extreme left in the canonical plot, away from all groups but closest to the domestics (Figure 54). Although collected in the wild in 1949, with no other data available, it is difficult to identify this specimen as anything other than a domestic piglet. It should be recalled, however, that the analyses reported here had a very limited ability to classify very young individuals.

The cranium of a subadult male from northern Taylor County (FSM 9191) plotted in an area of overlap between the feral and domestic groups (Figure 48) and was classified as a feral hog with a PP of 0.84.

The mandible of an adult female (FSM 3364) found on Cedar Key, Levy County, in 1948 was also classified as a feral hog (PP = 0.927; Figure 53).

Five specimens from Alachua County collected between 1955 and 1961 were mostly classified as feral hogs. Four of these—an adult male (FSM 3360), an adult female (FSM 6685), and two subadult males (FSM 5975 and 3282)—were all clearly classified as feral hogs in both cranial and mandibular analyses (Figures 43, 52, 53, 48, and 49). Another specimen, a yearling female (FSM 3340), fell within or close to both the feral and hybrid target groups for both cranial and mandibular analyses (Figures 46 and 47).

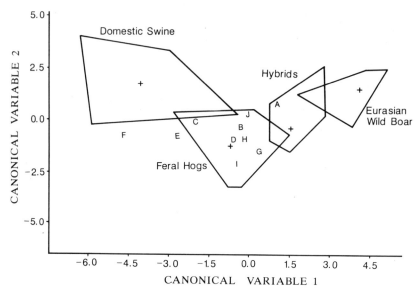

Figure 52. Canonical plot of adult female crania from the four target groups versus the following U.S. unknowns: A = Volusia Co., Fla.; B = Indian River Co., Fla.; C = Alachua Co., Fla.; D = Wheeler Co., Ga.; E = St. Martin Parish, La.; F = Lafayette Parish, La.; G = Travis Co., Tex.; H = San Patricio Co., Tex.; I = Chambers Co., Tex.; and J = McIntosh Co., Ga. Target group boundaries and means are taken from Figure 13.

No other data were available for these animals. Since no wild boar introductions have been reported for this area and the other samples from this county were clearly feral specimens, this last individual was likely also a feral hog.

A subadult female (FMNH 57173) was collected in Florida in 1950 by Colonel E. N. Wentworth, but no specific locality or other collection data were given. The animal was classified as a feral hog in the cranial analysis (PP = 0.78) and as a hybrid for the mandible (PP = 0.60; Figures 50 and 51). It did, however, also fall within the feral target group in this latter analysis.

Georgia

The largest sample of unknown specimens from a single locality in this state came from the Oconee River bottomlands in Baldwin County in the late 1970s. Fourteen specimens ranging from an adult male to neonate/infant were analyzed. As expected for a sample of this size, the results were mixed.

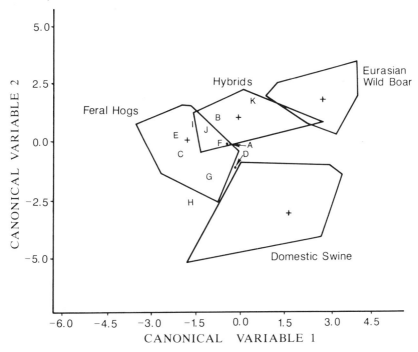

Figure 53. Canonical plot of adult female mandibles from the four target groups versus the following U.S. unknowns: A = Volusia Co., Fla.; B = Indian River Co., Fla.; C = Levy Co., Fla.; D = Alachua Co., Fla.; E = McIntosh Co., Ga.; F = Wheeler Co., Ga.; G = St. Martin Parish, La.; H = Lafayette Parish, La.; I = Travis Co., Tex.; and J and K = San Patricio Co., Tex. Target group boundaries and means are taken from Figure 13.

The collector of these animals claimed that all were feral hogs (J. Phelps, pers. comm.). The one adult male in the sample was classified as a feral hog in the cranial analysis (PP = 0.67) and as domestic in the mandibular analysis (PP = 0.69; Figure 43). In the canonical plot of the mandibular analysis, this specimen was within the domestic target group and adjacent to the hybrid target group but below the feral target group. This animal was all black and weighed 225 kg. Although almost all five yearling specimens fell within the feral target group, they were classified as either feral hogs or domestic swine depending on how close they were to either of the target group means (Figures 46 and 47). The only specimen that did not fall within the feral group was the mandible of one all-black yearling, which fell within the domestic target group. The seven juvenile specimens from this population all fell within the feral target groups in both the cranial and mandibular analyses (Figures 44 and 45). The single neonate/infant specimen was classified as a domestic animal (PP = 0.88). It was plotted outside but adjacent to the feral target group and immediately next to the domestic group mean (Figure 54). Coat coloration patterns in this sample included all black (6), all red/brown (1), all white (3), black-and-red/brown spotted (2), tricolored spotted (1), and black with white points (1). No curly, woollike underfur or light-tipped bristles were seen in any of these animals.

Two specimens were examined from Blackbeard Island, McIntosh County. No wild pigs have been reported recently on this island (Johnson et al., 1974), although Carter (1943) reported seeing six "wild hogs" in a slough on Blackbeard Island in April 1943. One specimen was the mandible of an adult sow (USNM, not catalogued), and the second was the cranium of another adult sow (CM 39985). These skeletal remains were collected on the island in 1940 and 1959, respectively. Both the cranium (PP = 0.88) and mandible (PP = 0.89) were classified as feral hogs (Figures 52 and 53, respectively).

An adult female (DLA 1) from near the Oconee River in Wheeler County was classified as a feral hog in both the cranial and mandibular analyses (PP = 0.76 and 0.62, respectively; Figures 52 and 53).

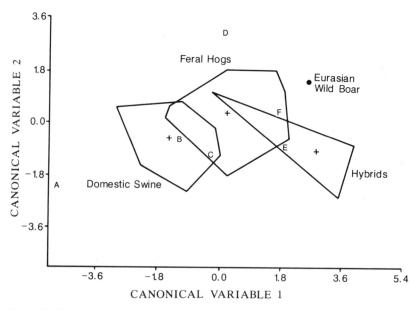

Figure 54. Canonical plot of neonate (sexes combined) crania from the four target groups versus the following U.S. unknowns: A = Putnam Co., Fla.; B = Baldwin Co., Ga.; C = Mitchell Co., N.C.; D = Travis Co., Tex.; E = Chatham Co., Ga.; and F = Beaufort Co., S.C. Target group boundaries and means are taken from Figure 18.

A juvenile specimen (AMNH 93138) collected on the Chinquapin Plantation in Thomas County in 1930 was similarly classified as a feral hog in both cranial (PP = 0.98) and mandibular (PP = 0.80) analyses (Figures 44 and 45).

One yearling specimen (SREL, not catalogued) from the BASF plant site on the Savannah River in Chatham County was also identified as a feral hog in the canonical analyses for both the cranium and mandible (PP = 0.61 and 0.54, respectively; Figures 46 and 47). A neonate/infant from the same locality (SREL, not catalogued), however, was classified as a hybrid in the cranial analysis (PP = 0.90; Figure 54). In the canonical plot, this animal, which was all black in color, fell close to an overlap area between the feral and hybrid target groups.

On 10 October 1981, an adult male wild pig (JJM 431) was killed in the town of Martinez, in the suburbs of Augusta, Columbia County, Georgia, after attacking several head of livestock and supposedly threatening the local populace. In both cranial and mandibular analyses this individual was classified as a feral hog (PP = 0.94 and 0.66, respectively; Figure 43). The origin of this animal was unknown. No populations of wild pigs are known to occur along that part of the Savannah River. However, as noted earlier, feral swine have been released by hunters along the Savannah River as far north as Augusta, Georgia, only a few kilometers from where this animal was killed. This animal was all black, had curly black underfur, and was estimated to weigh between 90 and 115 kg.

One subadult male specimen (FMNH 43099) from Georgia had no specific locality data. It was killed in the wild by Harry E. Mock before 1935. This animal was classified as a feral hog in both the cranial (PP = 0.66) and mandibular (PP = 0.50) analyses (Figures 48 and 49).

Louisiana

An adult female (USLBM 1204) found dead along a ditch bank in 1958 near St. Martinville, St. Martin Parish, was classified as a feral hog in both the cranial and mandibular analyses (PP = 0.902 and 0.910, respectively; Figures 52 and 53).

An adult female (USLBM 1672) and a yearling male (USLBM 1673) were collected from a heavily wooded area near Lafayette in Lafayette Parish. The sow was identified by the cranial analysis as domestic (PP = 1) and by the mandibular analysis as feral (PP = 0.63; Figures 52 and 53). The male was classified as feral by the cranial analysis (PP = 0.76) and as domestic by the mandibular analysis (PP = 0.95; Figures 46 and 47). The owner of the land where these animals were killed knew nothing of their origin but did state that they were part of a small group of hogs that had been living wild there for some time (M. B. Eyster, pers. comm.). No other data were available for these animals.

Mississippi

A subadult female (MMNS 6069) from north of Madison, Madison County, was classified as a hybrid by both the cranial and mandibular analyses (PP = 0.68 and 0.66, respectively; Figures 50 and 51). In both analyses, however, it fell either within the feral target group or between the feral and hybrid groups. It did not fall within the bounds of the hybrid target group in either analysis, although it was closer to the hybrid group means. No other data were available. Because it fell into an area of target group overlap and because hybrid wild pigs have been stocked into Madison County, the classification of this specimen as a hybrid may not be an error.

North Carolina

A neonate/infant specimen (USNM 47786) was collected by C. Hart Merriam on Roan Mountain in Mitchell County during a ground squirrel survey in 1892. The cranium fell on the canonical plot within the domestic target group (PP = 0.53) and also within the feral group (PP = 0.47) in an area of target group overlap (Figure 54). The skin of this animal had a faint but discernible pattern of juvenile striping over a pale tan body color, similar to those seen today in other populations where there are no clear records of wild boar or hybrid introductions (e.g., Hillsborough County, Florida, as described earlier). This animal was collected after Corbin's wild boar were brought to the United States but before any wild boar were introduced into the Tennessee–North Carolina mountains. Because there were supposedly no wild boar in this area at that time, and since juvenile striping has been reported in some purely feral populations, this animal was probably a feral animal.

The cranium of a juvenile animal of unknown sex (JJM, not catalogued) from one of the Outer Banks islands (Ocracoke Island, Dare County) was classified as a domestic animal (PP = 0.87) although it was also included within the boundaries of the feral target group (Figure 44). This skeletal specimen was found on the Ocracoke Sound side of the island by J. D. Lazell in the early 1970s. Although presumed at that time to be a wild-living animal, this was not known for certain. There has been a population of feral hogs on the Outer Banks in Carteret County, but not in Dare County. Because of this, it is likely that this was a domestic animal that had been butchered on the island. Webster (1988) does not list any extant feral hog populations in a recent survey of the mammalian fauna of the Outer Banks, including portions connected to the mainland in the north.

South Carolina

Four specimens were collected on Hilton Head Island in Beaufort County by W. L. Brown in December 1930. These consisted of two subadult females (USNM 256035 and 256036), one juvenile (USNM 256037), and one neonate/infant (USNM 256038). Crania of the two subadults were both classified as feral hogs (PP = 0.77 and 0.62), yet both mandibles were classified as hybrids (PP = 0.92 and 0.74; Figures 50 and 51). All four points were plotted in or near nonoverlapping areas of the respective target groups. The juvenile was classified as feral in both the cranial (PP = 0.896) and mandibular (PP = 0.84) analyses (Figures 44 and 45). The neonate/infant was equidistant from the wild boar, feral hog, and hybrid group means, but was included within the feral target group (PP = 0.81; Figure 54). Brown (1930) stated in his field notes only that "wild hogs" were found on the island. No introductions of wild boar or hybrids have ever been reported for Hilton Head Island.

Another subadult female (USNM 254667) collected by Brown in 1929 from Allendale County along the Savannah River was classified as a feral hog in the cranial analysis (PP = 0.89) and was between the feral and hybrid groups in the mandibular analysis (PP = 0.44 as feral and 0.56 as hybrid; Figures 50 and 51). The specimen fell adjacent to the feral target group, near an area of overlap with the hybrid group. Because target group resolution was better in the cranial canonical plot and it had a high PP in that analysis as belonging to the feral target group, this animal was probably a feral hog.

Texas

The mandible of a subadult male (KS 2) from the Webb Ranch in La Salle County had a PP of 0.75 as a hybrid animal but was plotted in an area of extensive overlap between the feral and hybrid target groups (Figure 49). Because of the poor resolution within this analysis and the common practice of introducing hybrids from the Edwards Plateau into the ranch country south of San Antonio, it was not possible to accurately classify this specimen without cranial measurements.

Thirty-four specimens from the Welder Wildlife Refuge in San Patricio County gave mixed results. These specimens consisted of matched and un-matched crania and mandibles and included specimens ranging in age from adult to juvenile. A cranial analysis of the single adult male in the sample (JJM 362) classified it between the feral and domestic groups (PP = 0.47

and 0.53, respectively) adjacent to an area of group overlap (Figure 43). On the other hand, the mandible was classified by the canonical plot below the hybrid target group and between the wild boar and domestic groups (Figure 43). It had a PP of 0.67 as belonging to the wild boar target group. A complete adult sow from this group (WWR H782) was classified as a feral hog in both the cranial and mandibular analyses (PP = 0.70 and 0.67, respectively; Figures 52 and 53). An adult female mandible (JJM 374), however, was classified as hybrid (PP = 0.91) and was plotted within the hybrid group near the wild boar target group (Figure 53). The three subadult female mandibles (WWR H766, H4, and not catalogued) were plotted between the feral and domestic target groups (Figure 51). The first two had PPs as feral hogs of 0.88 and 0.81, respectively, while the latter had a PP of 0.93 as domestic. The yearling crania were scattered on the canonical plot within the feral and hybrid target groups (Figure 46). The PPs ranged from 0.69 as belonging to the feral group to 0.98 as hybrid. The results of the mandibular analyses were similar, with PPs ranging from 0.44 for domestic to 0.93 for hybrid classifications (Figure 47). The juvenile cranial analyses resulted in the specimens being plotted within or near the feral group and in an area of overlap between that group and the domestics (Figure 44). Mandibular analyses gave the same results (Figure 45). Coat coloration patterns at the refuge included all black, all red/brown, black-and-white spotted, red/brown-and-white spotted, black-and-red/brown spotted, and tricolored spotted. None of the neonates exhibited the juvenile striped pattern. Light-tipped bristles and curly underfur were not present in this population. Average body weights for adults in this population ranged between 80 and 102 kg. Estimated maximum body weights for males on the refuge approach 159 kg. In conclusion, although morphologically variable, this population is almost certainly of feral origin.

One adult male (TCWC 1432) from south of Tilden in McMullen County was collected in 1922. In the cranial analysis it was classified as a feral animal, with a PP of 1 (Figure 43). In the mandibular analysis it fell near an overlap zone between ferals and hybrids, with a PP of 0.66 as hybrid and a PP of 0.33 as feral (Figure 43). This animal was estimated to weigh between 102 and 114 kg. No data on coat coloration or hair morphology were available. Three mounted adult male specimens killed near Tilden in the 1970s were also examined during this study. All three were completely black with no curly underfur. Other coat coloration patterns reported for the wild pigs in McMullen County included all red/brown, black-and-white spotted, black with white points, and black with white shoulder belt (W. Walker, pers. comm.; K. Schwarz, pers. comm.).

A subadult female (UIMNH 132985) from near Saratoga, Hardin County, in east Texas was classified as a hybrid in both cranial and mandibular analyses (PP = 0.78 and 0.89, respectively). In the cranial analysis it was plotted in a nonoverlap area of the hybrid group (Figure 50). In the mandibular analysis it was plotted in an overlap zone with the feral group (Figure 51). No additional data were available for this animal, although no introductions of wild boar or hybrids have been reported for this area. Based solely on the cranial analysis, however, this animal was indeed a hybrid.

The cranium of another subadult female (USLBM 2609) from a wooded area adjacent to a bayou near Cleveland in Liberty County in east Texas was classified as belonging to a feral hog (PP = 0.98; Figure 50).

A cranium of an adult female (UTEP 2379) from near Wallesville in Chambers County in east Texas was also classified as feral (PP = 0.98; Figure 52).

Three specimens were collected in march 1951 from the Norris Division of the King Ranch in Kenedy County. The adult male (OSU 3778) was plotted as a feral hog in both analyses (PP = 0.97 for the cranium and 0.78 for the mandible; Figure 43). Juvenile specimen OSU 3776 was classified as domestic in both analyses, and juvenile specimen OSU 3777 was classified as feral in both analyses (Figures 44 and 45). The cranial and mandibular PPs for these animals were 0.82 and 0.96 as a domestic for the first, and 0.94 and 0.80 as a feral for the second. No other data were available for these animals.

Although not well documented, the wild pig population in the Norris Division of the King Ranch is believed to have originated from domestic stock released by the early Spanish ranchers in the area, with subsequent releases also made by later ranchers (V. W. Lehman, pers. comm.). None of the piglets was reported to be striped, and most of the adult wild pigs in the Norris Division were solid black in coat coloration (V. W. Lehman, pers. comm.). Of 13 subadult/adult wild pigs observed by the senior author during a field survey of the King Ranch in 1982, 12 were solid black and 1 was red/brown. One solid black sow seen during the survey had a litter of six piglets, none of which was striped. All the subadult/adult animals appeared to have long snouts, erect to semierect ears, straight tails, long legs, and dense pelage. The feral hogs that inhabit the area around the Venado Hunt Camp on the King Ranch are also believed to be exclusively feral hogs (T. C. Urban, pers. comm.). Most wild pigs in that area are red/brown or spotted in color. Very few are solid black, and none of the piglets are reported to be striped (T. C. Urban, pers. comm.). The largest wild pig killed in this latter area of the ranch was a sow that weighed 113 kg (T. C. Urban, pers. comm.).

An adult sow (UTAVP M454) from near Austin in Travis County was also classified as a feral hog in both cranial (PP = 0.55) and mandibular (PP = 0.75) analyses (Figures 52 and 53).

An adult male (KS 1) from a ranch near Pearsall in Frio County was classified as a hybrid in the cranial analysis (PP = 1) and as feral in the mandibular analysis (PP = 0.77; Figure 43). The mandibular coordinates in the canonical plot were in an area of overlap with the hybrid target group. This animal was black-and-white spotted. Other coat coloration patterns from this area were all black, all red/brown, and black-and-red/brown spotted (R. T. Layman, pers. comm.; L. Penn, pers. comm.). However, hybrid animals are also known to occur in this area, and the results of the cranial analysis are probably correct.

A subadult female (MSB 46494) from the Laguna Atascosa National Wildlife Refuge in Cameron County in extreme southern Texas was classified as a hybrid in both the cranial and mandibular analyses (PP = 0.87 and 0.96, respectively). It also plotted in or closest to the hybrid target group in both cases (Figures 50 and 51). However, there have been no reports, recent or otherwise, of wild boar or wild boar × feral hog hybrids introduced into this area.

Discussion

Morphotype Determination

The identifying characteristics of the three different classifications of wild pigs in the United States, and in fact around the world, have been poorly studied on a comparative basis. Morphotypic distinctions have been further complicated by potential and actual hybridization between feral hogs and wild boar, and between both types and their hybrids or domestic swine. Stegeman (1938) discussed the physical characteristics of the introduced wild boar in Tennessee, giving some comparisons with local free-ranging domestic swine. Most of these characters were subjective descriptions. Both Hanson and Karstad (1959) and Golley (1964) described feral hogs as rangier and thinner in body conformation than domestic swine. Henry (1969) listed three basic characteristics indicative of pure wild boar ancestry: a striped pattern in the juvenile pelage, split tips on the guard hairs, and a diploid chromosome number of 36. Feral and domestic swine supposedly did not have striped young or split-tipped bristles and had a diploid chromosome number of 38. Henry was incorrect with respect to all three characters being absolute criteria. Juvenile striped coloration patterns have been reported to occur in both feral hogs and domestic swine (Darwin, 1867; Keller, 1902; Lydekker, 1908b; Hetzer, 1945; Crandall, 1964; Searle, 1968; this study). Splitting of the guard hairs or bristles has also been shown to exist in domestic swine, feral hogs, and wild boar × feral hog hybrids (Marchinton et al., 1974; this study). Rary et al. (1968) considered cytogenic differences more useful than morphological characters for identifying the three types of wild pigs. Table 21, however, shows the polymorphic variation expressed in the karyo-

Table 21. Variations in the diploid chromosome number of wild boar, domestic swine, feral hogs, and wild boar × feral hog/domestic swine hybrids.

Subspecies and Locality	Diploid Number			Reference
	36	37	38	
Wild Boar				
Sus scrofa				
Azerbaijan SSR		X		Tikhonov and Troshina, 1974
Sus scrofa ferus (?)				
Baltic States	X			Tikhonov and Troshina, 1974
Sus scrofa leucomystax				
Japan			X	Muramoto et al., 1965
Sus scrofa lybicus				
Israel			X	Epstein, 1971
Sus scrofa meridionalis				
Corsica			X	Popescu et al., 1980
Sus scrofa nigripes				
Kirghiz SSR	X	X		Tikhonov and Troshina, 1974
Sus scrofa reiseri				
Yugoslavia			X	Zivkovic et al., 1971
Sus scrofa scrofa				
France	X			Popescu et al., 1980
Germany	X			Grop et al., 1969
Netherlands	X	X	X	Bosma, 1976
Austria	X			Rittmannsperger, 1971
Sus scrofa ussuricus				
Pacific Maritime Prov.		X		Tikhonov and Troshina, 1974
Amur Region		X		Tikhonov and Troshina, 1974
Domestic Swine				
Sus scrofa domesticus				
Spain			X	Gimenez-Martin et al., 1962
Ohio			X	McConnell et al., 1963
Japan			X	Muramoto et al., 1965
Pennsylvania			X	Cornefert-Jensen et al., 1968
Austria			X	Rittmannsperger, 1971
Ghana		X	X	Marczynska and Pigon, 1973
Sweden			X	Hansen-Melander and Melander, 1974
Missouri			X	Pace et al., 1975
Brazil			X	Giannoni and Ferrari, 1977

Continued on next page

187

Table 21—*Continued*

Subspecies and Locality	Diploid Number			Reference
	36	37	38	
Feral Hogs				
Sus scrofa domesticus				
Dye Creek Ranch, Calif.			X	Barrett, 1971
Wild Boar × Feral/Domestic Swine Hybrids				
Sus scrofa spp.				
Tellico WMA, Tenn.	X	X	X	Rary et al., 1968
Monterey Co., Calif.	X		X	Barrett, 1971
Japan			X	Muramoto et al., 1965
Sweden		X		Gustavsson et al., 1973
France		X		Mauget, 1980
France		X	X	Popescu et al., 1980

type of this species. This variation has been the result of Robertsonian changes in one or two of the chromosome pairs (McFee et al., 1966). Based on the data in Table 21, an animal with 36 chromosomes could be a wild boar or a hybrid; one with 37 chromosomes could be a wild boar, hybrid, or domestic; and one with 38 could be any of the four types. Henry's (1969) characters, therefore, were either completely incorrect or not very accurate.

Since the above work, no attempt has been made to determine characteristics for classifying the four morphotypes of *Sus scrofa*. The results of this study show that the different morphotypes can be distinguished on the basis of a combination of morphological characters, including skull morphology, external body dimensions and proportions, coat coloration, and hair morphology. Used alone, however, any of these four morphological characters might be insufficient to determine the type of *Sus scrofa* in certain cases.

Skull Morphology

Skull characteristics, especially the size and shape of the cranial bones, are well recognized by taxonomists as one of the best means for classifying vertebrates (Lowe and Gardiner, 1974). It is not surprising, then, that the most successful method of morphotype classification utilized cranial comparisons. Various measurements, general shape, and the lacrimal index of the skull were the primary criteria from which most of the Eurasian wild boar subspecies were described and compared (Major, 1883; Nehring, 1889; Jentink; 1905; Miller, 1906; Thomas, 1912; Adlerberg, 1930; Heim de Balsac, 1937; Amon, 1938; Kelm, 1939). Several of these subspecies later turned out to be feral hogs. Unfortunately, most of the original descriptions were based on four or fewer specimens. Few if any cranial measurements were used in these descriptions, which did little to account for the potential variation exhibited by the morphotype. Cabon (1958) found a significant amount of morphological variation within all sex and age classes in a sample of 93 wild boar skulls from four localities in northeast Poland. Barrett (1971) compared Cabon's data with a sample of 113 feral hog skulls from the Dye Creek Ranch in California. He found that the best separation was obtained by doing a regression between the nasal length and zygomatic breadth of the adults. Barrett also found that the two forms could not be distinguished with any certainty in younger animals. This same decreasing resolution between the four morphotypes of *Sus scrofa* at younger ages was also found in this study. In addition, Barrett noted that feral hogs had "dished" dorsal profiles, while the wild boar skulls in Cabon's sample were mostly flat. Both authors, however, found variation in this character. Barrett found that individuals originating in the backcountry of the ranch where poor forage is found had flat profiles, while those from pastures with good forage had dished profiles. He postulated that profile depth was determined by both heredity and a nutritional factor related to the environment.

Pira (1909), Kelm (1938), Epstein (1971), and Bokonyi (1974) discussed the changes the *Sus scrofa* skull undergoes in domestication, including a shortening and broadening of the skull. The anterior premolars also become vestigial or disappear, and the dorsal profile becomes more dished. The shortening of the cranial bones is especially evident in the lacrimal bones (Epstein, 1971; Bokonyi, 1974). Pira (1909) also noted that the tooth row is proportionally longer in domestic swine, that the direction of the slope of the back of the cranium is anteriorly directed in domestics, and that the cross section of the lower male canines varies between wild and domestic swine.

No skull comparisons have previously been made between wild boar × feral hog hybrids and the other morphotypes of *Sus scrofa*. The only hybrid data in the literature were given by Henry (1970), who gave only the maximum measurements for "skull length" of males and females from the Tellico Wildlife Management Area in Tennessee. Since this measurement was not defined in that study, these data were not comparable with those of other studies.

On the basis of a single skull of either a primitive domestic pig or a feral hog from the Museo de Historia Natural in Buenos Aires, Pira (1909) claimed that domestic stock with broad skulls and dished profiles developed long narrow skulls and flat profiles after returning to a wild-living existence.

Von Nathusius (1864) showed that diet could have a marked effect on the cranium of littermates of the Berkshire breed of domestic swine. Those individuals reared on a low plane of nutrition had long, narrow skulls compared with those reared on a high plane of nutrition. The rostral length was also proportionally longer and the dorsal profile flatter in the malnourished siblings. McMeekan (1940a, 1940b, 1940c) showed the same results for domestic swine of the Large White breed. It appears, then, that the cranial morphology of at least domestic swine and feral hogs are plastic in their ontogeny, and that nutrition can be an important factor in determining the final outcome of any individual's cranial morphology. Deficiencies in calcium or phosphorus or a lack of vitamin D have been determined to retard the normal skeletal development in domestic swine (Ensminger, 1961).

Weaver and Ingram (1969) showed that different environmental temperatures had a variable effect on the size and shape of the skull of a growing domestic pig. Littermates reared at 5°C had significantly shorter skulls compared with siblings reared at 35°C. The piglets reared at the colder temperature also had slightly wider skulls, but these differences were not statistically significant. Environmental temperatures, then, can also have a marked effect on the gross morphology of a pig skull.

Morphological sexual dimorphism is characteristic of wild boar, feral hogs, hybrids, and domestic swine, with males both averaging and ranging larger than females. This dimorphism is exhibited in body measurements and weights and in size of the permanent canines (Nichols, 1962b; Harrison, 1968; Briedermann, 1970; Henry, 1970; Ensminger, 1961; Herring, 1972; Romic, 1975; Walker, 1975; Wood and Brenneman, 1977). Behavioral sex role differentiation is apparently the underlying reason for the morphological sexual dimorphism seen in suids (Herring, 1972).

With the exception of Mayer and Brisbin (1988), however, studies comparing the morphological sexual dimorphism in this species have yet to show that these differences are consistently statistically significant. Other studies have reported no statistically significant differences in body sizes between the sexes in either Polish wild boar or wild boar × domestic swine (*zlotnicka pstra* breed) hybrids (Zurowski et al., 1970). Brisbin et al. (1977a) reported similar findings for the body weights and external body measurements of feral swine. Barrett (1971), however, did find that the head lengths of male feral hogs from the Dye Creek Ranch, California, were significantly greater than those of females from the same population. Mansfield (1978) found that adult male hybrid wild pigs from Fort Hunter Liggett, California, were significantly heavier than adult females.

The discriminant function analysis and Student's *t*-test on the skulls of the four target groups in this study have shown that these differences are statistically significant for wild-living *Sus scrofa*, but not for domestic swine.

In general, these differences show that the skulls of wild-living males are significantly larger than wild-living females of the same morphotype. Selective breeding and a relaxation of selection for larger body size in male-male competition for females might explain the lack of a significant sexual difference in skull size in domestic swine.

External Body Dimensions and Proportions

Body measurements of Eurasian wild boar have been published by Fitzinger (1864), Miller (1903) Cabrera (1914), Phillips (1935), Amon (1938), Baumann (1949), Gaffrey (1961), Durov and Alexandrov (1968), Briedermann (1970), Diong (1973), Harrison (1974), Romic (1975), Lekagul and McNeely (1977), Roberts (1977), and Gunchak (1978). Body measurements for four feral hog populations in the United States were given by Nichols (1962a), Sweeney (1970), Barrett (1971), and Brisbin et al. (1977a). Henry (1970) gave similar data for the hybrid population in southeastern Tennessee. Of all of these studies, only Brisbin et al. (1977a) collected and compared data from more than one population. In that study, the long-term feral hog population from Ossabaw Island, Georgia, and the short-term feral hog population of the Savannah River Plant in South Carolina were compared.

Wild boar from populations in New Hampshire and the hybrid population in Tennessee and North Carolina are reported to be smaller than wild boar in Europe, the probable ancestors of both populations (Stegeman, 1938; Silver, 1957; Jones, 1959; Henry, 1970; Conley et al., 1972). Wild boar and hybrids from these two populations are reported to reach a maximum of 182 kg and 205 kg, respectively (Jones, 1959; A. Dobles, pers. comm.), while the maximum weight for wild boar in Europe is reported as 273 kg. This weight is exceptionally large for reported wild boar weights (Table 12). Snethlage (1950) reported having read and heard of wild boar approaching 180 kg but had never actually seen one. Normal maximum weights of male wild boar in Europe are around 200 kg. While hunting in the Federal Republic of Germany during a two-year period in the late 1960s, JJM heard of wild boar weighing around 90–114 kg but never saw any of these animals. One German hunter reported having heard of an adult male wild boar killed near Giessen in the state of Hesse that weighed 182 kg on the local post office scales. The hunter thought this was an exceptional animal (J. A. May, pers. comm.). These estimates for the comparisons between maximum weights of European wild boar and the introduced wild boar and hybrids in the United States were from exceptional animals and probably should not be used in such comparisons. In actuality, the averages for these morphotypes from the United States are probably not smaller than averages for the Eurasian wild boar.

Coat Coloration Patterns

No variation was seen in the coat coloration pattern of the wild boar sample during this study. All had the wild/grizzled coloration. The Corbin's Park wild boar population has had a white spot mutation since at least the mid 1950s (Silver, 1957). This mutation produces a piebald, or white-spotted, pattern. Recently, this mutation has even produced some all-white wild boar (E. P. Orff, pers. comm.). Most of these animals have been removed through selective harvest from the park's population. Because of this, the number of animals exhibiting this coloration has remained small, and generally they do not reach maturity (H. J. McCarthy, pers. comm.). An all-white phenotype similar to that reported for Corbin's Park has also been observed in a captive European wild boar population in France (Mauget, 1980). The spotting of the Corbin's Park wild boar is different from the spotty mutation described by Andrzejewski (1974) for the wild boar in the Kampinoski National Park, Poland. The animals in Poland had a cream-colored coat with black spots that was considered a naturally occurring mutation within that population. Some spotted coats in Eurasian wild boar are the result of female wild boar breeding with free-ranging male domestic swine. This interbreeding is rare and is generally prevented in Europe (Boback, 1957). Hybridization of this type would probably have a lasting effect on the phenotypic makeup of the coat coloration in a wild boar population because both all white and the white shoulder belt are dominant over the wild/grizzled pattern (Hetzer, 1945; Searle, 1968). In the New Hampshire; Calhoun County, Texas; and western European wild boar populations sampled, all neonates exhibited juvenile striping. This is a dominant color pattern, and no variation has been reported in natural populations.

Black appears to be the most common coat color in feral hogs. The results of Maynard (1872), Nichols (1962a), Sweeney (1970), Barrett (1971), Brisbin et al. (1977a), Belden and Frankenberger (1979), Duncan and Schwab (1986), and this study all found solid black to be the predominant coat color. This is also true for some populations in Mexico and Australia (Mearns, 1907; Pullar, 1953; McKnight, 1976). Nichols (1962a) reported that 88% of a sample of the population of mountain feral hogs on Mauna Kea on the island of Hawaii were solid black; the remainder were black with white feet. Spotted patterns were the next most common, followed by the belted patterns.

Researchers as far back as Darwin (1867) noted that all black was the most prevalent coat color in feral hog populations. The reasons for the high frequency of all-black coloration in this morphotype are unclear. Only a few theories have been developed to explain this phenomenon. Darwin quoted Professor Wyman, who stated that light-colored hogs contracted a disease from eating paint root (*Lachuanthes tinctoria*) that caused their hooves to drop off. Black-colored feral and domestic hogs apparently had a natural immunity to this plant's toxins. Maynard (1872) altered this slightly by saying that the coat color of the animal has no effect on this susceptibility, but the hooves must be black for the animal to be unaffected by eating paint root (he had seen black feral hogs with white feet that were lame from this disease). He also reported that white feral hogs were more numerous than black ones in some sections of Florida where paint root does not grow. Maynard also reported that settlers in Florida preferred black domestic swine to white because the former had a better chance of escaping predation by bears, especially at night. Thus the higher frequency of the black coat coloration in the southeastern feral populations may have been a result of founder effect as well as subsequent selection. Other authors, however, continue to identify selection as the sole cause of the predominance of black coat color in feral hog populations. McKnight (1976) stated that the dominance of dark shades among "well-established colonies" of feral hogs is at least an indication that most feral hog populations eventually become black in color, and that coat color frequencies may provide a clue to the relative length of time that a population has been feral. Brisbin et al. (1977a) suggested sun scalding as a possible factor selecting against white coats in feral hog populations that inhabit open areas, such as the *Spartina* salt marshes on Ossabaw Island, Georgia. Compared with these habitats, the shaded canopy forest of the Savannah River Plant had a greater incidence of all-white animals. However, since the coat coloration phenotypes of the stock that the early Spanish released on Ossabaw Island are unknown, it is impossible to determine whether it was founder effect or selection that resulted in the virtual absence of the all-white phenotype in the Ossabaw population. In one of the few long-term studies of color phenotype frequencies in a feral population, Mayer et al. (1989) showed that there was no general trend toward a higher frequency of black animals in the short-term feral population of the Savannah River Plant, South Carolina. These animals have inhabited a generally shaded area of closed-canopy bottomland swamps and have been free of directed selection for over 30 years.

All white was the most common coat coloration pattern among the sample of domestic swine. Searle (1968) stated that "self white" was common among domestic swine because it was deliberately selected by breeding programs and breed development. Because of the effect that artificial selection would have on the frequencies of domestic coat colorations, only the presence or absence of a coloration pattern is useful in comparisons with the wild-living morphotypes. With the exception of the wild type, or grizzled, coat coloration, only eight of the possible coat coloration categories defined by this study were observed in the wild-living swine. All colors, however, have been reported for domestic swine (Hetzer, 1945; Searle, 1968).

The juvenile striped pattern was reported for eight of the feral hog populations either observed by the authors or described in the literature and by correspondents in questionnaires. The occurrence of juvenile striping within these populations varied from very infrequent (only one or two out of hundreds of animals) to about 50% (Nichols, 1964; Barrett, 1971; Z. T. Kirkland, pers. comm.). Juvenile striping in strictly feral hog populations has been known since the mid 1800s (Darwin, 1867). Feral hog populations in Jamaica, New Granada, Australia, Andaman Islands, Nicobar Islands, Venezuela, and Sudan have also been reported to exhibit this phenotype (Darwin, 1867; Rolleston, 1879; Pullar, 1953; Ansell, 1971; T. Blohm, pers. comm.). This is not totally unexpected, because some breeds of domestic swine are known to have striped piglets (Rolleston, 1879; Keller, 1902; Hetzer, 1945; Crandall, 1964; Searle, 1968). This phenotype in feral populations has often been regarded as evidence of the introduction of wild boar stock (Henry, 1969; Barrett, 1971), but this is not always the case. Since juvenile striping is genetically dominant (Hetzer, 1945; Searle, 1968), it would be expected to be more common than it is in those populations where it does occur.

The association of juvenile striping with a slaty gray adult coat coloration is typical of the domestic Mangalitza hog of Europe. These color combinations found in feral populations with little or no documented history of hybrid introductions may represent primitive traits characteristic of European domestic swine introduced by early explorers or colonists. This may be particularly true in the southwestern portion of Florida in the area of the earliest Spanish introductions in the vicinity of Port Charlotte. These coat color patterns are found in feral hogs from the vicinity of the Peace and Caloosahatchee rivers; these animals show more of a general domestic than a long-term feral phenotype (R. M. Wright, pers. comm.). Additional reports of scattered occurrences of these patterns have also been received from Taylor, Jefferson, Wakulla, Lafayette, Hillsborough and Dixie counties, all located in the west-central–northwest region of the Florida peninsula (R. M. Wright, pers. comm.), all areas traversed by the earliest Spanish expeditions leading to the northwest out of the Port Charlotte area of colonization. In these Mangalitza-type animals, the slaty gray adult pelage is distinctively different from the wild/grizzled coat color of the wild boar–hybrid phenotypes, the only other morphotypes known to produce striping in the juvenile coat color (R. M. Wright, pers. comm.).

The wild/grizzled coat coloration is the one most commonly reported for wild boar × feral hog hybrid populations. Its frequency within these populations varies from 47.2% to 100% (Rary et al., 1968; Pine and Gerdes, 1969; Springer, 1977; this study). All black appears to be the next most common coat coloration. Rary et al. (1968) reported that 46.3% of the hybrid population in the Tellico River drainage in Tennessee had "blaze" faces. Spotting and belted patterns, however, were uncommon in all the above studies. Spotting was reported to occur in 4.4%–6.5% of these populations (Rary et al., 1968; this study), while belted patterns occurred in 5.6%–18.0% (Springer, 1977; this study). Juvenile striping has been reported for most of the hybrid populations in the literature. Fifty-two percent of the neonates from hybrid populations observed during this study were striped. In contrast to this, only 45.2% of the correspondents claimed to have seen striped piglets in hybrid populations. Most stated, however, that this striped pattern was seen in about half the piglets. Since juvenile striping is genetically dominant, its rarity in these reports may indicate inadequate field observations of appropriately aged animals. On the other hand, if the striped juvenile phenotype is indeed cryptically colored, it would be more difficult to observe under free-range conditions.

Hair Morphology

Differences between hair morphology of wild boar × feral hog hybrids and domestic swine have been studied extensively. Hansen et al. (1972) determined that captive hybrid piglets that originated from stock caught near Tellico Plains, Tennessee, had greater average pelage weight and density, a larger hair medulla, and more medullary vacuolation than domestic piglets. Foley et al. (1971) speculated that the greater pelage density of the hybrid piglets provided greater resistance to cold, which would have obvious selective advantages to a wild-living animal. Feder (1978) compared the bristles and underfur of European wild boar to those of two breeds of domestic swine and found that wild boar generally had longer, thicker hair. No quantitative comparisons have been made between any of these three morphotypes and feral hogs. Seasonal hair length differences (longer in winter, shorter in summer) have been noted in all four morphotypes (Boback, 1957; Mount, 1967; Barrett, 1971). Stegeman (1938) noted that the European wild boar had "heavier" pelage than the feral hog. Brisbin et al. (1977a) found that long-term feral hogs from Ossabaw Island, Georgia, had bristles significantly longer than those of short-term feral hogs from the Savannah River Plant, South Carolina. Jones (1959) also noted that wild boar had heavier pelage and that the distal tips of the bristles were split, in contrast to nonsplit bristles of feral and domestic swine. In addition, he noted that wild boar had silver-tipped bristles. Henry (1969) used the splitting of the bristle tips as an identifying character for wild boar. In a subsequent study, Marchinton et al. (1974) showed that a sample of wild boar × feral hog hybrid bristles had 94.4% split tips. They also showed that five breeds and one crossbred line of domestic swine had split-tipped bristles in frequencies varying from 92.1% to 96.0%. Light-tipped bristles, however, are characteristic of wild boar and hybrids (Springer, 1977; this study). Springer (1977) attributed the light-colored tips to the splitting of the distal tip, and not to a true color change. Feder (1978) noted that the white distal tip in wild boar bristles changed to a dark gray in the split region, but attributed the white tip to a pigment change. The splitting of a solid-colored bristle will produce the appearance of a slightly lighter-colored tip without an actual pigment change (Feder, 1978). Black bristles with dark brown split tips are common in feral hogs. Feder (1978) noted this same color change in the black bristles of Hanford miniature swine. However, this explanation does not explain the occurrence of white or cream tips on black bristles as reported by Springer (1977). In both morphotypes, bristles that had light-colored tips had some white tips that extended below the split tips, while others were much shorter than the split tips, with the basal portion of the prongs brown or black and the distal portion white. In these cases, the light tips must represent a pigment difference, and not a mechanical change.

Barrett (1971) stated that curly, woollike underfur was characteristic of European wild boar, but not feral hogs. This character has also been noted in hybrid populations (Table 21). As shown in this study, however, curly, woollike underfur occurs in all the wild-living morphotypes and is therefore not a useful characteristic in morphotypic determination.

Long-term versus Short-term Feral Hogs

A distinction in morphology of feral hogs that have long versus short feral histories has been reported previously. Pullar (1953) identified the two extremes as "early type" and "recent type" Australian feral hogs. "Early type" were described as smaller, black or dark red in color, narrow-backed with smaller hind legs, and having maximum weights approaching 32 and 64 kg for sows and boars, respectively. The "recent type" ferals were larger, had a higher proportion of light or mixed coat colors, a broader back, a less marked difference between fore and hind legs, and maximum weights of 68 and 136 kg for sows and boars. Pullar (1953) summed up this second type as resembling a "poorly developed modern domestic pig." He also reported intergrades between the two extremes, and the "early type" as being the scarcer of the two. Nichols (1962a) referred to the two types in Hawaii as "mountain pigs" (long isolated) and "forest pigs" (recent domestic escapees), referring to the primary type of habitat occupied by each form. The "mountain pigs" were solid black or black with white feet, had dense coats of bristles and underfur, high shoulders and long legs, and were normally smaller than 91 kg. "Forest pigs" were also mostly black, but they showed many mixed coat colorations, had sparse coats of bristles and rarely underfur, and a blocky body without either high shoulders or a sloping rump. "Forest pigs" also had somewhat shorter legs, and a maximum body weight of 260 kg. McKnight (1964, 1976) noted that the longer hogs in the United States and Australia had been feral, the more they developed leaner bodies with prominent backbone ridges. Animals with a longer feral history also tended to have longer snouts, longer hair, longer legs, and, frequently, longer and sharper tusks (McKnight, 1976). More recently, Brisbin et al. (1977a) compared external body measurements and coat coloration from a long-term (Ossabaw Island, Georgia) versus a short-term (Savannah River Plant, South Carolina) feral hog population. In general, their results agreed with those previously discussed.

In some cases in the present study, consistent misclassification of specimens from a "known" long- or short-term feral population suggested that the designation of the population itself might not have been entirely correct. In the case of the feral population on St. Catherines Island, Georgia, for example, historical records and personal interviews confirmed that the population there is indeed of recent feral origin. However, 13 of the 36 specimens misclassified as long-term ferals from known short-term populations were from St. Catherines Island. These findings must be interpreted in light of the fact that St. Catherines Island is separated from Ossabaw Island to the north by a channel of only several hundred meters of open water. The long-term feral hogs occupying Ossabaw are known to enter open water occasionally in attempts to escape pursuing hunting dogs by swimming, and it is not difficult to imagine that some of the individuals found within the short-term feral population on St. Catherines may actually be immigrants from the neighboring long-term feral population on Ossabaw Island. Furthermore, there is now a confirmed report of at least one hog captured on Ossabaw Island having been inadvertently released on St. Catherines Island as the result of a boating accident off the latter's coast in the early 1980s. Such dispersal events, whether man-made or natural, may occur frequently between long- and short-term feral populations and may have also been responsible, for example, for the sudden appearance of a generally long-term feral morphotype amongst the otherwise short-term feral population on the U.S. DOE Savannah River Plant in South Carolina (Figure 4; Plate 6). In any case, dispersal events between long- and short-term feral populations suggest that the "misclassification" of specimens by the statistical techniques employed here need not always result from limitations of the analytical techniques themselves. This would be particularly true when, as in the case of St. Catherines Island, there was a consistent tendency to misclassify a number of specimens, together with known geographical and historical factors supporting the possibility of interpopulation dispersal.

In many ways, the cranial characteristics of long-term feral hogs differed from those of short-term ferals in the same ways that wild boar differed from domestic pigs. Thus, for example, both long-term feral hogs and wild boar generally had shallower dorsal profiles (Figures 22 and 23) and larger angles of the occipital wall (Figures 24 and 25) than did their short-term feral and domestic counterparts, respectively. In these respects, then, it appears that longer periods of selection for survival in the feral state have indeed tended to reverse changes in cranial characteristics that have resulted from the domestication process. This is in contrast to the situation with coat color, where there has been no tendency for long-term feral populations to re-acquire the wild/grizzled adult color or striped juvenile pelage pattern of the wild boar. The difference in the ways that cranial characteristics versus coat colors have responded to a return to the feral state may be due to differences in the relative intensities of selection applied to these characteristics in most populations of feral swine. Nearly all feral swine populations have been strongly selected, as have those of their wild ancestors, to obtain food as effectively as possible. Accordingly, changes in cranial characteristics are almost certainly responsive to this need, conferring the ability for stronger and more effective rooting with long, straight snout profiles as opposed to the shorter and more upturned snouts of the domestic and short-term feral forms. Coat colors however, particularly to the degree that both the wild/grizzled and striped juvenile patterns provide cryptic coloration and concealment from predators, are probably not as essential for survival in the feral state—particularly in the United States. Larger predators such as panthers (*Felis concolor*) and wolves (*Canis lupus*) have been removed or greatly reduced in numbers in most areas currently inhabited by feral swine in this country. This, in turn, would greatly reduce the degree of selection for cryptic coloration. Although "predation" from human hunters has been sub-stituted for that of natural predators in many areas, such hunting generally does not involve any bias with respect to the color of adults taken (Mayer et al., 1989), and young juveniles are seldom hunted.

Non-U.S. Feral Hog Comparison

Feral hogs are probably the most widely distributed of the three wild-living morphotypes of swine. Feral populations occur in every major realm except Antarctica (Tisdell, 1982). Despite their varied origins and scattered dis-tributions, feral morphotypes appear to be consistently identifiable as such. Most of the morphometry conducted on non-U.S. hogs has been descriptive, with few examples of quantitative data and no comparative cranial morpho-metrics. Mearns (1907) stated that the feral hogs around the mouth of the Colorado River in Sonora, Mexico, were of "extraordinary size." He also gave skull measurements for a specimen (USNM 60356) from that locality, the same specimen also being used in this study. Bee (1957) reported whole body weights ranging in size from 27 to 82 kg for seven feral hogs from St. John Island, Virgin Islands. One adult female collected by Bee and used in this study (KU 74115) was estimated to weigh 68 kg. Husson (1960) reported that feral hogs on Curaçao had all the characteristics of domestic swine but were smaller than European wild boar, although no measurements were given. Daniel H. Janzen (pers. comm.) estimated that adult feral hogs in the Corcovado National Park in Costa Rica weighed 45–90 kg. Nizza (1966) reported that "half-free" Corsican swine averaged 65 kg in body weight at two years of age (sexes combined) when raised in captivity. He noted that this qualified them to be classified as "miniature swine." In comparison with the values presented previously, however, this is no smaller than the body weights reported for feral hogs from many other populations. Feral hogs in Australia have long legs, long snout, lean body, with shoulders and neck well developed, prominent backbone ridge, and conspicuous tusks (Pullar, 1953; McKnight, 1976).

Coat coloration patterns of non-U.S. feral hogs include black, rusty or reddish black, black-brown, blue-black, red, white, fawn, dun, roan, black-and-white spotted, and black with a white shoulder belt (Mearns, 1907; Pullar, 1953; Nizza, 1966; Bee, 1957; McKnight, 1976). Pullar (1953) also included two color phenotypes that he called "agouti colored" and "tiger markings." The first pattern is undefined but probably means banded bristles of alternating colors, as described for the wild type or grizzled coat color of the present study. The second is probably a variation of black-and-red/brown spotted. Mearns (1907), Pullar (1953), Husson (1960), Nizza (1966), and McKnight (1976) all cite black or very dark colored as the most common coat color. None of these descriptions of size, color, or external morphology suggest that any non-U.S. feral hog populations tend to differ in a significant way from their counterparts in the United States, with respect to these features.

Comparisons with Captive Wild Pigs

Nutritional effects on development and growth of the skeleton have been shown for domestic swine. The changes in the morphology of the skull in domestic swine as determined by Von Nathusius (1864) and McMeekan (1940a, 1940b, 1940c) were discussed earlier. Hammond (1962) noted that the limb bones of feral hogs in New Zealand resembled those of wild boar in thickness more closely than those of the domestic swine from which they originated. He attributed this difference to the fact that feral hogs and wild boar were reared on a low plane of nutrition while domestic swine were reared on a high plane of nutrition. McMeekan (1940b, 1940c) showed significant differences in limb length and thickness between domestic swine reared on high versus low planes of nutrition, with those on a higher plane of nutrition having longer and thicker bones. No comparisons were made with wild stock in that study. Barrett (1971) found that when caught very young and given adequate feed, captive-reared feral hogs from the Dye Creek Ranch, California, were capable of showing growth comparable to that of typical domestic swine. Barrett (1971) also determined that feral hogs from Dye Creek Ranch showed growth rates under wild conditions that were less than one-half that for domestic swine, and he attributed this difference to poor environmental conditions rather than genetic limitations.

The majority of the captive feral hogs whose morphology was analyzed in this study were from the long-term feral population of Ossabaw Island. In cranial analyses, these specimens were generally classified as or grouped with feral hogs. However, they were larger than their wild counterparts and had more dished dorsal profiles. In mandibular analyses, these captive feral hogs classified as either hybrids or wild boar, and not domestic swine, despite what one might expect after the studies with Dye Creek Ranch ferals. This may have been because the domestic specimens used to define that target group in this study were from animals collected from the late 1800s up to the present. Since Ossabaw Island animals originated from domestic swine that existed before intensive breed development began, the domestic ancestors of the Ossabaw feral hogs might have been very different morphologically from the domestic sample used in this study. It is known, for example, that unlike later domestic animals such as those used in this study, early domestic swine were smaller than wild boar (Zeuner, 1963). Colonial domestic swine were small to medium in size and had long legs, narrow backs, short bodies, long snouts, and "rough" hair (Towne and Wentworth, 1950). This description is more typical of an Ossabaw Island or long-term feral hog than of any short-term feral hog populations or recent domestic swine. Considering this and the previous findings on the morphological effects of nutrition in the short-term feral hogs from Dye Creek Ranch, feral hogs may be considered eco-types of domestic swine. Although they are morphologically distinct, these two morphotypes seem to be quite "plastic," even in skull characteristics, between generations, depending on the level of nutrition during rearing. More work needs to be done on the degree of phenotypic plasticity these forms may show in response to differing planes of nutrition.

Several species of wild mammals reared in captivity have been shown to differ morphologically from their wild counterparts (Hollister, 1917; Hediger, 1966; Scott, 1968; Berry, 1969; Corruccini and Beecher, 1982). In general, the captive-reared individuals tended to have broader, shorter, more massive, and bulkier bones. No studies have been done comparing the morphology of captive Eurasian wild boar with their wild-living counterparts. Conley et al. (1972) determined that captive hybrids from the Tellico Wildlife

Management Area in southeastern Tennessee had a growth rate twice that of wild-trapped animals from the same area. The growth rate leveled out at the same age for captive and wild-trapped animals, but the captive hybrids were significantly heavier at that time (approximately 160 kg versus 70–90 kg; Conley et al., 1972). Springer (1975) found that captive-reared hybrid animals from the Aransas National Wildlife Refuge, Texas, also had growth rates about twice that of their wild-living counterparts.

Identification of Unknown Wild Pigs

There have been no previous attempts to use morphometric methods to identify the racial origin of any specific wild pig population in the United States. As we noted earlier, this was due in part to the lack of a reliable means of identifying the three morphotypes. About 75% of the specimens from U.S. wild pig populations with little or no known history were identifiable in this study to one of the four morphotypes. The most successful morphotype determinations used multivariate skull analyses, in some cases supported by data on coat coloration and hair morphology. There were a number of reasons for failure to obtain a certain identification of some specimens from populations with unknown histories. The first of these was age related. Target group resolution decreased as age decreased, as reflected by the attempts to identify yearlings or younger animals. In some cases, such as the populations from Baldwin County, Georgia, or the Welder Wildlife Refuge, Texas, the older specimens were definitely of one type, while the younger specimens were classified as something else. This inconsistency complicated the exact identification of these populations. Since the older groups had the better original target group differentiation, these age classes should be used, where possible, in determination of the morphotypes of these populations.

In some cases, specimens were definitely identified to one morphotype while a sparse historical account indicated another morphotype. The primary examples here are two specimens from southern Alabama. Barrows et al. (1981) stated that this population was hybrid in composition. The results of the morphological workup, however, indicated that this population was feral. At least in this example, where resolution of the data seems reasonably conclusive, the validity of the statement by Barrows et al. (1981) concerning this population's origin seems to be questionable.

In still other cases, the identification of an animal was hindered by the poor or incomplete quality of the specimen available. For example, mandibles are morphologically conservative in swine, compared with crania. For this reason, some specimens that consisted of only a mandible were either difficult to identify or gave contradictory results. This was a particular problem in identifying specimens from areas that have highly interspersed small pockets of hybrid and feral populations, such as south Texas.

The success of this method of identification, then, depends primarily on the number, age, and quality of the specimens available for study from the population in question. The best specimen for morphotype identification is a cranium and skin from an adult male. In addition, a better understanding of the variation in the population in question would be obtained by examining about 25 specimens. This seems to be the most useful sample size for the identification of an unknown group using a stepwise discriminant function analysis.

4. Current Status

This chapter is an updated consensus of the current status of free-ranging *Sus scrofa* populations in the United States. Included are the present ranges, habitats, and morphotypic makeups of the various populations of each state. The degree of stability of these populations is also noted where these have been determined.

The information for this chapter was gathered through (a) a search of the recent literature, (b) letters and questionnaires sent to private individuals and federal and state personnel, (c) personal interviews with private individuals and federal and state personnel, and (d) the conclusions reached as a result of the morphological studies reported elsewhere in this volume.

In 1981, a detailed questionnaire (Appendix D) was sent to the responsible agency or individual in each state who was judged most likely to be able to supply information on the status of wild pigs in that state. The names of the individuals who eventually replied to these questionnaires are included among those listed in Appendix A. An identical questionnaire was sent out to these same state agencies in 1988, and the results of the 1988 status survey are included here and in Appendix D. The names of these respondents are also included in Appendix A.

Wild pigs presently occur in 19 of the United States (Table 22). The current geographic distribution of these wild populations, as determined in the present study, agrees closely with the distribution revealed by a survey by the Southeastern Cooperative Wildlife Disease Study (SCWDS, 1988; Appendix E). Discrepancies between the results of the present study and the distributions indicated in Appendix E are noted in the text that follows. Most of the animals are apparently of pure feral hog ancestry. Since 1890, however, the wild boar and especially the hybrid morphotypes have broadened their distribution in this country (Figures 55 and 56). Because of the present popularity of "Russian" and European wild boar among hunters in the United States (Springer, 1977; Barrett and Pine, 1980), the hybrid morphotype is likely to be introduced into most remaining areas that have persisting wild pig populations. Several animal dealers in Texas, Florida, South Carolina, and Tennessee cater specifically to persons interested in purchasing and stocking "wild boar." It is unlikely, however, that any of these animals sold for stocking are truly pure Eurasian wild boar.

Table 22. Free-living populations of wild boar, feral hogs, and wild boar × feral hog hybrids in the United States, as of 1988.

State	All Extirpated	Wild Boar Present	Wild Boar Only	Feral Hogs Present	Feral Hogs Only	Hybrids Present	Hybrids Only
Alabama				X		X	
Arizona				X	X		
Arkansas				X	X		
California				X		X	
Florida				X		X	
Georgia				X		X	
Hawaii				X	X		
Iowa	X						
Kentucky				X		X	
Louisiana				X		X	
Mississippi				X		X	
Missouri	X						
New Hampshire		X	X				
New Mexico				X	X		
North Carolina				X		X	
Oklahoma				X	X		
Oregon	X						
South Carolina				X		X	
Tennessee		X		X		X	
Texas				X			X
Virginia					X	X	
Washington	X						
West Virginia						X	X
Totals	4	2	1	16	6	12	2

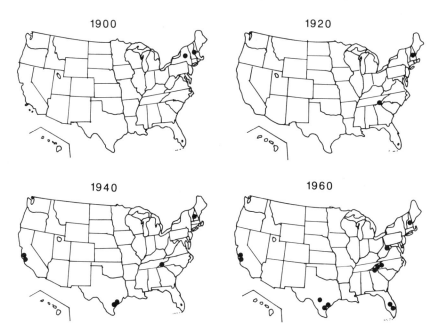

Figure 55. Localities in the United States where Eurasian wild boar or wild boar × feral hog hybrids were reported between 1900 and 1960.

Figure 56. Localities in the United States where Eurasian wild boar or wild boar × feral hog hybrids were reported in 1989.

Of the 19 states where wild pig populations currently occur, only two (New Hampshire and Texas) have animals that still morphologically resemble pure Eurasian wild boar. Only in Texas, in fact, is it possible that all three wild-living morphotypes are still present. Populations of both feral hogs and hybrids are found in 8 states, while 7 and 2 states, respectively, currently have populations of feral hogs and hybrids alone (Table 22). Of the states that had previously reported wild pig populations within their boundaries, four (Iowa, Missouri, Oregon, and Washington) now report that all these animals have been eradicated (Table 22).

Alabama

In 1981, wild pigs in Alabama were considered common in those areas where they occurred. The densest populations occurred in Choctaw, Marengo, Clarke, Washington, Baldwin, Monroe, and Mobile counties, in the areas surrounding the confluence of the Tombigbee and Alabama rivers (Appendix E-2). The 1981 survey, however, also noted the occurrence of populations at several locations that apparently no longer have wild pigs (SCWDS, 1988). These included locations along the Pea River in Geneva and Coffee counties, the Tallapoosa River in Montgomery and Macon counties, near White Oak in Barbour County, along the Tallapoosa River at the borders of Tallapoosa and Chambers counties, along the Tallapoosa and Little Tallapoosa Rivers in Randolph and Cleburne counties, around the east end of Weiss Lake in Cherokee County, and possibly along the Tennessee River in Marshall County.

These animals inhabit the lowland areas in Alabama, especially the bottomland hardwood forests and swamps. Annual flooding below the 10-m contour line in the southwestern corner of the state forces the hogs to high ridges and upland woodlands during the late winter and spring (J. R. Davis, pers. comm.).

Specimens examined and other information obtained on the populations in the southwestern corner of Alabama show that they are feral hogs with no wild boar or hybrid characteristics. The hybrid population reported to have occurred in Cherokee County would require an examination of specimens to confirm the presence of that morphotype. The remainder of the wild pig populations in the state are thought to be strictly feral in composition (J. R. Davis, pers. comm.).

As of 1988, the hunter harvest of wild pigs in Alabama was not governed by state regulations because the animals have been considered the property of the landowners in whose areas they occurred. During the 1980–81 hunting season, "feral swine and wild hogs" could be hunted only during the season set for white-tailed deer. By 1988, however, there were no longer any restrictions in either season or bag limit on hunting wild pigs in the state (J. Reid, pers. comm.). In 1982, a feral swine specialist position was added to the staff in the wildlife section of the Alabama Game and Fish Division. Several future studies of feral hogs in Alabama were planned at that time (J. R. Davis, pers. comm.). As of 1988, however, no such studies had been initiated (Appendix D; J. Reid, pers. comm.).

Arizona

The 1981 and 1988 surveys indicated that wild pigs in Arizona were common only in the Havasu National Wildlife Refuge (J. S. Phelps, pers. comm.). No information on other populations in the state was obtained during this study. However, SCWDS (1988) also reported additional low-density populations in northeastern Mohave County and in portions of Coconino, Pinal, Yuma, and Cochise counties, the latter representing a westward extension of the adjacent population in Hidalgo County, New Mexico (Appendix E-6). All these populations are apparently strictly feral hogs (J. S. Phelps, pers. comm.). The history of the Havasu population and the specimens and morphological data examined during this study confirm that these animals are indeed all true feral hogs. They are common on the refuge at present and are found in the cattail and bulrush stands along the edges of Topock Marsh. This population is hunted annually and does not appear to be expanding. It was recently estimated to number several hundred individuals (R. A. Gilbert, pers. comm.). Hunting of feral hogs in Arizona is not regulated by state law.

Arkansas

No specimens were examined from Arkansas during this study. All the histories and what little morphological data were available on these animals indicated that only feral hogs occur in this state; however, there have been reported releases of hybrids in Arkansas (B. L. Goatcher, pers. comm.). In general, feral hog populations in the state are stable to decreasing at present, but no population estimates are available. Populations are found along the Ouachita River watershed in Union, Ashley, Bradley, Drew, Cleveland, Calhoun, Dallas, Ouachita, Clark, Nevada, and Pike counties; along the Cossatot River in southern Polk County; and around the confluence of the Sulphur and Red rivers in Miller and Lafayette counties. They are also found in a number of other small, scattered locations around the southern half of the state and along the Mississippi River bottomlands in Desha and Chicot counties (SCWDS, 1988). The populations in these latter two counties are contiguous with high-density populations in Bolivar and Washington counties, Mississippi, to the west (Appendix E-4). These feral hogs are commonly found in the oak and hardwood bottomlands and swamps in Arkansas (F. Ward, pers. comm.; Sealander, 1979). Except for national wildlife refuges, the hunting of feral hogs in Arkansas is not regulated. Management is solely on a control basis and is designed to minimize interspecific competition with other game species (D. F. Urbston, pers. comm.).

California

Wild pig populations in California are made up of wild boar × feral hog hybrids and feral hogs. Although the 1988 survey indicated the presence of Eurasian wild boar in the state (Appendix D), evidence for the occurrence of this morphotype was not disclosed by any historical account or morphological analysis in this study. Recent estimates indicate that wild pig populations exist in almost half the 58 counties in the state (Appendix E-7; E. G. Hunt, pers. comm.), but a statewide survey is needed to better determine the complete distribution of these animals (Mansfield, 1978). In the late 1970s the statewide population estimate was 50,000–65,000 individuals (T. M. Mansfield, pers. comm.). This estimate increased to 100,000 animals in the mid 1980s (Anon., 1985). Based on known histories and morphotypic data determined by this study, hybrids occur only in the following counties: Monterey County from the Carmel Valley watershed just south of Monterey throughout the Los Padres National Forest to the southern edge of the county, extending eastward around King City; San Luis Obispo County in the northern half of the county, extending to as yet undetermined limits in the southern half; San Benito County in pockets of woodland habitat in most of the county, excluding the northwest and southeast corners; Mendocino County along the Elder Creek branch of the Eel River. In Tehama County, hybrid stocking on the Dye Creek Ranch between Antelope and Mill creeks has had an as yet undetermined impact on the morphology of the resident feral hog population there. The present status of hybrid animals introduced into Fresno and Shasta counties is unknown.

Feral hogs currently exist in the foothills of eastern Tehama County; in the Colusa National Wildlife Refuge, Colusa County; in the northwestern corner of Sonoma County; in pockets in most of San Benito County; in pockets in northern San Luis Obispo County; and in pockets in eastern Santa Barbara County. They also occur on Santa Cruz and Santa Rosa islands in Santa

Barbara County, in southern Ventura County, on Santa Catalina and San Clemente islands in Los Angeles County, and along the Colorado River south of Needles in San Bernardino County. The present status of feral hog populations in Humboldt, Shasta, Lake, and Santa Clara counties is unknown. Wild pigs in California have increased during the past decade due to their popularity as big game animals.

In general, wild pigs in California inhabit the oak woodlands and rolling oak grasslands with chamise or chaparral thickets (P. A. Dubsky, pers. comm.; P. Thomas, pers. comm.; J. Lopez; pers. comm.), but true chaparral, northern coastal sage, and streamside woodlands are the more commonly used habitats (Brown and Lawrence, 1965). On the Dye Creek Ranch in northern California, wild pig habitat ranges from sparse annual grasslands to dense foothill woodlands (Barrett, 1978). On the Channel Islands, these animals inhabit oak woodlands and coastal sage habitats (D. W. Baber, pers. comm.). On Santa Catalina Island, wild pigs move seasonally into cooler moist canyon bottoms during the dry months, with smaller home ranges during this period (Baber and Coblentz, 1986). Diets in this island population also vary seasonally, shifting abruptly to annual grasses and forbs during the winter months as opposed to fruits and forbs during the dry season (Baber and Coblentz, 1987). On Fort Hunter Liggett Military Reservation in Monterey County, common locations of wild pig activity are the oak grasslands around springs, creeks, and reservoirs (Mansfield, 1978).

In 1957, "wild pigs" were classified as big game animals in the state of California. In that same year, the hunting regulations were modified to provide for a shorter season, reducing the hunting pressure on the wild pig populations in the eastern portion of Tehama County. Before 1957, wild pigs had no game status in the state. The statewide hunting season has been variable over the years. For many years, the wild pig hunting season was year-round in California except in Monterey County, which had a six-month season. In the mid 1970s, wild pig hunting opened all year statewide (Pine and Gerdes, 1973; T. M. Mansfield, pers. comm.). At present, however, an even more liberalized bag limit is being considered for certain areas in the state in response to an increase in wild pig depredation (E. G. Hunt, pers. comm.). Statewide, the present bag limit is one pig per day and one in possession, except for Santa Catalina Island, which has a limit of two per day and four in possession.

Florida

Wild pigs in this state are almost all feral hogs. All but one of the specimens from Florida examined during this study were classified as feral animals. This individual, from Gilchrist County, consistently classified as a hybrid. Historical records and recent coat coloration data indicate that the hybrid morphotype is found only on Eglin Air Force Base and in parts of Alachua, Polk, Highlands, and Palm Beach counties. Of these sites, only the introduction in Alachua County and the one at Eglin Air Force Base are reported to have been successful in establishing the hybrid morphotype. Faint patterns of striping in juveniles from Hillsborough County area not necessarily indicative of hybrid morphotype introduction.

At present, wild pigs are reported to occur in all the counties in the state (F. H. Smith, Jr., pers. comm.; R. M. Wright, pers. comm.). The SCWDS (1988) survey, however, additionally indicates no substantial wild pig populations either in or near Seminole County (Appendix E-3). In general, wild pigs are found throughout the Florida panhandle, the east coast of the state, the everglades area south of Polk County to Florida Bay, and in the river systems of the west-central portion of the state. In the late 1980s, the population was estimated to exceed 500,000 (Stewart, 1989). These wild pigs live in a variety of habitats in the state, including hammocks surrounded by sawgrass marsh, hardwood bottomlands, river swamps, pine uplands, and pine flatwoods (Layne, 1974). Areas of high wild pig populations in Florida are generally large, forested tracts with dense understories that provide ample escape cover. These areas are usually managed for timber and pulpwood production (Belden, 1989). Dense wild pig populations are found in Levy, Dixie, Lafayette, and Taylor counties in areas dominated by coastal salt marsh and slash pine flatwoods interspersed with cypress swamps, hydric oak hammocks, and bayheads (Belden, 1989). Brackish and river marshes are the habitats most frequently used by feral hogs on the Chassahowitzka National Wildlife Refuge and Avon Park Air Force Range, respectively (E. Collinsworth, pers. comm.; R. L. Barker, pers. comm.). In Myakka River State Park, there is a seasonal shift between flatwoods and hammocks depending on the food availability (R. Dye, pers. comm.). Pine flatwoods are the preferred habitat in Hillsborough River State Park (R. Danser, pers. comm.).

The presence of hogs with unique morphological features in the southwestern counties of Florida in the vicinity of the Peace and Caloosahatchee rivers has focused the attention of biologists and historians on the feral populations in this part of the state. Interest has been heightened by the documentation of this region as perhaps the earliest in North America to have had domestic swine introduced as a result of the first Spanish explorations and settlements. This raises the possibility that some of the unique traits may be characteristic of the first domestic swine to have been brought to and bred within these settlements. Such speculation is particularly plausible in the case of the Mangalitza type found within feral populations of this region, as the combination of a slaty-gray adult with a striped juvenile color pattern is also characteristic of the domestic Mangalitza hog of Europe. The most interesting examples of hogs of this Mangalitza type are those that show more of a generally domestic/short-term feral appearance in combination with these color patterns, suggesting that the type of such individuals has been little changed from that of some of the earliest introductions. As a result, there is current interest in obtaining feral hogs of this Mangalitza type for use in the development of a captive-breeding population for use in such research and education programs as those featuring primitive breeds of livestock in various living-history farms or museums.

Similar interest is shown in the syndactylous, or mule-footed, hogs found in this same general region of the state. Although there is no specific documentation indicating any connection between either this trait or the presence of neck wattles, which often occur in the same individuals, and the earliest introductions of domestic swine by the Spanish, this may be the only region in the United States where such mule-footed hogs still occur in a feral population. Several hogs showing both wattles and the mule-footed trait have been captured from the feral population of this area for use in captive breeding programs, with progeny often released back into the same feral populations from which the breeding stock had been obtained. These activities may be responsible for the increases of these traits in these areas. Some of the earliest of the captive animals were all-black and, unlike the Mangalitza type, showed a typical long-term feral phenotype. A boar of this type was obtained from the Bradenton area by Ralph M. Wright of Plant City, Florida. An examination of this animal showed it to be of a generally short-term feral phenotype with a red coat with black spotting, but with both neck wattles and four syndactylous feet. Apparently no general statement can yet be made concerning the overall color or body phenotypes of animals showing the combined traits of wattles and syndactyly.

The history of the game status of the wild pig in Florida was reviewed by Belden and Frankenberger (1977). Briefly, these animals were first declared a game species in Florida in 1956 in the J. W. Corbett, Eglin Field, and Everglades wildlife management areas. Since that time, the species has been given game status in a total of 35 wildlife management areas, three state parks, two entire counties, parts of four other counties, one state wildlife refuge, and one private hunt club management area (Belden and Frankenberger, 1977). A statewide survey of hunters indicated that 43,536 wild pigs were harvested during the 1986–87 season. This represented a total of 58,967 hunters and 649,576 man-days of effort, ranking third behind deer and dove and second only to deer in these statistics (F. H. Smith, Jr., pers. comm.).

Georgia

Wild pigs are reported to occur in Georgia throughout the coastal plain and in extreme northern mountain areas, as well as in limited areas of the piedmont (J. Kurz, pers. comm.). The 1988 SCWDS survey, however, does not indicate significant populations in the latter, other than in Pike and Meriwether counties near the fall line (Appendix E-2). Histories and specimens examined indicate that feral hogs are the most common type of wild pig in the state, but hybrid stock may be currently increasing due to hunter interest and stocking activities. Two of the more recent hybrid introductions were into localities along a river system already populated by numerous scattered populations of feral hogs. The degree of gene flow between these feral populations is unknown, but it may be substantial enough to serve as a dispersal corridor for spreading the hybrid morphotype along river bottomlands. The hybrid morphotype has already increased in the areas where the hybrid stock was introduced along the Oconee River in Laurens County and along the Ocmulgee River in Houston County. In addition, the hybrid morphotype appears to be naturally increasing in both numbers and range in the mountains of northern Georgia. Hybrids in this area are found in the mountains of Murray, Fannin, Gilmer, and Union counties in north-central Georgia, and in Towns and Rabun counties in the northeastern corner of the state. Feral hog populations are found along the Savannah River from south of Augusta in Richmond County to Chatham County on the coast; along the Ogeechee River from Hancock and Warren counties eastward to the coast; in scattered swamps and pine forests along the Altamaha River system and in Camden, Glynn, McIntosh, Bryan, and Chatham counties along the coast; on the coastal islands of Cumberland, St. Simons, St. Catherines, Ossabaw, and Wassaw; and in small scattered populations in the extreme southern portion of the state, for example, in Baker, Clinch, Echols, and Thomas counties.

Swamps and marshes are the preferred habitats along the coastal plain. Less frequently, animals in these areas move into the uplands and pinewoods (Golley, 1964). Palmetto-oak forests and salt marshes are the preferred habitats on Cumberland Island (Z. T. Kirkland, pers. comm.).

Wild pigs do not have any legal or game status in Georgia. At present, they are considered domestic animals, and thus are the private property of the landowner on whose land they occur. There are, however, some license, weapon, and dog-use restrictions while hunting these animals on wildlife management areas in this state. All attempts to have this species reclassified as a game animal in Georgia have been defeated (J. Scharnagel, pers. comm.; C. Allen, pers. comm.; R. W. Whittington, pers. comm.), although no such attempts have been made in the past 17 years (J. Kurz, pers. comm.).

Table 23. Published research findings that document unique features of feral swine from Ossabaw Island (modified from Brisbin, 1989).

Trait	References
Body Size and Proportions	Brisbin et al. (1977a)
Hair Length	Brisbin et al. (1977a)
Genetic Isolation	Smith et al. (1980); Brisbin et al. (1977b)
Body Lipid Reserves	Hausman and Martin (1981); Scott et al. (1981b); Stribling et al. (1984)
Lipid Metabolism and Obesity	Martin et al. (1973); Martin and Herbein (1976); Buhlinger et al. (1978); Hoffman et al. (1983); Scott et al. (1981a)
Insulin and Growth Hormones	Wangsness et al. (1977)
Glucose Metabolism	Cote et al. (1982)
Reproductive Biology	Hagen and Kephart (1980); Hagen et al. (1980); Gilbertson-Beadling et al. (1988)
Muscle Development	Ezekwe and Martin (1975); Allen et al. (1982); Hausman et al. (1983); Hoffman et al. (1983)

Without doubt, one of the most scientifically important and best-studied populations of feral swine in the United States is the resident population of feral hogs on Ossabaw Island. From the mid 1970s through the mid 1980s, a number of published studies have documented the unique nature of many of the physiological and biochemical characteristics of animals from this population. These features have, in turn, made these swine important subjects for biomedical research in such areas as lipid metabolism, endocrinology, reproductive biology, muscle biology, and neonatal medicine (Table 23). Studies using Ossabaw Island swine have thus made important contributions to a number of areas relevant to both animal science and human health and medicine.

Because of the unique features of the feral swine of Ossabaw Island, concerns have been voiced over the vulnerability of these animals in the wild state to impacts from disease, hurricanes, and other catastrophic events. Consequently, beginning in the early 1980s, efforts were made to establish captive breeding herds of Ossabaw Island swine on the mainland. The first of these herds was established at the University of Georgia's Savannah River Ecology Laboratory (Brisbin et al., 1977b) with four males and five females caught as neonates on the island in 1978 and 1979. These animals and their progeny were subsequently transferred in 1983 to the Claude Moore Colonial Farm at Turkey Run, McLean, Virginia (one male and three females), and to the Sedgwick County Zoo in Wichita, Kansas (two males and four females). The breeding line was not maintained at the Turkey Run Farm, and the Ossabaw Island animals hybridized with other swine in that collection. However, the animals at the Sedgwick County Zoo continue to be maintained in a pure line (Baker, 1984). These animals are registered under the International Species Inventory System (ISIS), and their progeny have been transferred to other zoo collections (e.g., the Brookfield Zoo, Chicago, Illinois; the Cincinnati Zoo, Cincinnati, Ohio; and the Living History Farm, Des Moines, Iowa).

Concomitant with the activities described above, breeding herds of Ossabaw Island swine were also established at the University of Georgia (R. J. Martin, pers. comm.) and Pennsylvania State University (D. R. Hagen, pers. comm.). Both herds were maintained to produce animals for use in research at these institutions, and both herds have now been terminated, with no animals from either contributing to the present captive gene pool.

In the winter of 1986, the American Minor Breeds Conservancy (AMBC; P.O. Box 477, Pittsboro, NC 27312) established and assumed responsibility for maintenance of the Ossabaw Island Pig studbook (Heise and Christman, 1989). At the time of its initiation, the studbook listed three living foundation males and three living foundation females that had all been captured in the wild, and two males and four females that had been born in captivity. Because of the high degree of inbreeding that would inevitably result from such a small number of captives, an expedition sponsored by the AMBC and Sedgwick County Zoo was sent to Ossabaw Island in February 1988. This expedition resulted in the capture of eight additional neonates (two males and six females), which have now been used to establish another registered breeding herd in Blacksburg, Virginia, under the direction of P. Sponenburg of the Virginia-Maryland Regional School of Veterinary Medicine. Plans call for the transfer of some of these wild-caught animals to the Sedgwick County Zoo and for another expedition to capture still more breeding stock from the islands in the near future (D. Bixby, pers. comm.).

In the meantime, field research on the Ossabaw Island swine has continued. During the past several years, most of this research has been aimed at studying swine diseases and disease vectors under natural conditions and has been conducted by the Southeastern Cooperative Wildlife Disease Study at the University of Georgia (e.g., Nettles, 1989). Ownership of

Ossabaw Island has now been acquired by the state of Georgia, which manages the island and its wildlife through the Game and Fish Division of the Department of Natural Resources. Public hunts are held for deer, wild turkey, and small game, but no hogs may be taken. E. T. West, the island's former owner, has a life estate in her residence on the island and ownership of the island's feral animals, including all swine, cattle, horses, and donkeys. Under her direction, a program of live-trapping feral hogs continues, with captured animals periodically removed from the island for sale on the mainland (E. T. West, pers. comm.).

Hawaii

Surveys conducted in this study indicate that wild pigs are currently found on the islands of Niihau, Kauai, Oahu, Molokai, Maui, and Hawaii. The SCWDS (1988) survey, however, did not indicate any substantial populations of wild pigs on Niihau (Appendix E-8). The densest populations occur on the island of Hawaii. Although often called wild boar, these animals are strictly feral hogs. The response to the 1988 questionnaire indicated the occurrence of Eurasian wild boar and feral hog × wild boar hybrids (Appendix D), but there is no credible evidence to indicate either the past or present occurrence of either morphotype in the islands. The histories and the morphology of specimens examined from the various populations support this position.

The preferred habitats of wild pigs in Hawaii are dense rain forest and parkland-forest ranchlands, but they are not restricted to these areas and are present in all areas of the islands that are not intensively used by humans. This includes all available habitat from lowland tropical forests up to 3,400 meters on volcanic slopes above the timberline (Nichols, 1962a; Kramer, 1971).

Feral hogs in Hawaii are classified as big game animals. There are varying seasons and bag limits depending on the island and the specific area being hunted (T. Sutterfield, pers. comm.). Currently, the U.S. Park Service is conducting research aimed at developing more effective means of removing pigs from native ecosystems. Removal methods suggested by this work are also being implemented by the Nature Conservancy on lands under its control. The Hawaii Division of Fish and Game is also planning similar removal programs, with an emphasis on areas containing unique native ecosystems. Eradication programs have a lower priority in areas where most of the vegetation consists of introduced or exotic species. Such areas are most often managed for public hunting on a sustained yield basis (T. M. Lum, pers. comm.).

Iowa

All indications are that wild-living pig populations no longer exist in this state (T. W. Little, pers. comm.). The populations that were in the state appear to have been short-lived, and the persistence of the animals along the Raccoon River is doubtful (L. Gladfelter, pers. comm.). These animals have no legal game status in Iowa and are considered domestic livestock and the property of the landowner where they occur (R. Bishop, pers. comm.).

Kentucky

As in Iowa, the future of feral hogs in Kentucky is uncertain at best. There is a possibility, however, that hybrid animals from the Cumberland Plateau in Tennessee may have spread into south-central Kentucky. Hines (1988) reports wild hog populations in "only a few counties in the southeastern part of the state." The 1988 SCWDS survey, however, does not indicate any significant populations in this southeastern portion of the state, and notes scattered populations only in Marion, Estill, Lee, Wayne, and McCreary counties (Appendix E-1). A photograph accompanying Hines's (1988) article suggests a wild boar–hybrid morphotype.

Wild hogs in Kentucky have no game status and in the past have been legally considered private property (J. Durrell, pers. comm.). Hines (1988), however, reports that the importation or possession of "wild hogs" is now illegal and that hunting regulations are geared toward reducing the numbers of any populations present. Persons hunting for feral hogs on lands other than their own are often arrested or cited as suspected deer poachers because Kentucky feral hog populations are so tenuous (J. Durrell, pers. comm.).

Louisiana

The histories and specimens examined from the Louisiana wild pig populations indicate that these animals include both feral hogs and hybrids. They are now found mostly in the central and southwestern portions of the state (Appendix E-4), concentrated in two major areas. One is along and around the Calcasieu River system, including areas around Whiskey Chitto Creek, Bundick Creek, and Beckwith Creek. The other concentration is between the Little and Red rivers north of Alexandria. Another large population is found in the Atchafalaya floodway between St. Martin and Iberville parishes. Some feral hogs are scattered along the Pearl River bottomlands in Washington and St. Tammany parishes. In addition, small scattered pockets of feral hogs are found in the northern counties of the state, especially along the Ouachita River system and Bodcan Creek impoundment (Appendix E-4). Hybrid populations are found along the Mississippi between St. Joseph and Ferriday, in Caldwell and Winn parishes, and throughout the Mississippi River delta and the Lake Pontchartrain region of southeastern Louisiana. Although no populations of wild pigs were reported for the western portions of Caddo and DeSoto parishes, the existence of abundant populations in the immediately adjacent portions of east Texas (Appendix E-5) suggests that some animals in these parts of northwestern Louisiana may have been overlooked. Feral hogs prefer hardwood bottomlands and pine-hardwood uplands in Louisiana (J. D. Newsom, pers. comm.).

Wild pigs are not considered game animals in Louisiana. Unmarked free-ranging hogs may be taken by hunters on certain wildlife management areas only during the open seasons for the game species (e.g., white-tailed deer) prescribed for the area. These persons must have the proper licenses, permits, and other requirements for hunting the game species for which the area is open at that time (J. W. Farrar, pers. comm.).

Mississippi

Wild pigs in Mississippi are found mostly in the counties bordering the Mississippi and Pearl rivers in the southeastern corner of the state (Appendix E-4). The densest population is along the Pearl River in Pearl River and Hancock counties. Several small populations of feral hogs also exist on the offshore islands between Mississippi Sound and the Gulf of Mexico. Other scattered populations are found along the Tallahatchie River between Lafayette and Marshall counties, along the Noxubee River in Noxubee County, along the Mississippi River in Bolivar County, between the Bayou Pierre and Big Black River in Claiborne County, and along the Pearl River in Madison County.

All evidence suggests that all the populations in the state are feral hogs except for ones on the Pearl River Game Management Area in Madison County, on the Copiah County Game Management Area in Copiah County, and along the Mississippi River between Port Gibson and Fayette, which are of hybrid composition. The response to the 1988 questionnaire (Appendix D) indicated that Eurasian wild boar occurred in the state in addition to hybrid populations, but no historical or morphological data exist to support this claim.

At present, the wild pig populations in Mississippi appear to be restricted primarily to river systems and their associated bottomlands (R. K. Wells, pers. comm.).

Wild pigs are not considered game animals in Mississippi except on state game management areas, where they can be hunted during the open season on specified game species (E. Cliburn, pers. comm.). Hybrid animals on the Pearl River Game Management Area are considered big game by the Mississippi Department of Wildlife Conservation but are not officially classified as such (J. H. Phares, pers. comm.).

Missouri

The state of Missouri has no unfenced populations of wild pigs at this time (G. P. Dellinger, pers. comm.).

New Hampshire

Historically, the captive and free-ranging wild pig populations in New Hampshire have been either Eurasian wild boar or hybrid in their makeup. All the specimens examined from the captive population in Corbin's Park were morphologically pure Eurasian wild boar, although the population presently inside the park is now probably no longer pure because of the apparent introduction of hybrid animals by park management. No specimens, however, have been collected recently from the park. It is likely that individuals or small groups of wild boar will continue to escape from Corbin's Park. The small pocket of outside wild boar in Grafton County seems to be persisting. Hunting may eliminate these animals, but more will undoubtedly continue to escape from the park, at least until such time as the park is enclosed by the proposed chain-link fence.

No habitat preferences have ever been reported for this population. Personal field observations have indicated that these animals are found in mixed farmland and forest habitat of mixed white pine, hemlock, and hardwoods such as beech, birch, and maple.

The wild boar and hybrids have no legal game status in New Hampshire. There are no set seasons or bag limits. However, a state hunting license is required to pursue and kill these animals except for state residents hunting on their own land (H. C. Nowell, pers. comm.). It is also illegal to hunt them at night (E. P. Orff, pers. comm.).

New Mexico

The population of feral hogs in Hidalgo County is stable in distribution and number (Appendix E-6). At least one of the ranches in this county, the Gray Ranch, is presently planning to implement a game management program for these animals (S. J. Dobrott, pers. comm.). Because of the U.S. Forest Service policy concerning them, the future of the feral hogs in the Coronado National Forest in Hidalgo County is uncertain.

Their range in New Mexico is between 1,500 and 1,900 meters in elevation, and varies from open grassland to desert arroyos bordered by willow and hackberry trees to piñon, juniper, and oak woodlands (B. Donaldson, pers. comm.). The animals on the Gray Ranch seem to prefer moist bottomlands in the summer and oak woodlands in the fall and winter (S. J. Dobrott, pers. comm.).

Feral hogs have no legal game status in the state of New Mexico. They are classified as strayed livestock and are under the jurisdiction of the New Mexico Livestock Sanitary Board (W. A. Snyder, pers. comm.).

North Carolina

Wild pigs in western North Carolina are mostly hybrids. These hybrid populations are found in the mountains in the southwestern corner of the state, perhaps reflecting a greater cold-hardiness as a result of wild boar ancestry (Foley et al., 1971) and thus greater tolerance of colder winter temperatures. These animals are most numerous in Cherokee, Clay, Graham, Macon, Swain, and Jackson counties. Other wild pig populations have been reported in Burke, Caldwell, Rutherford, and Cleveland counties (G. L. Bames, pers. comm.).

In the eastern part of the state, feral hog populations have previously been reported in the bottomlands of the Green Swamp in Bladen County (Rutledge, 1965) and on the McKay National Wildlife Refuge in Currituck County. The SCWDS survey (1988), however, fails to note the persistence of significant numbers of wild pigs in either county (Appendix E-1). Responses to the 1988 questionnaire similarly indicate that feral hogs have now been eliminated from these locations on the lower coastal plain (J. M. Collins, pers. comm.). Webster (1988) does not list the feral hog as part of the mammalian fauna of the northern Outer Banks. Nevertheless, SCWDS (1988) indicates a high-density wild pig population still persisting in southeastern Virginia immediately adjacent to Currituck County, and it is possible that some feral hogs may still exist in this area (Appendix E-1).

In the western part of the state, wild pigs live in oak-hickory-beech forests but seem to prefer grassy balds seasonally (D. Rhodes, pers. comm.). In the McKay National Wildlife Refuge, these animals were reported to frequent wax myrtle thickets (I. W. Ailes, pers. comm.).

"Wild boar" are a big game species in North Carolina with scheduled seasons and bag limits (G. L. Bames, pers. comm.). However, only "wild boar" have been given legal status. The identification of "wild boar" as compared with "feral hogs" has "proven most difficult in court" (R. B. Hamilton, pers. comm.). State big-game tag reports indicate that in the 1986–87 and 1987–88 seasons, respectively, a total of 72 and 89 "wild boar" were harvested by hunters in the state, with 50 and 73 of these taken in Graham County alone.

Oklahoma

Historically, only feral hogs have existed in this state, but no specimens were available for examination. Feral hogs in Oklahoma are a remnant of previously larger population numbers. Only a few of these animals are left in Pushmataha and McCurtain counties in the southeastern corner of the state (Appendix E-5). There have also been reports of feral hogs in Carter County, in the area of the Arbuckle Mountains, and in Muskogee County (R. Masters, pers. comm.), although no such occurrences were noted in these areas by the SCWDS survey (1988). Aside from the occasional domestic stock that are released or escape from time to time, the future of established feral hog populations in this state is doubtful (R. E. Thackston, pers. comm.; G. Bukenhofer, pers. comm.). R. E. Thackston (pers. comm.) reported in 1981 that feral hogs in the southeastern corner of the state were restricted to narrow hardwood bottomlands. However, R. Masters (pers. comm.) reported in 1988 that feral hogs have also been observed in upland oak-pine habitats in the Ouachita Mountains and that in winter months they may prefer areas such as hardwood corridors or north slopes, where oaks predominate and mast production is good.

These animals are not classified as game animals in Oklahoma, but hunters are required to have a state hunting license when pursuing them (R. Umber, pers. comm.; R. E. Thackston, pers. comm.).

Oregon

Based on historical accounts and the one specimen examined, only feral hogs seem to have occurred in this state. At present, Oregon has no wild pig populations. It is possible, however, that feral hogs might become established in Siskiyou County, California, and then spread north into Oregon in the future. Any wild pigs that exist in Oregon at present have no legal game status (F. Newton, pers. comm.; A. Polenz, pers. comm.).

South Carolina

Wild pig populations in South Carolina are found mainly along river systems and in the lower coastal plain of the state (Appendix E-2). A few of the populations in the state are thought to have some wild boar ancestry, although this claim has yet to be proven conclusively either with historical or morphological data in all cases. D. Shipes (pers. comm.), in responding to the 1988 questionnaire (Appendix D) indicated that Eurasian wild boar had been released in the Congaree Swamp in Richland and Calhoun counties and on Groton Plantation in Allendale County. However, these animals were descendants of pigs originally obtained from a wild pig research program at the University of Georgia (Hansen et al., 1972), and these are definitely of hybrid origin (JJM and ILB, pers. comm.), having been obtained from Tennessee–North Carolina sources.

Feral hog populations are found along the Savannah River from Aiken County to the coast, along the Pee Dee and Little Pee Dee rivers, along the southern portions of the Edisto River between Colleton County and Dorchester and Charleston counties, along the Wateree River between Richland and Sumter counties, in the pine forests and bottomlands between and including the Santee and Cooper rivers in Berkeley and Charleston counties, and in the Chattooga Ridge region of Oconee County.

River swamps, hardwood bottomland, and some pine flatwoods appear to be the preferred habitat types for this species in South Carolina (Golley, 1966; Holt, 1970; Wood and Lynn, 1977). In the Francis Marion National Forest, they prefer bottomland hardwoods around streams and swamps (R. Tyler, pers. comm.). The animals in the Santee Coastal Reserve are found in marsh bordering the upland areas of the reserve (T. Strange, pers. comm.).

Wild pigs have no legal game status in South Carolina. A state hunting license is required to hunt them, and on state game management areas they may only be hunted during the deer season or during special designated hunts (W. B. Conrad, pers. comm.).

Tennessee

Wild pig populations in Tennessee (most if not all of them hybrids) are presently found in three general areas of the state. J. Murrey (pers. comm.) indicated in response to the 1988 questionnaire that only Eurasian wild boar and hybrids were currently found in the state (Appendix D) and noted that "in certain counties and specific management areas, feral hogs were hunted out back in the 1950s." Again, cold-hardiness has undoubtedly been a factor in the ability of these hybrid populations to colonize areas with relatively cold winters. The majority of the animals are concentrated in the south-eastern corner of the state in Polk, Monroe, Blount, and Sevier counties (Appendix E-1). The next largest population is in the Cumberland Plateau in Pickett, Scott, Fentress, Overton, Morgan, Putnam, Cumberland, Bledsoe, and Rhea counties. Wild pigs in Bledsoe and Rhea counties, however, have been indicated by SCWDS (1988) as no longer extant. The last of these populations is found in and around the Anderson-Tully Wildlife Management Area along the Mississippi River in Lauderdale County.

In the eastern portion of the state, this species inhabits mixed hardwood forests and mountain laurel and rhododendron thickets (Conley et al., 1972). In the Anderson-Tully Wildlife Management Area, hybrid animals are found in the hardwood bottomlands of the Mississippi River (Lawrence, 1982).

This species is classified as a big game animal in Tennessee, and their harvest has been regulated since around 1936 (J. Murrey, pers. comm.). Scheduled seasons and bag limits are enforced in all areas (C. J. Whitehead, pers. comm.). It is interesting to note that the 1988 hunting regulations provide for the taking of "wild boar" in Tipton County, immediately to the south of Lauderdale County, although no survey has ever indicated the presence of any significant populations of wild pigs in that county (e.g., SCWDS, 1988; Appendix E-1).

Texas

The Texas wild pig populations are found mostly in the eastern and southern quarters of the state (Appendix E-5) and include wild pigs that range from morphotypically pure wild boar to feral hogs. As in California, recent hunter interest in wild boar has contributed to the spread of this morphotype through private introductions into various areas of Texas. From histories and specimens examined, wild boar × feral hog hybrids occur in Kerr, Gillespie, Kendall, Kimble, Edwards, Bandera, Bexar, Medina, Real, Uvalde, and Frio counties in and around the Edwards Plateau; in Webb, LaSalle, Dimmit, Zavala, and Maverick counties in the Rio Grande valley; in Calhoun, Aransas, Refugio, and Victoria counties in central coastal Texas; and in Throckmorton, Haskell, Cottle, Foard, King, Knox, Young, Shackelford, and Stephens counties in north Texas. Feral hogs are found in almost all the east Texas counties from the Red River plains in the north to the coastal marshes around Galveston in the south. Ferals also are found in Brown, McCulloch, San Saba, Mason, Llano, Burnet, and Travis counties in central Texas; in Val Verde, Kinney, and Zapata counties along the Rio Grande; and in Atascosa, McMullen, Live Oak, Duval, Jim Wells, Kleburg, Kenedy, San Patricio, Gonzales, De Witt, and Goliad counties in south Texas. Although a specimen examined in this study was collected from the Laguna Atascosa National

Wildlife Refuge, Cameron County, and wild pigs were reported to be abundant in that area in the late 1970s (N. Scott, pers. comm.), SCWDS (1988) indicated that there were no longer any substantial numbers of wild pigs present there (Appendix E-5).

Wild pig habitat varies in Texas. In general, wild pigs occupy white-tailed deer habitat in Texas; however, these swine have much lower population densities in the more arid portion of the state's white-tailed deer range (Cox, 1981). In south Texas, they are found in river bottomlands with heavy brush (J. B. Wise, Jr., pers. comm.). On the Powderhorn Ranch in Calhoun County, the wild boar prefer areas with at least 25% brush with mowed uplands and extensive flats with scattered waterholes (J. C. Smith, pers. comm.). Lynn Drawe (pers. comm.), director for big game research at the Welder Wildlife Refuge, observed that the wild pigs on the refuge were found in the chaparral–mixed grass community, paspalum–aquatic plant community, woodland–spring aster community, spring aster–longtom community, and around lakes and ponds. On the Aransas National Wildlife Refuge, the habitat preference of the resident hybrid population shifts depending on the seasonal presence of certain desirable foods (Springer, 1975). The habitat types occupied on the refuge include grass-sedge flats, cordgrass flats, marshes, live oak brush, and live oak mottes. The latter two are the only types selected year-round for cover (Springer, 1975).

This species is not classified as a game animal in Texas. In the early 1980s, an estimated 2,500 to 3,000 wild pigs were being harvested annually by hunters in Texas (Cox, 1981). There are no statewide set seasons or bag limits. All hunters, however, are required to have a hunting license to hunt wild pigs in Texas. Several state areas have regulated annual hunts for feral hogs. There are restricted seasons and bag limits on some areas, and special permits may be required (C. K. Winkler, pers. comm.).

Virginia

The feral hog population in Princess Anne County in extreme southeastern Virginia (Appendix E-1) is established, and its survival for the near future appears certain. Feral hogs on both the Back Bay National Wildlife Refuge and the adjoining False Cape State Park occur throughout this area, with as much as 75%–80% of the population occurring on the state lands, and a total population estimate of 400–500 for both areas in 1986 (Duncan and Schwab, 1986). Hogs are most commonly found in the dense wax myrtle and oak areas, which they seem to prefer (I. W. Ailes, pers. comm.). The animals seem to move seasonally between freshwater marshes, which they occupy in summer and fall, and upland areas where they are found in late fall and winter when acorn mast is available (Duncan and Schwab, 1986).

The feral hogs in Virginia receive protection but are not listed in the game laws, thus there is no legal hunting season prescribed by law. The staff of the Back Bay National Wildlife Refuge was attempting to exterminate the resident feral hog population through controlled shooting in 1981 (R. H. Cross, Jr., pers. comm.; R. Duncan, pers. comm.). In 1986, plans were being made to eventually include the taking of hogs in deer hunts on both state and federal lands in that area (Duncan and Schwab, 1986).

Washington

Present reports indicate that the introduced wild boar population in the north-central portion of the state is either extinct or near extinction. All sightings of boar have resulted in attempts to eliminate them (R. Johnson, pers. comm.). At present, it is unlawful to import or possess live specimens of "wild boar, *Sus scrofa*, and hybrids involving the species *Sus scrofa*" in the state of Washington (King and Schrock, 1985).

West Virginia

The hybrid population in Boone, Logan, and Wyoming counties introduced into southern West Virginia is established and has spread throughout the mountains in these three counties (Appendix E-1). Litters have been produced annually, and this population has expanded its range from 13,000 ha in 1981 to 66,000 ha in 1988 (T. Dotson, pers. comm.).

Although the response to the 1988 questionnaire indicated that only Eurasian wild boar were found in the state (Appendix D), the earlier history of the animals from which this population was established clearly indicates a hybrid origin, as discussed earlier. Nevertheless, there is no evidence of any subsequent hybridization after release, and no boar harvested in the state "has shown physical characteristics common to boar–feral hog hybrids" (R. L. Miles, pers. comm.).

This population occupies the rugged, hilly terrain in the southwestern corner of the state. The area where the hybrids exist is 90%–95% mixed hardwood forest (Igo et al., 1979).

From the release date in 1971 until 1979, these animals were given complete legal protection from harvest. A 3-day season was established in 1979 and the population was classified as harvestable big game (Igo et al., 1979). This status has continued until the present. In 1988, a 15-day archery season and a 3-day gun season were held for wild boar (R. L. Miles, pers. comm.). Hunter kill in the state between 1979–86 has ranged from 3 (1979) to 76 (1984) animals per year (Dotson, 1986; 1987).

5. Conclusions

Feral hogs were the only type of free-ranging wild pig found in North America and the Hawaiian Islands between about A.D. 750 and 1890. Since that time, Eurasian wild boar have been stocked or escaped into at least five locations in the continental United States. Hybrid descendants from these wild boar populations have been subsequently stocked into eight other states, as well as numerous additional localities in the states where they were initially established. Because of their popularity as a big game species, wild boar and their hybrid crosses with feral hogs are now being stocked into new areas and are increasing in numbers in areas where they already existed. If this trend continues at its present rate, the majority of the wild pig populations in the United States in the near future will have at least an element of, if not be dominated by, this mixed wild boar–feral hog ancestry.

The morphotypes of the wild boar, feral hog, wild boar × feral hog hybrid, and domestic swine can be determined through an examination of cranial measurements and a consensus of coat coloration pattern and hair morphology. In the skull analysis, adult males were the sex and age group that gave the best intergroup separation using the stepwise discriminant function and canonical variates analyses, with cranial measurements giving superior results as compared with mandibles. The remainder of the cranial and all the mandibular analyses had poorer differentiation between groups and were therefore less useful. Separation of the skulls of long-term and short-term feral hogs was clearest in the older age classes. Significant differences between the sexes were found in the skulls of the three wild morphotypes, with those of males being larger than those of females. Differences in coat coloration and hair morphology were useful in separating wild boar and hybrids from feral hogs and domestic swine. Eurasian wild boar were monomorphic in coat coloration, while the other morphotypes were variable. The wild boar coat coloration was a white or tan grizzled coat of dark brown or black basally colored bristles, with the light-tipped bristles especially evident in the facial region. The undersides were lighter in color, and the ears and distal portions of the limbs, snout, and tail were solid dark brown to black. The feral and domestic morphotypes consisted of various combinations of solid color patterns, solid color patterns with white points, spotted, and belted patterns. The hybrids included various combinations of all the above patterns. The hybrids included various combinations of all the above patterns. Juvenile striping was not useful for distinguishing among the four morphotypes. Hair morphological differences were based on the presence or absence of light-tipped bristles. Wild boar and hybrids showed grizzled or solid guard hairs, while feral and domestic swine had only solid-colored bristles. The presence of curly, woollike underfur was not a useful character for separating morphotypes. The results of the body measurements and weights were not useful in these comparisons. In body measurements, however, domestic swine were the largest overall except in snout length. Feral hogs were the most variable and the largest of the free-living forms in most parameters. An examination of non-U.S. feral hog specimens showed that the feral morphotype was relatively consistent throughout the world. In general, captive wild pigs resembled but were larger than their wild-living counterparts. Because of the morphometric overlap, it is difficult in some cases to distinguish between the wild-living and captive ecotypes. Approximately three quarters of the unknown specimens examined could be identified to one of the known morphotypes. The success of such determinations depended on the age class and quality of the specimen.

Wild pigs currently exist in 19 of the United States. Transitional populations may occasionally persist for short periods in 2 or 3 additional states. Feral hogs, the most common morphotype of free-living pig, are found in 16 states, 6 of which have feral hogs exclusively. Wild boar × feral hog hybrids are the second most numerous wild-living morphotype. They are found in 12 states, 2 of which have only hybrid populations. Pure, unhybridized populations of Eurasian wild boar probably no longer exist in this country, with the possible exception of an area within a large fenced hunting preserve in New Hampshire. In general, wild pigs are found in a large variety of habitats in the United States. Where populations persist, there has recently been an increase in the number of states instituting some form of control over the hunting of these animals.

Appendix A

Personal Communications

The following is an alphabetical list of persons who provided information for this study through either personal correspondence or interviews with the authors. Titles and addresses were current at the time these individuals provided information used in this book. Although some individuals may no longer hold the positions indicated, the addresses given should provide contact with their successors. These references are all cited in the text as personal communications.

Ailes, I. W. (refuge staff), Back Bay National Wildlife Refuge, Pembroke Office Park, Pembroke Building 2, Suite 218, Virginia Beach, VA 23462.

Allen, C. (wildlife biologist), Game and Fish Division, Georgia Department of Natural Resources, 270 Washington St., S.W., Atlanta, GA 30334.

Andrews, C. H. III (sportsman), 3455 McCrays Mill Rd., Sumter, SC 29150.

Arnold, H. W., Sr. (sportsman), Rte. 1, Box 870, Jackson, SC 29831.

Atkeson, T. Z. (refuge manager), Wheeler National Wildlife Refuge, P.O. Box 1643, Decatur, AL 35602.

Baber, D. W. (graduate student), Department of Fisheries and Wildlife, Oregon State University, Corvallis, OR 97331.

Baldwin, W. P. (land management consultant), P.O. Box 818, Summerville, SC 29483.

Bames, G. L. (wildlife biologist), North Carolina Wildlife Resources Commission, P.O. Box 2919, Raleigh, NC 27602.

Barker, R. L. (wildlife biologist), Avon Park Air Force Range, Avon Park, FL 33825.

Barrett, R. H. (associate professor), Department of Forestry and Conservation, University of California, Berkeley, CA 94720.

Belden, R. C. (biological scientist supervisor), Florida Game and Fresh Water Fish Commission, Wildlife Research Lab., 4005 South Main St., Gainesville, FL 32601.

Bishop, R. (wildlife biologist), Iowa Conservation Department, 300 Fourth St., Des Moines, IA 50319.

Bixby, D. (executive director), American Minor Breeds Conservancy, Box 477, Pittsboro, NC 27312.

Blohm, T. (rancher and conservationist), Apartado 69, Caracas 1010-A, Venezuela.

Bruna, J. (wildlife biologist), Kentucky Department of Fish and Wildlife Resources, State Office Building Annex, Frankfort, KY 40601.

Bukenhofer, G. (USFS staff), Choctaw Ranger District, Ouachita National Forest, P.O. Box B, Heavener, OK 74937.

Butts, K. (assistant refuge manager), Aransas National Wildlife Refuge, P.O. Box 100, Austwell, TX 77950.

Clary, T. (sportsman), Rte. 2, Box 572, Batesburg, SC 29006.

Cliburn, E. (wildlife biologist), Mississippi State Game and Fish Commission, Box 451, Jackson, MS 39205.

Collins, C. T. (professor of biology), California State University, Long Beach, CA 90840.

Collins, J. M. North Carolina Wildlife Resources Commission, 512 N. Salisbury Street, Raleigh, NC 27611.

Collins, P. W. (associate curator), Santa Barbara Museum of Natural History, 2559 Puesta Del Sol Rd., Santa Barbara, CA 93105.

Collinsworth, E. (refuge manager), Chassahowitzka National Wildlife Refuge, Rte. 2, Box 44, Homosassa, FL 32646.

Conley, B. (game biologist), District VI, Arkansas Game and Fish Commission, Little Rock, AR 72201.

Conley, R. H. (game biologist), Tennessee Game and Fish Commission, Rte. 2, Tellico Plains, TN 37385.

Conley, W. (associate professor), Department of Fishery and Wildlife Sciences, Box 4901, New Mexico State University, Las Cruces, NM 88003.

Conrad, W. B. (chief of wildlife), South Carolina Wildlife and Marine Resources Department, P.O. Box 163, 1015 Main St., Columbia, SC 29202.

Convers, I. (businessman), 9002 State Rd. 39 N., Plant City, FL 33566.

Cornelison, S. (land owner), 18 Silver Bluff Ct., Aiken, SC 29801.

Craft, J. (USFS staff), Osceola National Forest, P.O. Box 1649, Lake City, FL 32055.

Crawford, B. T. (research chief), Game Division, Missouri Department of Conservation, 2901 N. Ten Mile Dr., Jefferson City, MO 65101.

Cross, R. H., Jr. (biologist), Virginia Commission of Game and Inland Fisheries, P.O. Box 11104, Richmond, VA 23230.

Danser, R. (park superintendent), Hillsborough River State Park, Rte. 4, Box 250-L, Zephyrhills, FL 33599.

Davis, J. R. (district game biologist), Alabama Department of Conservation, P.O. Box 352, Jackson, AL 36545.

Davis, R. (sportsman), Rte. 1, Box 134, Douglas, GA 31533.

Dearinger, J. (secretary), The American Kennel Club, 51 Madison Ave., New York, NY 10010.

Dellinger, G. P. (superintendent), Planning, Coordination and Administration Section, Wildlife Division, Missouri Department of Conservation, P.O. Box 180, Jefferson, MO 65101.

Denman, L. G., Jr. (rancher), 503 Terrell Rd., San Antonio, TX 78209.

Dobles, A. (president), Blue Mountain Forest Association, 1250 S. Willow, Manchester, NH 03103.

Dobrott, S. J. (range and wildlife specialist), Gray Ranch, Rural Rte. Box 165, Animas, NM 88020.

Doherty, J. G. (general curator of mammals), New York Zoological Society, Bronx Zoo, Bronx NY 10460.

Donaldson, B. (wildlife biologist), New Mexico Department of Game and Fish, 1480 N. Main, Las Cruces, NM 88001.

Dotson, T. L. (district game biologist), Rte. 1, Box 484, Point Pleasant, WV 25550.

Drawe, L. W. (range and wildlife biologist), Welder Wildlife Refuge, P.O. Drawer 1400, Sinton, TX 78387.

Dubsky, P. A. (fish and wildlife administrator), Department of the Army, Fort Ord Complex, CA 93941.

Dufour, R. F. (district wildlife officer), New Hampshire Department of Fish and Game, RFD 1, Littleton, NH 03561.

Duncan, R. (refuge staff), Back Bay National Wildlife Refuge, Pembroke Office Park, Pembroke Building 2, Suite 218, Virginia Beach, VA 23462.

Durell, J. (wildlife biologist), Kentucky Department of Fish and Wildlife Resources, State Office Building Annex, Frankfort, KY 40601.

Dye, R. (park superintendent), Myakka River State Park, Rte. 1, Box 72, Sarasota, FL 33577.

Ebert, P. (wildlife biologist), Oregon State Game Commission, P.O. Box 3503, Portland, OR 97208.

Ehrlich, F. T. (taxidermist/sportsman), 119 Gadsden St., Chester, SC 29706.

Eyster, M. B. (professor), Department of Biology, Box 545, University of Southwestern Louisiana, Lafayette, LA 70501.

Farrar, J. W. (wildlife biologist), Louisiana Wildlife and Fisheries Commission, 400 Royal St., New Orleans, LA 70130.

Fouts, J. (animal dealer), Tanganyika Wildlife Co., P.O. Box 12084, Wichita, KS 67201.

Fowler, H. (refuge staff), Wheeler National Wildlife Refuge, P.O. Box 1643, Decatur, AL 35602.

Frankenberger, W. B. (wildlife biologist), Florida Game and Fresh Water Fish Commission, 4005 South Main St., Gainesville, FL 32601.

Garabedian, M. S. (wildlife biologist), Blue Mountain Forest Association, Star Route H1, Box 4, East Lempster, NH 03606.

Garcelon, D. K. (president), Institute for Wildlife Studies, P.O. Box 127, Arcata, CA 95521.

Gibbons, J. W. (ecologist), Savannah River Ecology Laboratory, P.O. Drawer E, Aiken, SC 29801.

Gidden, C. S. (refuge manager), St. Marks National Wildlife Refuge, P.O. Box 68, St. Marks, FL 32355.

Gilbert, R. A. (refuge manager), Havasu National Wildlife Refuge, P.O. Box A, Needles, CA 92363.

Gladfelter, L. (wildlife research biologist), Iowa Department of Natural Resources, Wallace State Office Building, Des Moines, IA 50319.

Gore, H. G. (director of wildlife statewide projects), Texas Parks and Wildlife Department, 4200 Smith School Rd., Austin, TX 78744.

Hagen, D. R. (associate professor), Department of Animal Science, Pennsylvania State University, University Park, PA 16802.

Hamilton, D. (USFS staff), Osceola National Forest, P.O. Box 1649, Lake City, FL 32055.

Hamilton, R. B. (assistant chief of big game), Game Division, North Carolina Wildlife Resources Commission, P.O. Box 2919, Raleigh, NC 27602.

Hill, L. (refuge manager), Colusa National Wildlife Refuge, Rte. 1, Box 311, Willows, CA 95988.

Hillestad, H. O. (ecologist), Law Environmental, Inc., 112 Town Park Dr., Kennesaw, GA 30144.

Hogsed, G. (sportsman), Hwy 19, Aiken, SC 29801.

Hubbard, J. (sportsman), Aiken, SC 29801.

Hughes, T. W. (graduate student), Department of Aquaculture, Fisheries, and Wildlife, Clemson University, Clemson, SC 29631.

Hunt, E. G. (chief), Wildlife Management Division, California Department of Fish and Game, 1416 Ninth St., Sacramento, CA 95814.

Jackson, J. A. (professor), Department of Biological Sciences, Mississippi State University, P.O. Box GY, Mississippi State, MS 39762.

Janzen, D. H. (professor), Department of Biology, University of Pennsylvania, Philadelphia, PA 19104.

Joachim, A. (biologist), Division of Fish and Wildlife, New York Department of Environmental Conservation, Wildlife Research Laboratory, Delmar, NY 12054.

John, D. (wildlife biologist), Louisiana Wildlife and Fisheries Commission, 400 Royal St., New Orleans, LA 70130.

Johns, P. E. (wildlife biologist), Savannah River Ecology Laboratory, P.O. Drawer E, Aiken, SC 29801.

Johnson, F. (refuge manager), Aransas National Wildlife Refuge, P.O. Box 100, Austwell, TX 77950.

Johnson, G. (agricultural inspector), Bureau of Contagious and Infectious Diseases,, Florida Department of Agriculture, 5501 N. Bailey Rd., Plant City, FL 33566.

Johnson, R. (big game manager), Department of Game, 600 N. Capitol Way, Olympia, WA 98540.

Johnston, L. (biologist II), Arkansas Game and Fish Commission, State Capitol Grounds, Little Rock, AR 72201.

Jones, M. L. (registrar), The San Diego Zoo, Balboa Park, P.O. Box 551, San Diego, CA 97112-0551.

Joyner, S. K. (refuge manager), Catahoula National Wildlife Refuge, P.O. Drawer LL, Jena, LA 71342.

Kammermeyer, K. (wildlife biologist), Georgia Department of Natural Resources, Wildlife Division, Gainesville, GA 30501.

King, L. L. (refuge manager), Felsenthal National Wildlife Refuge, P.O. Box 279, Crossett, AR 71635.

Kirkland, Z. T. (chief ranger), Cumberland Island National Seashore, P.O. Box 806, St. Marys, GA 31558.

Knowles, J. R. (wildlife biologist), Natural Resources Division, Eglin Forestry Branch, Jackson Guard Station, Eglin Air Force Base, Niceville, FL 32542.

Kurz, J. (assistant chief), Game and Fish Division, Georgia Department of Natural Resources, 205 Butler St., S.E., Suite 1362, Atlanta, GA 30334.

Laramie, H. A. (chief of game division), New Hampshire Fish and Game Department, 34 Bridge St., Concord, NH 03301.

Layton, R. T. (taxidermist/sportsman), Layton's Northside Taxidermy, 103 N.E. Loop 410, San Antonio, TX 78216.

Lehmann, V. W., Sr. (wildlife biologist), Caesar Kleberg Wildlife Research Institute, College of Agriculture, Texas A&I University, Kingsville, TX 78363.

Lihoude, B. (refuge manager), Lake Woodruff National Wildlife Refuge, P.O. Box 488, DeLeon Springs, FL 32028.

Lipe, J. (wildlife biologist), Mississippi Division of Wildlife Conservation, Box 451, Jackson, MS 39205.

Little, T. W. (wildlife research supervisor), Iowa Department of Natural Resources, Wallace State Office Building, Des Moines, IA 50319.

Lopez, J. (range technician), Los Padres National Forest, 406 S. Mildred Ave., King City, CA 93930.

Lum, T. M. (staff wildlife biologist), Hawaii Division of Fish and Game, 1151 Punchbowl St., Honolulu, HI 96813.

Maloy, C. (wildlife biologist), Natural Resources Division, Eglin Forestry Branch, Jackson Guard Station, Eglin Air Force Base, Niceville, FL 32542.

Mancke, R. E. (South Carolina state naturalist), South Carolina State Museum Commission, P.O. Box 11296, Columbia, SC 29211.

Mansfield, T. M. (associate wildlife biologist), California Department of Fish and Game, 1416 Ninth St., Sacramento, CA 95814.

Martin, R. J. (professor), College of Home Economics, University of Georgia, Athens, GA 30602.

Masters, R. (regional biologist), Oklahoma Department of Wildlife Conservation, 1605 E. Osage, McAlester, OK 74501.

May, J. A. (sportsman), Giessen Army Depot, APO New York, NY 09169.

McAnally, B. (game biologist), District V, Arkansas Game and Fish Commission, Little Rock, AR 72201.

McDaniels, R. S. (superintendent), Congaree Swamp National Monument, 200 Caroline Sims Rd., Hopkins, SC 29061.

McKenzie, J. H., Jr. (veterinarian), 1350 E. Derenne Ave., Savannah, GA 31406.

McMaster, R. R. (refuge manager), White River National Wildlife Refuge, P.O. Box 308, DeWitt, AR 72042.

Miles, R. L. (chief), Wildlife Resources Division, Department of Natural Resources, Charleston, WV 25305.

Morgan, K. O. (superintendent), Cumberland Island National Seashore, P.O. Box 806, St. Marys, GA 31558.

Morris, J. G. (assistant professor), Department of Biology, Florida Institute of Technology, Melbourne, FL 32901.

Murphy, D. A. (assistant chief), Game Division, Missouri Department of Conservation, 2901 N. Ten Mile Dr., Jefferson City, MO 65101.

Murrey, J. (wildlife biologist), Tennessee Wildlife Resources Agency, P.O. Box 40747, Nashville, TN 37204.

Newsom, J. D. (professor), School of Forestry and Wildlife Management, 212 Forestry Building, Louisiana State University, Baton Rouge, LA 70803.

Newton, F. (wildlife biologist), Oregon State Game Commission, P.O. Box 3503, Portland, OR 97208.

Nowell, H. C. (chief), Game Management and Research, New Hampshire Department of Fish and Game, 34 Bridge St., Concord, NH 03301.

O'Neal, D. (refuge staff), Savannah National Wildlife Refuge, P.O. Box 8487, Savannah, GA 31412.

Orff, E. P. (game biologist), New Hampshire Fish and Game Department, 34 Bridge St., Concord, NH 03301.

Owens, L. (game biologist II), District X, Arkansas Game and Fish Commission, Little Rock, AR 72201.

Payne, P. W. (sportsman), RFD, Boerne, TX 78006.

Penn, L. (taxidermist/sportsman), Penn's Taxidermy Studio, 23645 Hwy 281 N., San Antonio, TX 78216.

Perry, M. D. (refuge manager), St. Vincent National Wildlife Refuge, Apalachicola, FL 32320.

Phares, J. H. (wildlife biologist), Mississippi State Game and Fish Commission, Box 451, Jackson, MS 39205.

Phelps, J. (sportsman), Camak, GA 30807.

Phelps, J. S. (predator/furbearer specialist), Arizona Game and Fish Department, 2222 W. Greenway Rd., Phoenix, AZ 85023.

Polenz, A. (big game staff biologist), Oregon Department of Fish and Wildlife, P.O. Box 59, Portland, OR 97207.

Reid, J. (area manager), Alabama Department of Conservation and Natural Resources, Rte. 1, Box 74, Coffeeville, AL 36524.

Reiner, J. W. (sportsman), 502 Cherokee Dr., North Augusta, SC 29841.

Rhodes, D. (wildlife biologist), Shining Rock Wilderness Area, Pisgah Ranger District, Pisgah National Forest, P.O. Box 8, Pisgah Forest, NC 28768.

Roest, A. I. (associate professor), Biological Sciences Department, California Polytechnic Institute, San Luis Obispo, CA 93407.

Rogers, M. J. (biologist III), District VII, Arkansas Game and Fish Commission, Calico Rock, AR 72519.

Rolley, R. E. (graduate student), Oklahoma Cooperative Wildlife Research Unit, Oklahoma State University, Stillwater, OK 74078.

Scharnagel, J. (wildlife biologist), Game and Fish Division, Georgia Department of Natural Resources, 270 Washington St., S.W., Atlanta, GA 30334.

Schwarz, K. (taxidermist/sportsman), Schwarz Taxidermy, 1851 Rigsby, San Antonio, TX 78210.

Scott, N. (wildlife ecologist), U.S. Fish and Wildlife Service, P.O. Box 1306, Albuquerque, NM 87103.

Shipes, D. (deer project leader), South Carolina Department of Wildlife and Marine Resources, P.O. Box 167, Columbia, SC 29202.

Siegler, H. R. (chief of game management and research), New Hampshire Fish and Game Department, 34 Bridge St., Concord, NH 03301.

Smart, R. F. (veterinarian/sportsman), P.O. Box 686, Boerne, TX 78006.

Smith, F. H., Jr. (chief), Bureau of Wildlife Management, Florida Game and Fresh Water Fish Commission, 620 S. Meridian St., Tallahassee, FL 32399-1600.

Smith, J. C. (wildlife biologist), Texas Parks and Wildlife Department, 715 S. Bronte, Rockport, TX 78382.

Smith, R. (general curator), San Antonio Zoological Society, 3903 N. St. Mary's St., San Antonio, TX 78212.

Snow, R. (taxidermist/sportsman), Snow's Taxidermy Studio, Broxton, GA 31519.

Snyder, W. A. (wildlife biologist), New Mexico Department of Game and Fish, State Capitol, Santa Fe, NM 87501.

Springer, M. D. (wildlife biologist), 1006 Dexter, College Station, TX 77840.

Stanley, M. E. (farm advisor), Animal Science and Range Management, University of California Agriculture Extension Service, P.O. Box 815, 625 Division St., King City, CA 93930.

Stanton, C. (president), Santa Cruz Island Company, P.O. Box 435, Port Hueneme, CA 93041.

Stewart, J. D. (wildlife biologist), Vermont Fish and Game Department, Division of Wildlife Research and Management, Roxbury, VT 05669.

Strange, T. (waterfowl biologist), South Carolina Wildlife and Marine Resources Department, Box 37, McClellanville, SC 29458.

Stribling, H. L., Jr. (assistant professor), Department of Zoology and Wildlife Science, Auburn University, Auburn, AL 36849-5414.

Stye, R. (wildlife biologist), Cheoah Ranger District, Rte. 1, Box 16-A, Nantahala National Forest, Robbinsville, NC 28771.

Sutterfield, T. (wildlife biologist), Division of Forestry and Wildlife, Hawaii Department of Land and Natural Resources, 1151 Punchbowl St., Honolulu, HI 96813.

Tatum, E. (animal dealer), Holly Springs, AR.

Taylor, T. (biologist II), District IV, Arkansas Game and Fish Commission, Perrytown, AR 72126.

Thackston, R. E. (S.E. regional biologist), Oklahoma Department of Wildlife Conservation, P.O. Box 705, Wilburton, OK 74578.

Thomas, P. (range technician), Los Padres National Forest, 406 S. Mildred Ave., King City, CA 93930.

Tyler, R. (wildlife biologist), U.S. Forest Service, Francis Marion National Forest, Rte. 3, Box 630, Moncks Corner, SC 29461.

Umber, R. (wildlife biologist), Oklahoma Department of Wildlife Conservation, 1801 N. Lincoln, Oklahoma City, OK 83105.

Urban, T. C. (manager), Venado Hunt Camp, Encino, TX 78353.

Urbston, D. F. (assistant chief, research), Arkansas Game and Fish Commission, 2 Natural Resources Dr., Little Rock, AR 72205.

Walker, R. L. (chief), Division of Forestry and Wildlife, Hawaii Department of Land and Natural Resources, 1151 Punchbowl St., Honolulu, HI 96813.

Walker, W. (rancher), P.O. Box 453, Fowlerton, TX 78021.

Walther, J. R. (refuge manager), Sabine National Wildlife Refuge, M.R.H. Box 107, Hackberry, LA 70645.

Ward, F. (biologist II), District II, Arkansas Game and Fish Commission, Little Rock, AR 72201.

Waters, V. (teacher and conservationist), 1621 Apple Valley Dr., Augusta, GA 30906.

Wells, R. K. (wildlife biologist), Mississippi Division of Wildlife Conservation, Box 451, Jackson, MS 39205.

Wetzel, R. M. (professor), Biological Sciences Group, Box U-42, University of Connecticut, Storrs, CT 06268.

West, E. T. (conservationist), P.O. Box 13397, Savannah, GA 31416-0397.

Whitehead, C. J. (wildlife biologist), Tennessee Game and Fish Commission, P.O. Box 40747, Ellington Center, Nashville, TN 37220.

Whittington, R. W. (wildlife biologist), Game and Fish Division, Georgia Department of Natural Resources, 270 Washington St., S.W., Atlanta, GA 30334.

Whitmore, F. C. (curator), Department of Vertebrate Paleontology, Smithsonian Institution, Washington, DC 20560.

Williams, L. E., Jr. (wildlife biologist), Wildlife Research Projects, Florida Game and Fresh Water Fish Commission, 2606 N.E. Terrace, Gainesville, FL 32601.

Winkler, C. K. (wildlife biologist), Texas Parks and Wildlife Department, 4200 Smith School Rd., Austin, TX 78744.

Wise, J. B., Jr. (sportsman), 402 E. Palfrey, San Antonio, TX 78223.

Woods, J. T., Jr. (caretaker/foreman), Noble Foundation, St. Catherines Island, Sunbury, GA.

Wright, R. M. (conservationist and historian), 4402 Bruton Rd., Plant City, FL 33566.

Appendix B

*Survey of the Subspecies of
Eurasian Wild Boar*

Since the early 1900s, the subspecific taxonomic status of the Eurasian wild boar has been dealt with only sparingly. In addition, many of the original descriptions of these subspecies were based on restricted samples of from one to four skulls (Swinhoe, 1863; Gray, 1868; Milne-Edwards, 1872; Blanford, 1875; Heude, 1888; Nehring, 1889; Jentink, 1905; Miller 1906; Thomas, 1912; Adlerberg, 1930; Heim de Balsac, 1937). This did little to account for local variability, much less the broader geographic variation present within the species. Antonius (1922) considered the Eurasian subspecies of *Sus scrofa* to be the result of hybridization between the eastward migrating *Sus scrofa scrofa* and the westward migrating *Sus scrofa cristatus*. Adlerberg (1930) synonymized all the Eurasian wild boar under *Sus scrofa* Linnaeus and described a new subspecies, *Sus scrofa raddeanus*. Amon (1938) followed Antonius (1922) and explained the geographic variation seen in wild boar as a distributional mixture of the western species, *Sus scrofa ferus*, with the eastern *Sus vittatus vittatus*. Kelm (1939) noted that the skull morphology of Eurasian wild boar underwent a clinal change from west to east, with the skull, especially the lacrimal bones, becoming shorter and higher. Ellerman and Morrison-Scott (1966) presented only the basic subspecific synonymies and ranges and did little to clarify the vagaries among the various recognized geographic races. Epstein (1971) reviewed the literature on the Eurasian wild boar subspecies, discussing the theories on the causes of geographic differentiation and comparing in detail only those wild-living subspecies of *Sus scrofa* thought to be the most direct ancestral forms of the domestic pig. Hemmer (1978) determined that the insular forms of *Sus scrofa nicobaricus* and *Sus scrofa andamanensis* were feral hog populations, although this had been stated earlier by both Blyth (1858) and Miller (1902) when these subspecies were first described, and later by Abdulali (1962) with respect to the wild pigs on the Andaman Islands. Corbet (1978) simplified the problem by stating that the subspecies of Eurasian wild boar have been based on differences in size and coat coloration, but that the subspecific range boundaries were poorly defined. Groves (1981), in a revision of the entire genus *Sus*, accepted only 14 of the 23 normally recognized subspecies, synonymized 7 other subspecies, provisionally revived 1 previously described subspecies, and provisionally described 2 new subspecies, naming only 1 of them, *Sus scrofa davidi.*

The following is a taxonomic review of the normally recognized geographic subspecies of Eurasian wild boar. Included for each subspecies is a synonymy, identification of the type of specimen(s) and type locality when possible, delineation of the range, description of morphological characters, and other relevant comments. Morphological characters are based on descriptions in the literature and specimens examined by the authors. Color names used in the coat coloration descriptions of the various Eurasian wild boar specimens were based on comparisons of color plates and descriptions in Smith (1974). Color name and number cross-reference information between the colors described by Smith (1974) and those of Ridgeway (1912) is provided in Smith (1974).

Sus scrofa algira Loche
North African Wild Boar

Sus scrofa, var. *barbarus* Sclater. 1860. Proc. Zool. Soc. London, p. 443 (nomen nudum).
Sus scrofa var. *algira* Loche. 1867. Expl. Sci. de l'Algerie, Zool. Mamm., p. 59.
Sus scrofa barbarus Gray. 1868. Proc. Zool. Soc. London, p. 31.
Sus scrofa sahariensis Heim de Balsac. 1937. Bull. Soc. Zool. France, 62:333.
Sus scrofa algirus Heim de Balsac. 1937. Bull. Soc. Zool. France, 62:333 (lapsus calumi).

Holotype: adult male; from type locality.

Type locality: country of Beni Sliman, Algeria.

Range: Tunisia, Algeria, Morocco, and Spanish Sahara.

Morphological characters: The adult pelage is fuscous to black with white hair tips over most of the body. The ears, distal rostrum, limbs, and tail are solid fuscous to black. The facial area has white-tipped bristles in a facial saddle. Some specimens have been described as having a reddish middorsal stripe (Heim de Balsac, 1937). The presence of underfur is variable (Heim de Balsac, 1937; Epstein, 1971). When present, the underfur has been described as dark brown (Heim de Balsac, 1937). Juvenile animals exhibit a stripped phenotype (Loche, 1867).

External body measurements (in mm) of one adult male from Morocco (USNM 476855) were: head-body length = 1,440; tail length = 190; hind foot length = 270; ear length = 116.

Skull measurements (in mm) of two adult males (USNM 476855 and ANSP 2341, respectively) were: condylobasal length = 320, 372; nasal length = 184, 213; zygomatic breadth = 137, 164; mandibular length = 263, 303; lower 3d molar length = 36.6, 41.3; width of first cusp row of lower 3d molar = 15.8, 18.8. The lacrimal indices were 1.21 and 1.20. Both cranial profiles were flat/concave.

Comments: According to Trouessart (1905), this subspecies does not differ morphologically from *Sus scrofa scrofa*. The isolation of the distribution of this subspecies is probably the best argument for its validity. Populations of this subspecies became extinct in Libya between 1910 and 1920 (De Beaux, 1928). Trouessart (1905) included Egypt in the distribution of this geographic race.

Specimens examined (2): MOROCCO—Oujda Prov., 3 km NE of Taforalt (1, USNM). TUNISIA—Tunis (1, ANSP).

Sus scrofa attila Thomas
Middle East Wild Boar

Sus attila Thomas. 1912. Proc. Zool. Soc. London, 105:1.
Sus scrofa attila Lydekker. 1914. Ward's Records of Big Game, 7th ed., p. 452.

Holotype: BM 12.1.23.1; adult male; skin and skull; presented by Fraulein Sarolta von Wertheimstein; 8 Dec. 1911; from type locality.

Type locality: Kolozsvar (= Cluj-Napoca), Transylvania, Romania.

Range: Romania and Moldavian SSR east to Georgian SSR, Iran, Iraq, and extreme eastern Turkey.

Morphological characters: Based on the examination of 20 adult skins, the coat coloration pattern in this subspecies is highly variable dark grayish brown/fuscous to black with antique brown, buff, cream, fawn, cinnamon brown, or burnt umber hair tips. The distal limbs, rostrum, tail, and ears are burnt umber to black. The facial area has white- to buff-tipped bristles in the lateral portions. One specimen (MCZ 29674) from Iraq had a mouth streak of white-tipped bristles. The juveniles are striped (FMNH 84477). Bristle lengths range from 30 to 145 mm middorsally and from 25 to 85 mm laterally. The middorsal bristles increase in length anteriorly. The presence of underfur seems to vary with the season, being absent in summer (n = 2) and densely present in fall (n = 9), winter (n = 1), and spring (n = 8). The underfur colors include drab, fawn, smoke gray, and cinnamon brown.

The means and ranges of external body measurements (in mm) from four male and seven female adult specimens were: head-body length, males = 1,363.8 (1,290–1,465)/females = 1,364.7 (1,234–1,480); tail length, males = 242.5 (230–250)/females = 223.5 (190–256); hind foot length, males = 326.3 (310–360)/females = 314.7 (295–333); ear length, males = 119.3 (103–130)/females = 133.6 (120–160). Adult body mass for this subspecies is reported to range from 136 to 318 kg (Banoglu, 1952; Page, 1954; Hatt, 1959; Heptner et al., 1966; Harrison, 1968; Bauer, 1970a; Harrington, 1977).

Skull measurements (in mm) for the type specimen were: condylobasal length = 405; nasal length = 250; zygomatic breadth = 176; mandibular length = 363; lower 3d molar length = 42.0; width of first cusp row of lower 3d molar = 19.5 (Miller, 1912). Mean skull measurements (in mm) for adult male (n = 10) and female (n = 8) specimens were: condylobasal length, males = 364.5 (333–395)/females = 349.6 (327–361); nasal length, males = 216.2 (199–234)/females = 198.6 (169–212); zygomatic breadth, males = 160.9 (145–170)/females = 151.8 (146–156); mandibular length, males = 300.0 (279–323)/females = 281.6 (274–291); lower 3d molar length, males = 47.5 (39.8–50.2)/females = 43.1 (38.0–48.1); width of first cusp row of lower 3d molar, males = 19.1 (17.1–20.8/females = 19.4 (17.9–20.1). The lacrimal indices for these specimens were: males = 1.46 (1.07–1.93)/females = 1.39 (1.14–1.69). Koslo (1975) gave means and ranges for condylobasal lengths (in mm) of wild boar from two areas within the distribution of this subspecies as: Beresin Reserve, U.S.S.R., males = 403 (397–414)/females = 353 (347–357); Caucasus, U.S.S.R., males = 388 (362–414)/females = 347 (317–378). Kelm (1939) reported the range of lacrimal indices for this geographic race as 1.3–1.9. The cranial profiles were all of the flat/concave phenotype.

The diploid number of chromosomes for this subspecies is 37 (Tikhonov and Troshina, 1974).

Comments: Groves (1981) noted that the geographic boundary between *Sus scrofa attila* and *Sus scrofa scrofa* was very distinct based on skull size.

Specimens examined (26): IRAN—Turkman Desert, Dar Kaleh (3, AMNH); vicinity of Hamadan (1, USNM); Fars, 17.4 km SW of Yasoodj (1, FMNH); Fars, 2.4 km E of Yasoodj (1, FMNH); Isfahan, Kuhrang (1, FMNH); Kerman, 32 km SW of Zabol (2, FMNH); Gorgan, 16 km ESE of Gorgan (2, FMNH); Khuzistan, E of Kermanshah (2, FMNH). IRAQ–Jebel Baradost, Ishkufti, Khurwatan (1, MCZ); 4.8 km S of Baghdad (1, FMNH); Darband area between Kirkuk and Sulimaniyah Liwa (1, FMNH); Khanagin (2, FMNH); Place Ramlla, 16 km from Khanaquin (3, FMNH); Kikuk Liwa,

Wadi Hostocki, 0.8 km S of Towaka Zhari (1, FMNH); Chahala, near Amhara (2, FMNH); Baradost (1, FMNH); 30 km N of Hilla (1, UMMZ).

Sus scrofa baeticus Thomas
Andalusian Wild Boar

Sus scrofa baeticus Thomas. 1912. Proc. Zool. Soc. London, pp. 391–393.
Sus scrofa castilinaus Ellerman and Morrison-Scott. 1951. Checklist Palearctic & Indian Mamm., p. 347.
Sus scrofa boeticus Epstein. 1971. Origin Dom. Anim. Africa, 2:314 (lapsus calumi).

Holotype: BMNH 8.3.8.12; male; presented by Abel Chapman, Esq.; collected on 6 Feb. 1908; from type locality.

Type locality: Coto Donana, Huelva, Spain.

Range: Southern Iberian Peninsula, from the Andalucia region to the river valley of the Bajo Buadalquivir and the coast of the Huelva Provinces.

Morphological characters: The adult pelage is fuscous to black with light-tipped bristles over most of the body. The ears, distal limbs, and tail are black. The facial area has white-tipped bristles on the rostrum (Cabrera, 1914). Juveniles are striped (Cabrera, 1914). Curly, woollike underfur is not present in this subspecies, even in winter skins (Thomas, 1912; Cabrera, 1914).
　　Body measurements (in mm) of an adult male topotype were: head-body length = 1,190; tail length = 152; hind foot length = 250; ear length = 126; shoulder height = 600 (Cabrera, 1914).
　　Skull measurements (in mm) of the type specimen and an adult female topotype (BMNH 8.3.8.13) were, respectively: condylobasal length = 305, 285; nasal length = 177, 158; zygomatic breadth = 142, 127; mandibular length = 268, 250; lower 3d molar length = 35.0, 29.0; width of first cusp row of lower 3d molar = 17.6, 13.2 (Miller, 1912). The cranial profile is flat/concave.

Comments: Thomas (1912) described this subspecies as "very much smaller" than *Sus scrofa scrofa*. Cabrera (1914) noted that other than a reduced size, there were no distinguishing skull characters that would separate it from *Sus scrofa castilianus*. Ellerman and Morrison-Scott (1966) tentatively synonymized this subspecies with *Sus scrofa castilianus*. Groves (1981) synonymized it under *Sus scrofa meridionalis*.

Specimens examined (1): SPAIN—Virgen de la Cabeza, Sierra Morena (1, AMNH).

Sus scrofa castilianus Thomas
Castilian Wild Boar

Sus scrofa castilianus Thomas. 1912. Proc. Zool. Soc. London, pp. 391–393.

Holotype: BM 11.10.5.3.; male; skin and skull; collected by the Reverend Saturio Gonzalez; presented by the Honorable N. Charles Rothschild; 1911; from type locality.

Type locality: Quintanar de la Sierra, SE of Burgos, Prov. of Burgos, Spain.

Range: Central and northern Iberian Peninsula, from the Pyrenees and Cantabrica mountains to the Spanish province of Galicia in the north, and from the Sierra Morena to the Portuguese province of Alemtejo in the central peninsula.

Morphological characters: The coat coloration of a subadult male (USNM 154146) was fuscous to black with cream to buff hair tips on the back, sides, and undersides of the body. The ears, distal rostrum, limbs, and tail were black. The facial area was fuscous with white- to cream-tipped bristles in a facial saddle. Juveniles animals are striped (Cabrera, 1914). Bristles on the subadult male described measured 62–113 mm middorsally, and 34–80 mm laterally. The underfur on that animal was dense and cinnamon colored.

External body measurements (in mm) from an adult male from Monte Negre, Barcelona, were: head-body length = 1,400; tail length = 153; hind foot length = 280; ear length = 130; shoulder height = 625 (Cabrera, 1914).

Skull measurements (in mm) of the type specimen, an adult female topo-type (BNMH 8.7.7.33), and another adult male were, respectively: condy-lobasal length = 336, 306, 340; nasal length = 191, 166, 191; zygomatic breadth = 137, 131, 141; mandibular length = 295, 269, 285; lower 3d molar length = 42.0, 36.0, 34.0; width of first cusp row of lower 3d molar = 18.0, 16.4, — (Miller, 1912; Cabrera, 1914).

Comments: Thomas (1912) stated that this subspecies was an intermediate form between *Sus scrofa scrofa* and *Sus scrofa baeticus*. Groves (1981) synon-ymized this subspecies under *Sus scrofa scrofa*.

Specimens examined (1): SPAIN—Palacio de la Sierra, Prov. Burgos (1, USNM).

Sus scrofa chirodontus Heude
Southern Chinese Wild Boar

Sus leucomystax Swinhow. 1870. Proc. Zool. Soc. London, p. 639.
Sus chirodontus Heude. 1888. Mem. concern. l'Hist. Nat. de l'Emp. Chinois,
2:54.
Sus palustris Heude. 1888. Mem. concern. l'Hist. Nat. de l'Emp. Chinois,
2:54.
Sus leucorhinus Heude. 1892. Mem. concern. l'Hist. Nat. de l'Emp. Chinois,
2:114.
Sus paludosus Heude. 1892. Mem. concern. l'Hist. Nat. de l'Emp. Chinois,
2:114.
Sus melas Heude. 1892. Mem. concern. l'Hist. Nat. de l'Emp. Chinois, 2:114.
Sus flavescens Heude. 1899. Mem. concern. l'Hist. Nat. de l'Emp. Chinois,
4:130.
Sus chirodonticus Heude. 1899. Mem. concern. l'Hist. Nat. de l'Emp.
Chinois, 4:130.
Sus meles Sowerby. 1917. Proc. Zool. Soc. London, p. 15 (lapsus calumi).
Sus scrofa chirodontus Allen. 1930. Amer. Mus. Novitates no. 430, p. 4.
Sus scrofa chirodonta Allen. 1940. Mann. China & Mongolia, p. 1123 (lapsus
calumi).

Syntypes: One of the skulls in the Heude collection at the Sikawei Museum,
Shanghai, China. No numbers given; four male specimens; all skulls alone;
all from Kiente, NE of Poyang Lanke, south of Anking-fu, Anhui Province,
China (Sowerby, 1917).

Type locality: Poyang Lake, Kiangsi, China.

Range: Provinces of Hupeh, Anhwei, Chekiang, Hunan, Kiangsi, Fukien,
Kwangsi, Kwangtung, and Hianan in China.

Morphological characters: The adult coat coloration is variable. Four of the
five skins examined were burnt umber to black with buff to cinnamon brown
hair tips over the body; solid burnt umber to black ears, distal rostrum, limbs,
and tail; and with a mouth streak of cream- to white-tipped bristles on the
face. One of these skins (AMNH 41472) had black eye rings. A fifth skin
(FMNH 39329), collected in July, was solid fuscous in color with solid cream-
colored bristles on the mouth streak. Juveniles are striped (AMNH, 59966).
Winter bristle lengths were 40–100 mm middorsally and 25–45 mm laterally. The bristle lengths on the summer skin were 45–50 mm middorsally
and 15–25 mm laterally. The presence of underfur was also variable. The

summer skin and three of the winter skins had none. The other winter skin (AMNH 47859) had sparse smoke gray underfur.

Marshall (1967) reported the following measurements for this subspecies: head-body length = 1,219 mm; tail length = 152 mm; body mass = 159–182 kg. Duff (1934) reported body masses of two large adult males collected in the Kuling Foothills as 157 and 171 kg. Allen (1940) reported the body mass of a large male from Kwangtung Province of southern China as 134 kg. An adult male specimen (CAS 5082) collected in Sike, China, weighed 102 kg.

Mean skull measurements (in mm) of five adult male and five adult female specimens were: condylobasal length, males = 334.6 (324–352)/females = 313.4 (294–331); nasal length, males = 188.6 (177–200)/females = 168.4 (160–179); zygomatic breadth, males = 156.4 (147–164)/females = 139.0 (131–145); mandibular length, males = 274.2 (260–291)/females = 248.4 (235–259); lower 3d molar length, males = 38.8 (35.7–44.4)/females = 36.7 (34.0–40.7); width of first cusp row of lower 3d molar, males = 18.2 (17.0–19.0)/females = 18.0 (16.1–21.3). The lacrimal indices for these specimens were: males = 0.97 (0.80–1.13)/females = 0.88 (0.64–1.00). All specimens exhibited flat to convex cranial profile.

Comments: Sowerby (1943) noted that this subspecies was smaller and redder than the Yangtze Valley form. Allen (1940) reported that specimens of *Sus scrofa chirodontus* were darker than *Sus scrofa moupinensis*. Groves (1981) synonymized this subspecies under *Sus scrofa moupinensis*.

Specimens examined (34): CHINA—Fukien Prov., Yenping (8, AMNH; 1, MCZ; 1, FMNH); Fukien Prov., Chungen Hsien (2, AMNH; 1, FMNH); Fukien Prov., Hsien, Fuching (1, AMNH); Kiang-su, Nanking (1, UMMZ); Hainan, Nodoa (13, AMNH; 3, FMNH); Hainan, Nam Fong (1, AMNH); Chekiang, near Tung-Lu, Sike (1, CAS); southern China, no specific locality (1, AMNH).

Sus scrofa coreanus Heude
Korean Wild Boar

Sus coreanus Heude. 1897. Mem. concern. l'Hist. Nat. de l'Emp. Chinois, 3:191.

Sus leucomystax coreanus Tate. 1947. Mamm. East. Asia, p. 313.

Sus scrofa coreanus Ellerman and Morrison-Scott. 1951. Checklist Palearctic & Indian Mamm., p. 347.

Sus scrofa koreanus Heptner, Nasimovic, and Bannikov. 1966. Säuget. Sowjetunion. Band I: Parrhufer und Unpaarhufer, p. 53 (lapsus calumi).

Holotype: In the Heude collection at the Sikawei Museum, Shanghai, China; no number given; male; skull only; from type locality (Sowerby, 1917).

Type locality: Fusan (= Pusan), Korea.

Range: Korean peninsula.

Morphological characters: The coat coloration of an adult male (MCZ 30234) was dark grayish brown with buff hair tips over most of the body. The undersides were dark grayish brown with cream hair tips. The ears were fuscous to dark grayish brown with a few buff bristles mixed in. The distal limbs, rostrum, and tail were fuscous. The facial area was dark grayish brown with short buff-tipped hairs forming a mouth streak. The bristle lengths on this animal were 60–110 mm middorsally and 60–80 mm laterally. The underfur present was dense and smoke gray.

The skull measurements (in mm) of two adult males (CAS 4651 and MCZ 30234, respectively) were: condylobasal length = 355, 388; nasal length = 177, 230; zygomatic breadth = 161, 178; mandibular length = 290, 317; lower 3d molar length = 38.6, 51.9; width of first cusp row of lower 3d molar = 18.1, 24.5. The lacrimal indices from these two specimens were 0.92 and 1.29. Both skulls exhibited a flat/concave dorsal profile. Mori (1922) gave the following skull measurements (in mm) for an adult specimen belonging to this subspecies: condylobasal length = 365; nasal length = 225; zygomatic breadth = 155.

Comments: Sowerby (1943) stated that the skull of the type specimen of this subspecies was smaller than those of the Ussuri and Manchurian specimens. Groves (1981) synonymized this subspecies under *Sus scrofa ussuricus.*

Specimens examined (6): KOREA—Seishin (1, MCZ); Kosan, Chosen (5, CAS).

Sus scrofa cristatus Wagner
Indian Wild Boar

Sus cristatus Wagner. 1839. Akad. Wiss. Munch. Gelehrt. Anz., 9:435.
Sus aper var. *apiomus* Hodgson. 1841. J. Asiat. Soc. Bengal 10:911.
Sus aper var. *isonotus* Hodgson. 1841. J. Asiat. Soc. Bengal 10:911.
Sus indicus Gray. 1843. List. Mamm. British Mus., p. 185.
Sus affinis Gray. 1847. Cat. Osteol. British Mus., p. 71.
Sus bengalensis Blyth. 1860. J. Asiat. Soc. Bengal 29:105.

Sus vittatus affinis Major. 1883. Zool. Anz., 6:296.
Sus vittatus bengalensis Major. 1883. Zool. Anz., 6:296.
Sus vittatus cristatus Major. 1883. Zool. Anz., 6:296.
Sus vittatus indicus Major. 1883. Zool. Anz., 6:296.
Sus cristatus typicus Lydekker. 1900. Great & Small Game of India, p. 261.
Sus cristatus cristatus Lydekker. 1915. Cat. Ung. Mamm. British Mus., 4:319.
Sus scrofa cristatus Ellerman and Morrison-Scott. 1951. Checklist Palearctic
   & Indian Mamm., p. 345.

Holotype: In Naturhistor. Staatsmuseum in Vienna, Austria; skull and skin;
juvenile animal; collected by Frenherrn von Hügel.

Type locality: Probably the Malabar Coast of India (Lydekker, 1915).

Range: Pakistan, India, Nepal, East Pakistan, and Bangladesh.

Morphological characters: The adult coat coloration (n = 3) of the back,
sides, and undersides is fuscous to drab with cream or buff bristle tips. The
ears, distal limbs, rostrum, and tail are black to fuscous. The facial area
is fuscous to drab with cream or white bristle tips. These tips are lighter
posterior to the corner of the mouth, forming a distinct mouth streak. One
individual (FMNH 27436) had a solid-colored coat of mixed buff and clay
bristles, with fuscous-colored ears and distal rostrum, and patches of solid
cream-colored bristles on the sides of the face. The neonates have the striped
juvenile coloration pattern (Tate, 1947; Roberts, 1977). Bristle lengths in
adults (n = 4) range from 35 to 130 mm middorsally and from 15 to 75 mm
laterally. These lengths increase anteriorly. Epstein (1971) described the
nuchal crest, or mane, as "well developed," but noted that the remainder of
the coat was "scanty." The presence of underfur is variable from absent to
very dense. Some individuals (n = 2) lacked it even in winter. Underfur
colors observed include drab, fawn, and smoke gray. A few authors (e.g.,
Blanford, 1888; Lydekker, 1907; Epstein, 1977) have reported that this subspecies lacks underfur.

   Ranges of external body measurements in adults (sexes combined) were:
head-body length = 1,150–1,524 mm; tail length = 180–305 mm; ear
length = 100–140 mm; shoulder height = 700–1,016 mm; body mass =
91–230 kg (Blanford, 1888; Garga, 1945; Prater, 1965; Roberts, 1977). An
adult male (FMNH 25715) and female (ANSP 650) from India had the following measurements (in mm) respectively: head-body length = 1,530, 1,438;
tail length = 250, 312; hind foot length = —, 208; ear length = —, 106;
shoulder height = 900, —.

Means and ranges of skull measurements (in mm) of adult male (*n* = 18) and female (*n* = 9) specimens were: condylobasal length, males = 335.2 (300–380)/females = 302.9 (285–324); nasal length, males = 185.2 (145–200)/females = 161.8 (147–176); zygomatic breadth, males = 152.6 (136–178)/females = 139.0 (130–147); mandibular length, males = 273.1 (231–307)/females = 242.2 (222–262); lower 3d molar length, males = 42.4 (34.5–49.0)/females = 38.2 (35.6–40.4); width of first cusp row of lower 3d molar, males = 18.3 (15.8–20.8)/females = 16.7 (14.6–18.4). The lacrimal indices from these specimens were: males = 0.96 (0.75–1.37)/females = 0.85 (0.75–1.02). The cranial profiles were evenly divided between the two phenotypes; approximately 50% showed a flat/concave phenotype and the other 50% were convex.

The diploid number of chromosomes in this subspecies is 38 (Groves, 1981).

Comments: Groves (1981) separated the wild boar from southern India into *Sus scrofa affinis.*

Specimens examined (47): INDIA—Hardware (1, AMNH); Sonaipur, N Kheri Forest (1, AMNH); Surguja, Ambikapur (1, USNM); Kurrachee (1, USNM); Calcutta (1, MCZ); Pegu (1, MCZ); Punjab (1, MCZ); Amballa (2, MCZ); Oudh, Palia (2, FMNH); Agra, United Prov. (2, FMNH); Rajputana (5, FMNH); Madhyn Pradesh Prov., Itarsi (7, MSU); Madhyn Pradesh Prov., 112 km NE of Bilaspur (1, MSU); Kashmir, Lake Woolar (2, USNM); Kashmir, Valley of Kashmir (1, USNM); Assam, 19.2 km E of 3d Tank Area (1, USNM); Ganges Delta, Sandarbans (2, ANSP); no specific locality (2, AMNH; 3, ANSP; 1, USNM; 6, MCZ; 1, FMNH). NEPAL—Dharan Bazar (1, USNM). PAKISTAN—Sialkot District (1, USNM).

Sus scrofa falzfeini Matschie
Baltic Wild Boar

Sus falzfeini Matschie. 1918. Sitz. Ber. Ges. Naturf. Fr. Berlin, 8:5.
Sus scrofa attila Adlerberg. 1930. Compt. Rend. Acad. Sci. USSR, p. 93.
Sus scrofa falzfeini Ellerman and Morrison-Scott. 1951. Checklist Palearctic & Indian Mamm., p. 347.
Sus scrofa scrofa Heptner, Nasimovic, and Bannikov. 1966. Säuget. Sowjetunion. Band I: Parrhufer und Unpaarhufer, p. 52.

Holotype: In the Zoologischen Museum der Universitat in Berlin; no. 29453; adult male; collected by Friedrich von Falz-Fein in 1912; from type locality.

Type locality: Drainage of the Neman River, Naliboki, between Grodno and Minsk, White Russia or Russian SSR.

Range: Eastern Poland and Russian SSR, Latvian SSR, Lithuanian SSR, and Estonian SSR.

Morphological characters: The skin of one large adult male from Bialowieza National Park (FMNH 46401) was fuscous to black with antique brown to russet tips over the back and sides; cream on the undersides; burnt umber to black on the ears, distal rostrum, tail, and limbs; with fuscous to black bristles on the face and black eye rings and cream-tipped bristles in a facial saddle. The bristle lengths were 35–140 mm middorsally and 27–80 mm laterally. The underfur was dense and varied from cream to smoke gray in color.

The body mass for adult specimens of this subspecies ranges from 150 to 286 kg (Heptner et al., 1966; Amato, 1976).

The skull measurements (in mm) from FMNH 46401 were: condylobasal length = 437; nasal length = 255; zygomatic breadth = 175; mandibular length = 359; lower 3d molar length = 47.3; width of first cusp row of lower 3d molar = 20.6. Cabon (1958) gave skull measurements (in mm) for 33 adult wild boar from Poland. These measurements were summarized as means and ranges from her data table as follows: condylobasal length, males (n = 12) = 380.2 (357–410)/females (n = 9) = 353.0 (336–368); nasal length, males (n = 16) = 223.0 (201–255)/females (n = 10) = 210.2 (195–231). The lacrimal indices calculated from Cabon's measurements were: males (n = 21) = 1.61 (1.33–1.92)/females (n = 11) = 1.52 (1.18–2.18).

Comments: Groves (1981) synonymized this subspecies under *Sus scrofa attila.*

Specimens examined (1): POLAND—Bialowieza National Park (1, FMNH).

Sus scrofa jubatus Miller
Southeast Asian Wild Boar

Sus jubatus Miller. 1906. Proc. U.S. Natl. Mus., 30:745.
Sus jubatulus Miller. 1906. Proc. U.S. Natl. Mus., 30:746.
Sus cristatus jubatus Lydekker. 1907. Game Animals of India, Burma, Malaya, and Tibet, p. 283.
Sus cristatus jubatulus Lydekker. 1907. Game Animals of India, Burma, Malaya, and Tibet, p. 283.

Sus scrofa jubatus Ellerman and Morrison-Scott. 1951. Checklist Palearctic & Indian Mamm., p. 347.

Holotype: USNM 83518; adult male; skin and skull; collected by Dr. W. L. Abbott; original number "B"; 1896; from type locality.

Type locality: Trong (also Trang or Tarang), lower Siam, Thailand.

Range: Vietnam, Thailand, Laos, Cambodia, Burma, and Malayan Peninsula.

Morphological characters: Most skins (n = 4) examined were solid black with a mouth streak of cream-tipped bristles. One (FMNH 31791) was fuscous with buff- to cream-tipped bristles over the body; black to fuscous ears, distal rostrum, legs, and tail; and a mouth streak of cream-tipped bristles. The juveniles are striped (Lekagul and McNeely, 1977). Bristles in adults (n = 5) ranged from 55 to 115 mm middorsally and from 25 to 50 mm laterally. The middorsal bristles increased in length from posterior to anterior. None of the skins (winter, n = 4; spring, n = 1) had underfur.

Ranges of adult external body measurements and weights were as follows: head-body length = 1,350–1,500 mm; tail length = 200–300 mm; ear length = 95–105 mm; shoulder height = 600–750 mm; body mass = 75–200 kg (Van Peenan et al., 1969; Lekagul and McNeely, 1977). One adult male (AMNH 87595) from Vietnam had the following measurements (in mm): head-body length = 1,294; tail length = 229; hind foot length = 298.

Means and ranges of skull measurements (in mm) of adult male (n = 7) and female (n = 7) specimens were: condylobasal length, males = 326.0 (302–340)/females = 303.7 (276–328); nasal length, males = 176.0 (159–192)/females = 160.3 (147–176); zygomatic breadth, males = 157.2 (130–170)/females = 139.8 (131–153); mandibular length, males = 261.0 (256–270)/females = 242.6 (231–264); lower 3d molar length, males = 39.9 (35.0–42.7)/females = 37.5 (34.6–40.0); width of first cusp row of lower 3d molar, males = 19.8 (17.6–21.1)/females = 17.5 (16.3–18.9). Lacrimal indices from these were: males = 0.93 (0.78–1.17)/females = 0.92 (0.76–1.20). Thirteen specimens had convex dorsal profiles, and one had a flat/concave profile.

Comments: Groves (1981) synonymized the western portion of this subspecies under *Sus scrofa cristatus,* the eastern portion under *Sus scrofa moupinensis,* and the southern portion under *Sus scrofa vittatus.*

Specimens examined (27): VIETNAM—Cochin China, Lagna River (1, AMNH); southern Annam (1, MCZ); Annam, Dong-Me (1, AMNH);

Annam, Quangtri, Phone Mon (1, FMNH); no specific locality (5, AMNH). LAOS—Lao Fu Chay (1, FMNH). THAILAND—Mt. Angka (1, MCZ); Trang (1, AMNH); Kam Pang Pet Prov., Upper Klong Klung (2, FMNH); no specific locality (2, AMNH). BURMA—Sathiana (1, AMNH); Tenasserim, Tanjong Budak (4, USNM); Tenasserim, Buk Pyin (1, USNM); no specific locality (1, AMNH). MALAYSIA—near Pahang River (1, AMNH); Johore (1, USNM); Johore, Gunong Pulai (1, USNM); Johore, Muar District, near Tangak (1, FMNH).

Sus scrofa leucomystax Temminck
Japanese Wild Boar

Sus leucomystax Temminck. 1842. Siebold's Fauna Japon., Mamm., p. 6.
Sus vittatus leucomystax Major. 1883. Zool. Anz., 6:296.
Sus vittatus japonica Nehring. 1885. Zool. Garten, 26:336.
Sus leucomystax japonicus Nehring. 1889 Sitz. Ber. Gessel. Nat. Freunde Berlin, 7:143.
Sus leucomystax leucomystax Lydekker. 1915. Cat. Ung. Mamm. British Mus., 4:322.
Sus scrofa leucomystax Sowerby. 1943. Mamm. Japanese Isl. Musée Heude, Notes de Mamm., no. 1, p. 57.

Holotype: Possibly a subadult female illustrated as a skull in Temminck (1842).

Type locality: Japan.

Range: Japan, including the islands of Honshu, Shikoku, and Kyushu.

Morphological characters: The adult coat in this subspecies is blackish brown on the back and sides, with white undersides and a mouth streak of white-tipped bristles (Gray, 1868; Sowerby, 1943). Underfur is present and dense (Gray, 1868). One adult male killed in the Kyoto Prefecture was devoid of bristles (Asahi, 1971). Juvenile animals are striped (Imaizumi, 1960).

Imaizumi (1960) gave the external body measurements and weights for *Sus scrofa leucomystax* as: head-body length = 1,370 mm; tail length = 230 mm; hind foot length = 265 mm; ear length = 120 mm; body mass = 75–190 kg. Asahi (1971) reported the external measurements and weight for an adult male from the Kyoto Prefecture as: head-body length = 1,605 mm; tail length = 195 mm; ear length = 110 mm; body mass = 130 kg.

Ranges of skull measurements (in mm) from a variable number of specimens of each sex were: condylobasal length, males = 266–327/females =

269–309; nasal length, males = 164–186; zygomatic breadth, males = 130–165; mandibular length, males = 232–292/females = 229–269 (Kelm, 1939; Asahi, 1971; Imaizumi, 1973). Nehring (1889) gave the condylobasal length and zygomatic breadth for an adult male from Japan as 307 mm and 155 mm, respectively. Kelm (1939) gave the range for the lacrimal indices as 0.8–1.1.

The diploid number of chromosomes is 38 (Muramoto et al., 1965).

Comments: Groves (1981) noted that the skulls of *Sus scrofa leucomystax* were similar to small skulls of *Sus scrofa scrofa* but broader.

Specimens examined (2): JAPAN—Honshu, Okubt, Chiba Prt. (2, USNM).

Sus scrofa lybicus Gray
Turkish Wild Boar

Sus scrofa Tristram. 1866. Proc. Zool. Soc. London, p. 84.
Sus lybicus Gray. 1868. Proc. Zool. Soc. London, p. 31.
Sus vittatus lybicus Major. 1883. Zool. Anz., 6:296.
Sus scrofa lybicus Lydekker. 1915. Cat. Ung. Mamm. British Mus., 4:316.

Holotype: BM 44.7.13.7; adult female; skull only; presented by Sir Charles Fellows; original number 713a; 1844; from type locality.

Type locality: Xanthus, near Gunek, Turkey.

Range: Central and western Turkey, Syria, Lebanon, Jordan, and Israel.

Morphological characters: Based on the skins of two subadult males (FMNH 44722 and 44723) from Syria, the coloration pattern is fuscous with cream to buff hair tips on the back and sides; fuscous on the undersides; fuscous to burnt umber on the distal rostrum, tail, and limbs; cinnamon and buff on the ears; and fuscous on the facial area, with a buff-tipped facial saddle. Juvenile animals are striped (Harrison, 1968). The bristle lengths on the two subadult males ranged from 60 to 90 mm middorsally and from 10 to 60 mm laterally. Underfur was densely present and drab.

The external body measurements (in mm) of an adult male (AMNH 327738) from Turkey were: head-body length = 1,310; tail length = 90; hind foot length = 290; ear length = 135. Shoulder heights of adults ranged from 749 to 950 mm (Banoglu, 1952; Harrison, 1968). Body mass ranged from 125 to 185 kg (Banoglu, 1952).

Mean skull measurements (in mm) of six adult males and two adult females were: condylobasal length, males = 362.7 (340–401)/females = 340.7 (320–355); nasal length, males = 215.8 (194–250)/females = 186.7 (175–195); zygomatic breadth, males = 157.0 (145–160)/females = 148.0 (136–157); mandibular length, males = 297.3 (282–323)/females = 268.3 (253–279); lower 3d molar length, males = 38.5 (36.8–39.8)/females = 39.3 (38.5–40.3); width of first cusp row of lower 3d molar, males = 18.1 (16.9–19.5)/females = 18.1 (17.9–18.5). The lacrimal indices from these skulls were: males = 1.27 (0.92–1.62)/females = 1.34 (1.28–1.41). All specimens had flat/concave dorsal profiles.

The diploid number of chromosomes in the karyotype was 38 (Epstein, 1971).

Comments: Groves (1981) included a specimen from the Nile Delta in this subspecies and also expanded the range of this geographic race to include populations in Yugoslavia.

Specimens examined (12): TURKEY—Yeni Foca, N of Izmir (1, AMNH); near Mazgirt (1, MCZ); hills W of Elazig (1, MCZ); Metsia (Igel), Tarsus Forest (2, FMNH); no specific locality (3, YPM). ISRAEL—no specific locality (2, MCZ). SYRIA—Amouk plains (2, FMNH).

Sus scrofa majori de Beaux and Festa
Italian Wild Boar

Sus scrofa majori de Beaux and Festa. 1927. Mem. Soc. Ital. Sci. Nat. Milano, 9:270.

Lectotype: In the Museo di Geologia e Paleontologia della R. Universita di Fidenza; adult male; skull only; from type locality.

Type locality: Mount Pescali, Tuscany Maremma, Italy.

Range: Italian peninsula and Sicily.

Morphological characters: The general coat coloration was described as "snuff brown" and "tany olive" [*sic*], with the mane of the anterior half of the body black and the muzzle strongly grizzled with white (de Beaux and Festa, 1927). The underfur was described as dense and "smoke gray" and "drab," with longitudinal stripes in the juvenile pelage (de Beaux and Festa, 1927).

The skulls of two adult male specimens (Turin 4820; Turin 4821) from San Rossore, Italy, had the following measurements (in mm): condylobasal length

= 334, 332; nasal length = 196, 191; zygomatic breadth = 148, 152; mandibular length = 295, 290; lower 3d molar length = 36.6, 32.8; width of first cusp row of lower 3d molar = 16.0, 15.0 (Miller, 1912). Ranges of skull measurements (in mm) of five subadult/adult specimens were given by de Beaux and Festa (1927) as: condylobasal length = 289–339; and zygomatic breadth = 126.5–147.

Comments: Groves (1981) synonymized this subspecies under *Sus scrofa scrofa*. On the basis of craniometric and electrophoretic data, Randi et al. (1989) also suggested synonymizing *Sus scrofa majori* under *Sus scrofa scrofa*.

Specimens examined (0).

Sus scrofa meridionalis Major
Mediterranean Wild Boar

Sus scrofa meridionalis Major. 1882. Atti Soc. Tosc. Sci. Nat. Pisa, Proc. Verb., 3:119.
Sus scrofa var. *sardous* Strobel. 1882. Atti Soc. Ital. Sci. Nat. Milano, 25:221.
  (Later identified as a domestic pig and not a wild boar.)
Sus meridionalis Miller. 1912. Cat. Mamm. West. Europe, p. 960.

Holotype: No information available.

Type locality: Sardinia.

Range: Sardinia and Corsica.

Morphological characters: Based on the skin of an adult male (USNM 198012) from Sardinia, the coat coloration is fuscous with cream to buff-tipped bristles over the back, sides, and undersides. The ears, distal rostrum, limbs, and tail are fuscous to black. The face has a facial saddle of white-tipped bristles, which change to a cream color in the posteriolateral portion of the head. Underfur is present, dense, and smoke gray. The bristle lengths are 60–122 mm middorsally and 25–65 mm laterally.
  The skull measurements (in mm) of an adult male (USNM 198012) were: condylobasal length = 274; nasal length = 158; zygomatic breadth = 120; mandibular length = 221; lower 3d molar length = 30.1; width of first cusp row of lower 3d molar = 13.4. The lacrimal index for that specimen was 1.11. The cranium had a flat/concave dorsal profile. Miller (1912) listed the skull measurements (in mm) of four adult males and one adult female from Sardinia, for which the means and ranges were as follows: condylobasal length,

males = 265 (260–270)/female = 242; nasal length, males = 152.0 (139–158)/female = 128; zygomatic breadth, males = 121 (119–124)/female = 103; mandibular length, males = 232.5 (227–238)/female = 197; lower 3d molar length, males = 29.1 (27.6–30.4)/female = 27.0; width of first cusp row of lower 3d molar, males = 13.2 (13.0–13.4)/female = 13.0.

The diploid number of chromosomes for wild boar from Corsica is 38 (Popescu et al., 1980).

Comments: Groves (1981) expanded the range of this subspecies to include populations from Andalusia in the southern Iberian Peninsula. Groves (1981) also described specimens of *Sus scrofa meridionalis* as significantly smaller than *Sus scrofa scrofa*. On the basis of biochemical analyses, Apollonio et al. (1985) concluded that Sardinian wild boar were either entirely descended from introduced feral hogs or the product of considerable hybridization between the native wild boar population and introduced feral hogs.

Specimens examined (1): SARDINIA—Sinuai, Cagliari (1, USNM).

Sus scrofa moupinensis Milne-Edwards
Northern Chinese Wild Boar

Sus moupinensis Milne-Edwards. 1872. Nouv. Arch. Mus. l'Hist. Nat. Paris, 7:93.

Sus vittatus moupinensis Major. 1883. Zool. Anz., 6:296.

Sus oxyodontus Heude. 1888. Mem. concern. l'Hist. Nat. de l'Emp. Chinois, 2:54 (nomen nudum).

Sus dicrurus Heude. 1888. Mem. concern. l'Hist. Nat. de l'Emp. Chinois, 2:54.

Sus curtidens Heude. 1892. Mem. concern. l'Hist. Nat. de l'Emp. Chinois, 2:114.

Sus laticeps Heude. 1892. Mem. concern. l'Hist. Nat. de l'Emp. Chinois, 2:114.

Sus collinus Heude. 1892. Mem. concern. l'Hist. Nat. de l'Emp. Chinois, 2:114.

Sus acrocranius Heude. 1892. Mem. concern. l'Hist. Nat. de l'Emp. Chinois, 2:114.

Sus planiceps Heude. 1899. Mem. concern. l'Hist. Nat. de l'Emp. Chinois, 4:132.

Sus cristatus moupinensis Lydekker. 1900. Great and Small Game of India, Burma, Malaya, and Tibet, p. 266.

Sus scrofa moupinensis Allen. 1930. Amer. Mus. Novitates no. 430, p. 3.

Sus scrofa moupinensis Heptner, Nasimovic, and Bannikov. 1966. Säuget. Sowjetunion. Band I: Parrhufer und Unpaarhufer, p. 52 (lapsus calumi).

Holotype: In Museum d'Histoire Naturelle de Paris; adult; skin and skull; collected by Abbe Pere Armand David about 1873 from the type locality.

Type locality: Moupin, Sze-Chwan, China.

Range: Provinces of Sze-Chwan, Shensi, Honan, Shantung, Shansi, Kansu, and Hopei, China.

Morphological characters: The adult coat coloration (n = 8) is fuscous to dusky brown with cream to buff bristle tips on the back, sides, and undersides. The ears, distal rostrum, limbs, and tail are fuscous, dusky brown, or burnt umber. The facial area has a basal bristle color of black fuscous or dusky brown with white to cream in a variable pattern. Six skins had white- to cream-tipped bristles in the mouth streak. The remaining two skins had no mouth streak. Two specimens had fuscous eye rings. Sowerby (1930) noted that old males were occasionally almost completely black. Juveniles are striped (ANSP 15124). The adult bristle lengths (winter, n = 5; fall, n = 1) ranged from 50 to 110 mm middorsally and from 30 to 60 mm laterally. Smoke gray–colored underfur was present in all skins.

The external body measurements and weight of an adult male collected near Kolan Shan in western Shansi were: total length = 1,880 mm; ear length = 114 mm; shoulder height = 940 mm; body mass = 146 kg (Sowerby, 1923). The body masses of four adult males from Shansi averaged 153.4 kg with a range of 177–182 kg (Sowerby, 1918; 1923). Sowerby (1923) stated that large boars may weigh up to 182 kg.

The skull measurements (in mm) of three adult male and five adult female specimens had means and ranges as follows: condylobasal length, males = 330.7 (306–344)/females = 337.2 (320–360); nasal length, males = 180.0 (152–196)/females = 185.0 (171–204); zygomatic breadth, males = 152.0 (139–162)/females = 150.4 (143–158); mandibular length, males = 264.7 (242–276)/females = 272.8 (261–296); lower 3d molar length, males = 37.8 (33.9–40.3)/females = 38.0 (35.5–40.0); width of first cusp row of lower 3d molar, males = 18.2 (16.5–19.2)/females = 18.7 (16.4–20.0). The lacrimal indices of these specimens were: males = 0.82 (0.67–0.95)/females = 0.95 (0.75–1.22). All specimens had convex cranial profiles.

Comments: Groves (1981) expanded the range of *Sus scrofa moupinensis* to include populations found in Indochina.

Specimens examined (11): CHINA—Sze-Chwan, Schulingkon (1, ANSP); Wenchwan, Tsoo Po, 24 km SW of Wenchwan (1, AMNH); Shansi Prov., He-Shwin (2, AMNH; 1, MCZ; 1, FMNH); Sze-Chwan, Wanshien, Yen-

Ching-Kon (1, AMNH); Sze-Chwan, Wanshien (1, AMNH; 1, FMNH); Shansi Prov., Tsing Ting Shan, Tai Pei Shan (1, AMNH); Sze-Chwan, Tachien-Lu (1, MCZ).

Sus scrofa nigripes Blanford
Central Asian Wild Boar

Sus scrofa var. *nigripes* Blanford. 1875. J. Asiat. Soc. Bengal, 44(2):112.
Sus scrofa nigripes Lydekker. 1901. Great and Small Game of Europe, Western and Northern Asia and America, their Distribution, Habits, and Structure, p. 279.

Co-types: In the Indian Museum, Calcutta, India; male; skin and skull; female; skin and skull; both collected from the type locality by Capt. F. Stoliczka in 1874.

Type locality: Tien Shan Mtns., Kashgar District, Sinkiang, China.

Range: Southern Kazakh SSR, Afghanistan, and Sinkiang, China; may include northern Pakistan.

Morphological characters: Adult skins examined (n = 2) were fuscous with cream to buff tips on most body bristles. Stubbe and Chotolchu (1968) described this coloration as black-brown to black with yellow-brown bristle tips. The ears, distal rostrum, tail, and limbs were burnt umber to fuscous. The face was fuscous with cream-tipped bristles in a facial saddle. One specimen (MCZ 29675) had fuscous eye rings. The two skins, collected in the fall, had bristle lengths 70–105 mm middorsally and 45–55 mm laterally. Underfur was present and dense in both skins. The color varied from smoke gray to burnt umber. Stubbe and Chotolchu (1968) described the color as yellow-brown.

Pfeffer (1960) gave the maximum record of adult male body mass as 300 kg. Sloudsky (1956) reported the following means and ranges for adult body masses (in kg): males = 106.4 (80–183)/females = 68.7 (49–80).

Skull measurements (in mm) of an adult male (FMNH 25705) and an adult female (AMNH 97793) were, respectively: condylobasal length = 338, 338; nasal length = 19, 19; zygomatic breadth = 157, 157; mandibular length = 279, 283; lower 3d molar length = 45.6, 37.7; width of first cusp row of lower 3d molar = 21.0, 18.0. Kelm (1939) reported ranges of the adult condylobasal length for each sex as: males = 345–372/females = 322–344 mm. The respective lacrimal indices for FMNH 25705 and AMNH 97793 were 1.10 and 1.38. Kelm (1939) gave the range of the lacrimal indices of this subspecies as 1.1–1.7. Both crania had flat/concave profiles.

The diploid number of chromosomes in this subspecies was reported as 36 and 37 by Tikhonov and Troshina (1974).

Comments: Kelm (1939) noted that the skull of this race was similar to that of *Sus scrofa scrofa* and *Sus scrofa attila.*

Specimens examined (3): AFGHANISTAN—Rabut Kashan, Mourghal (Murghab) Valley, 30 km S of the Russian border (1, AMNH). CHINA—Sinkiang, Agijas (Agiaz River Valley), Agyas at 2000 m (1, FMNH). USSR—Uzbekistan SSR, E Bokhara, Pjandge River (1, MCZ).

Sus scrofa riukiuanus Kuroda
Ryukyu Wild Boar

Sus leucomystax riukiuanus Kuroda. 1924. On New Mammals from Riu Kiu Islands (Tokyo), p. 11.
Sus scrofa riukiuanus Sowerby. 1943. Mamm. Japanese Isl. Musée Heude, Notes de Mamm. no. 1, p. 57.
Sus riukiuanus Imaizumi. 1973. Mem. Nat. Sci. Mus., Tokyo, 6:119.

Holotype: N. Kuroda coll. no. 920; adult male; skull only; collected by H. Orii; 4 Dec. 1921; from type locality.

Type locality: Kabira, Ishigakijima, Ryukyu Islands.

Range: Ryukyu Islands

Morphological characters: Sowerby (1943) noted that this subspecies exhibited a white mouth streak. The pelage is somewhat darker than *Sus scrofa leucomystax* (Kuroda, 1924). Juvenile animals are striped (Bangs, 1901; Kuroda, 1924).

External body measurements (in mm) of the type specimen were: head-body length = 1,220; tail length = 175; ear length = 85; (Kuroda, 1924).

The ranges of adult skull measurements (in mm) of this subspecies as compiled from Imaizumi (1973) were: condylobasal length, males (n = 5) = 238–266/female (n = 1) = 237; nasal length, males (n = 3) = 123–137/female (n = 1) = 130; mandibular length, males (n = 10) = 201–241/females (n = 3) = 195–201. Skull measurements (in mm) of the type specimen were: condylobasal length = 269; nasal length = 153; zygomatic breadth = 133; mandibular length = 232; lower 3d molar length = 27; width of first cusp row of lower 3d molar = 13.2 (Kuroda, 1924).

Comments: Imaizumi (1973) has argued that this form was not of feral origin and because of its isolation should be given species status. In comparing it with *Sus scrofa leucomystax,* Imaizumi (1973) stated that *Sus scrofa riukiuanus* had simpler and smaller second and third molars, smaller auditory bullae, and less developed condylar and angular portions of the mandible. Sowerby (1943) described this race as darker in color than *Sus scrofa leucomystax.*

Specimens examined (0).

Sus scrofa scrofa Linnaeus
European Wild Boar

Sus scrofa Linnaeus. 1758. Syst. Nat., 10th ed., 1:49.
Sus scrofa aper Erxleben. 1777. Syst. Regn. Anim., 1:176.
Sus setosus aper Boddaert. 1785. Elench. Anim., 1:157.
Sus scrofa ferus Gmelin. 1788. Linn. Systm. Nat., 13th ed., 1:217.
Sus scrofa fasciatus Wagner. 1844. Schreber's Säugetiere, suppl., 4:322.
Sus scrofa var. *celtica* Strobel. 1882. Atti Sco. Ital. Sci. Nat. Milano, 25:79.
Sus vittatus fasciatus Major. 1883. Zool. Anz. 6:296.
Sus scrofa scrofa Thomas. 1912. Proc. Zool. Soc. London, p. 392.

Holotype: Unknown.

Type locality: Germany.

Range: Germany, France, Belgium, Netherlands, Switzerland, Austria, western Czechoslovakia, western Poland, and southern Denmark.

Morphological characters: Based on five adult skins, the adult coat coloration is burnt umber to black with buff to cinnamon tips over the back and sides of the body. The underside is variable, some are colored the same as on the back, while others have a solid smoke gray–colored belly, and others have mixed cream, buff, and white undersides. The ears, distal rostrum, limbs, and tail are solid burnt umber to black. The facial area has the same base color as the back and sides, with white- to cream-colored bristle tips in a facial saddle pattern. Some had burnt umber to black eye rings. Groves (1981) described the coat coloration as "dull to dark brown or olive-grey" with hair tips of "golden-brown to red-brown." Juvenile animals are striped (pers. observ. by JJM in West Germany). The bristle lengths of the five winter skins ranged from 70 to 105 mm middorsally and from 45 to 90 mm laterally. Underfur was present on all skins examined and was very dense and smoke gray to burnt umber. Groves (1981) described the underfur as "copious" and dark gray to red-brown in color.

External body measurements (in mm) reported for this subspecies were: head-body length = 1,500–1,850; tail length = 170–260; hind foot length = 250–300; ear length = 96–110; shoulder height = 850–1,000 (Baumann, 1946; Gaffrey, 1961). Gaffrey (1961) reported body mass of adult males to range from 100 to 230 kg. Heptner et al. (1966) stated that the largest wild boars from central Europe ranged from 153 to 230 kg. Epstein (1971) reported the maximum body mass for this subspecies as 350 kg. Briedermann (1970) determined the following means for 11 male and 58 female adult wild boar from East Germany: head-body length, males = 1,498 mm/females = 1,362 mm; shoulder height, males = 794 mm/females = 711 mm; body mass, males = 89.6 kg/females = 61.1 kg.

Mean skull measurements (in mm) for adult male (n = 15) and female (n = 4) specimens of this geographic race were: condylobasal length, males = 353.5 (305–398)/females = 328.3 (310–348); nasal length, males = 201.3 (161–230)/females = 183.5 (162–196); zygomatic breadth, males = 149.7 (141–165)/females = 146.8 (141–158); mandibular length, males = 288.2 (253–330))/females = 263.5 (256–272); lower 3d molar length, males = 38.4 (34.5–45.5)/females = 36.2 (34.3–37.1); width of first cusp row of lower 3d molar, males = 17.7 (16.5–19.8)/females = 16.9 (16.2–18.8). The lacrimal indices of these specimens were: males = 1.34 (0.70–1.95)/females = 1.34 (1.16–1.63). Kelm (1939) reported the ranges of condylobasal lengths as: males = 320–368 mm; females = 290–351 mm. Baumann (1949) stated that 370 mm was the maximum condylobasal length known for this subspecies. Kelm (1939) determined the lacrimal index range as 1.3–2.1.

The diploid number of chromosomes for this geographic race is polymorphic and has been reported as 36, 37, and 38 (Gropp et al., 1969; Rittmannsperger, 1971; Bosma, 1976; Popescu et al., 1980).

Comments: Groves (1981) expanded the range of *Sus scrofa scrofa* to include populations found in the Italian and Iberian peninsulas.

Specimens examined (19): FRANCE—Ardennes (1, AMNH); Lamargelle (1, USNM; 1, AMNH). GERMANY—Harz Mountains (1, USNM); Sichenbergen (1, AMNH); Sendichirfi (1, AMNH); Sachenwald (1, AMNH); Friedrichsruh, Sachsenwald (1, AMNH); Berlin, Grunewald (1, AMNH); 14.4 km N of Wildflecken (1, FSM); Brandenburg, Potsdam (1, UCONN); Bavaria (1, MSU); no specific locality (1, USMN; 3, AMNH; 3, MCZ).

Sus scrofa sennaariensis Fitzinger
Nile River Wild Boar

Sus larvatus Fitzinger. 1853. Sitz. Ak. Wiss. Wien, 10:362. (Synonym for *Potamochoerus porcus.* Specimens misidentified as African bush pigs.)
Sus sennaariensis Fitzinger. 1858. Sitz. Ak. Wiss. Wien, 29:365 (nomen nudum).
Sus sennaariensis Fitzinger. 1864. Sitz. Ak. Wiss. Wien, 50:388.
Sus sennaarensis Gray. 1868. Proc. Zool. Soc. London, p. 32 (lapsus calumi).
Sus scrofa var. *africanus* Rolleston. 1879. Trans. Linn. Soc., ser. 2, 1:257.
Sus vittatus sennaariensis Major. 1883. Zool. Anz., 6:296.
Sus scrofa Anderson and de Winton. 1902. Mamm. Egypt, p. 354.
Sus scrofa sennaarensis Lydekker. 1908a. Game Animals of Africa, p. 390 (lapsus calumi).
Sus scrofa sennaariensis Setzer. 1956. Proc. U.S. Natl. Mus., 106:570.
Sus scrofa barbarus Heptner, Nasimovic, and Bannikov. 1966. Säuget. Sowjetunion. Band I: Parrhufer und Unpaarhufer, p. 53.

Holotype: In Naturhistor. Staatsmuseum, Vienna, Austria; subadult female; skull only; from type locality.

Type locality: Sennar, Kordofan, Sudan.

Range: Nile Valley from Sudan to Egypt.

Morphological characters: The coat coloration of this poorly known subspecies has been described as blackish brown to olive black bristles with light brown or yellow tips (Fitzinger, 1864; Gray, 1868; Epstein, 1971). The bristles are approximately 40 mm in length and elongated into a mane on the neck (Epstein, 1971). The piglets have been reported both to have stripes (Osborn and Helmy, 1980) and to be devoid of stripes (Epstein, 1971).

Epstein (1971) described external body measurements (in mm) as reaching the following: "body length" = 750; shoulder height = 600; tail length = 70–80.

As illustrated in Antonius (1922), the skull of a subadult female specimen from Sennar had a flat/concave dorsal profile. The lacrimal bone has been described as short and high (Epstein, 1971).

Comments: The validity of this subspecies as a truly wild race is uncertain. Rutimeyer (1878) believed that the wild pig in Sennar was a feral descendant of domestic stock. Ellerman and Morrison-Scott (1966) accepted the Sudan as a valid component of the native distribution of *Sus scrofa.* Allen (1954) accepted it as valid and included the extinct populations of wild boar from Egypt in this subspecies. Setzer (1956) listed this race only on the basis of observations of feral hogs in several parts of the Sudan. Ansell (1971) concluded that this subspecies was based on a feral or primitive domestic form and that *Sus scrofa* was probably never an indigenous wild species in the Sudan. Epstein (1971) noted that although this race was in doubt, from a zoogeographic point of view the occurrence of true wild boar in Sennar could not be ruled out.

Specimens examined (0).

Sus scrofa sibiricus Staffe
Mongolian Wild Boar

Sus canescens Heude. 1888. Mem. concern. l'Hist. Nat. de l'Emp. Chinois, 2:54 (nomen nudum).
Sus leucomystax sibiricus Staffe. 1922. Arb. Lehr. Tierz. Hochs. Boden., Wien, 1:51.
Sus scrofa raddeanus Adlerberg. 1930. Compt. Rend. Acad. Sci. USSR, p. 95.
Sus scrofa ferus Adlerberg. 1930. Compt. Rend. Acad. Sci. USSR, p. 95.
Sus scrofa ussuricus Allen. 1930. Amer. Mus. Novitates no. 430, p. 2.
Sus scrofa sibiricus Kelm. 1939. Zeit. Tier. Zucht., 43:366.
Sus scrofa raddeana Allen. 1940. Mamm. China & Mongolia, p. 1115 (lapsus calumi).

Holotype: In Naturhist. Staatsmuseum in Vienna, Austria.

Type locality: Tunkinsk Mountains, southern Siberia.

Range: Mongolia and the Tunkinskiye Gol'tsy and Sayan mountains in southern Siberia.

Morphological characters: Based on the skin of an adult female (USNM 123456), the back and sides are burnt umber to fuscous with cream- to buff-colored hair tips. The underside is a mixed cream, white, and burnt umber. The ears, distal limbs, tail, and rostrum are fuscous to black. The facial area is burnt umber in color, with buff-tipped bristles on the crown and white-tipped bristles in the mouth streak. Epstein (1971) also noted the presence of a mouth streak from the corners of the snout to the neck. Epstein (1971) further described this subspecies as having a profusely developed coat. Underfur is present, dense, and a smoke gray to drab. Adlerberg (1930) reported the underfur to be vinaceous buff.

An adult male specimen had the following skull measurements (in mm): condylobasal length = 336; zygomatic breadth = 143 (Adlerberg, 1930). Kelm (1939) reported the ranges of the condylobasal lengths (in mm) as: males = 310–324; females = 282–308. The aforementioned adult specimen had a flat/concave dorsal profile. Kelm (1939) reported the lacrimal index as ranging from 0.6 to 1.1 in this subspecies.

Comments: Kelm (1939) stated that specimens of *Sus scrofa sibiricus* were markedly smaller than neighboring races.

Specimens examined (1): MONGOLIA—96 km NE of Urga (= Ulan Bator) (1, AMNH).

Sus scrofa taivanus Swinhoe
Taiwanese Wild Boar

Porcula taivana Swinhoe. 1863. Proc. Zool. Soc. London, p. 360.
Sus taivanus Swinhoe. 1864. Proc. Zool. Soc. London, p. 383.
Sus leucomystax Gray. 1868. Proc. Zool. Soc. London, p. 26.
Sus vittatus taivanus Major. 1883. Zool. Anz., 6:296.
Sus leucomystax taivanus Lydekker. 1915. Cat. Ung. Mamm. British Mus., 4:322.
Sus scrofa taivanus Sowerby. 1943. Mamm. Japanese Isl. Musée Heude, Notes de Mamm. no. 1, p. 57.

Syntypes: "One or more of these are now in the British Museum" (Swinhoe, 1863). A neonatal specimen in this series (BM 62.12.24.8) was destroyed in 1873 (Jones, 1975). The syntype series includes the following specimens from Formosa, purchased by Swinhoe in 1870: BM 70.2.10.39 skin; BM 70.2.10.40 skin, imm.; BM 70.2.10.41 skin, imm.; BM 70.2.10.42 skin, imm.; BM 70.2.10.43 skin; BM 70.2.10.84 skull, female; BM 70.2.10.85 skull; BM 70.2.10.86 skull, female, imm.; BM 70.2.10.87 skull, female; BM 70.2.10.88 skin and skull, female, imm. (Lydekker, 1915).

Type locality: Taiwan.

Range: Taiwan.

Morphological characters: The coat of a juvenile male was described as black with buff or gray bristle tips and white-tipped bristles in the mouth streak (Kuroda, 1935). The presence of this white mouth streak is characteristic of this subspecies (Sowerby, 1943). The middorsal bristles reached 100 mm on this animal (Kuroda, 1935).

The external body measurements (in mm) on the above-mentioned juvenile male were: head-body length = 930; tail length = 135; hind foot length = 200; ear length = 80 (Kuroda, 1935).

The skull measurements (in mm) of an adult male (USNM 358646) were: condylobasal length = 288; nasal length = 151; zygomatic breadth = 135; mandibular length = 230; lower 3d molar length = 33.0; width of first cusp row of lower 3d molar = 17.0. The lacrimal index on this animal was 0.94. The dorsal profile was flat/concave. Mori (1922) gave the following adult skull measurements: condylobasal length = 305 mm; zygomatic breadth = 150 mm; lacrimal index (calculated from lacrimal measurements) = 1.04.

Comments: Sowerby (1943) described this subspecies as more reddish than *Sus scrofa leucomystax.* Allen (1940) mistakenly reported that the name of *Sus scrofa taivanus* referred to the domestic swine found on Formosa.

Specimens examined (8): TAIWAN—Nan-Tou, Hsien, Hwa-Chi Village, Jen-Ai (1, USNM); Taipei, Hsien, Tamsui (1, USNM); Ilan Hsien (1, USMN); Hualien, Hualiun (2, USNM); Wu-Lai (2, USNM); I-Lan, I-Lan Hsien (1, USNM).

Sus scrofa ussuricus Heude
Siberian Wild Boar

Sus ussuricus Heude. 1888. Mem. concern. l'Hist. Nat. de l'Emp. Chinois, 2:54.
Sus leucomystax var. *continentalis* Nehring. 1889. Sitz. Ber. Ges. Nat. Freunde Berlin, 7:141.
Sus gigas Heude. 1892. Mem. concern. l'Hist. Nat. de l'Emp. Chinois, 2:114.
Sus songaricus Heude. 1897. Mem. concern. l'Hist. Nat. de l'Emp. Chinois, 3:191.
Sus mandchuricus Heude. 1897. Mem. concern. l'Hist. Nat. de l'Emp. Chinois, 3:192.
Sus scrofa continentalis Adlerberg. 1930. Compt. Rend. Acad. Sci. USSR, p. 95.
Sus scrofa ussuricus Allen. 1930. Amer. Mus. Novitates no. 430, p. 2.

Holotype: In the Heude Collection at the Sikawei Museum, Shanghai, China; no number given; subadult male; skull only with cranium missing; from type locality (Sowerby, 1917).

Type locality: Ussuri Valley, eastern Siberia.

Range: Manchuria, China, and Ussuri and Amur river regions of eastern Siberia, USSR.

Morphological characters: The coloration of two adult skins collected in winter was fuscous to black with buff hair tips on the back and sides. The undersides were burnt umber to fuscous with cream hair tips, with a few scattered solid white bristles mixed in. The ears, distal rostrum, limbs, and tail were burnt umber to fuscous. The facial area was burnt umber to fuscous with cream to white bristle tips, lighter in the cheek, crown, and mouth streak. The bristle lengths were 70–100 mm middorsally and 40–60 mm laterally. Curly underfur was present and dense on both skins. The color varied from smoke gray to burnt umber.

The skull measurements (in mm) of one adult male and one adult female were, respectively: condylobasal length = 400, 379; nasal length = 225, 199; zygomatic breadth = 171, 174; mandibular length = 313, 302; lower 3d molar length = 45.4, 46.2; width of first cusp row of lower 3d molar = 23.0, 21.5. The lacrimal indices were, respectively, 1.14 and 1.21. Both crania exhibited a flat/concave dorsal profile. Kelm (1939) gave the ranges of the adult condylobasal lengths (in mm) as: males = 361–393; females = 332–365. The collective range of the lacrimal indices was 0.8–1.3 (Kelm, 1939). Mori (1922) gave the following adult cranial measurements (in mm): condylobasal length = 393; zygomatic breadth = 166. Lacrimal index (calculated from lacrimal measurements) = 1.16. Adlerberg (1930) reported that the condylobasal length of this subspecies reached 490 mm.

The diploid number of chromosomes is 37 (Tikhonov and Troshina, 1974).

Comments: Groves (1981) expanded the range of this subspecies to include populations found in the Korean peninsula.

Specimens examined (5): CHINA—I-Mien-Po, N. Kirin (1, USNM); Barim Station, on Chinese Eastern Railway, Great Khingam Mountains, N Manchuria (1, CAS); Shitouhotze Station, on Chinese Eastern Railway, Kirin Prov., N Manchuria (1, CAS). USSR—Mukhen and Alchi rivers, Maratime Prov., Siberia (1, AMNH); Amur Land, Amur River, 128 km N of Khabarovsk (1, AMNH).

Sus scrofa vittatus Müller and Schlegel
Banded Pig or Indonesian Wild Boar

Sus vittatus Müller and Schlegel. 1842. Verh. Zoogh. Indoneische Archip., 1:172.

Sus vitatus Fitzinger. 1864. Sitz. Akad. Wiss. Wien, 50:393 (lapsus calumi).

Aulacochoerus vittatus Gray. 1873. Hand-List Thick-Skinned Mamm. British Mus., p. 58.

Sus vittatus vittatus Major. 1883. Zool. Anz., 6:296.

Syntypes: In Leyden Museum; four mounted specimens: two adult females, one adult male, and one immature male; all collected by S. Müller in 1836 from the type locality.

Type locality: Padang, Sumatra.

Range: Sumatra and Java.

Morphological characters: The adult coat coloration (*n* = 4) is yellowish to fuscous with cream to buff hair tips over most of the body; brown to black distal limbs, rostrum, tail, and ears; a black middorsal stripe; and white to reddish brown–tipped bristles in a broad mouth streak. Immature animals are striped (Jentink, 1905; Diong, 1973). Adult bristle lengths were 27–127 mm middorsally and 19–64 mm laterally. None of the skins examined (*n* = 4) had underfur.

Harrison (1974) gave the following range of body measurements (in mm) and weights (in kg) for this subspecies: head-body length = 1,000–1,500; tail length = 200–300; shoulder height = 700; body mass = 50–100. Adult male (*n* = 4) and female (*n* = 6) specimens from Sumatra had the following means and ranges for external body measurements (in mm): head-body length, males = 1,171 (1,060–1,250)/females = 1,083 (995–1,170); tail length, males = 233 (226–240)/females = 190 (130–268); hind foot length, males = 253 (244–265)/females = 230 (212–255). Two adult females had these additional measurements: ear length = 95, 96 mm; body mass = 55, 56 kg.

Adult skull measurements (means and ranges, in mm) of five male and seven female specimens were: condylobasal length, males = 282.5 (258–308)/females = 265.7 (230–282); nasal length, males = 145.3 (129–159)/females = 136.6 (117–145); zygomatic breadth, males = 131.3 (120–137)/females = 124.3 (118–134); mandibular length, males = 240.5 (213–255)/females = 226.1 (203–245); lower 3d molar length, males = 30.1 (28.4–33.5)/females = 31.2 (27.0–37.6); width of first cusp row of lower 3d molar, males = 16.3 (15.0–18.5)/females = 16.6 (15.0–19.0). The lacrimal indices of this series of skulls were: males = 0.89 (0.53–1.01)/females = 0.85 (0.74–0.96). All specimens had convex dorsal profiles. Jentink (1905) gave the skull measurements (in mm) of two adult male specimens as: condylobasal length = 327, 325; nasal length = 185, 183; zygomatic breadth = 150, 146; lower 3d molar length = 40.0, 37.0.

Comments: This subspecies is variable in overall skull size and in the complexity of the third molar, with specimens from the mainland and large islands being larger and having more complex molars than those from small islands.

Specimens examined (15): JAVA—no specific locality (1, AMNH). SUMATRA —Atjeh, Blangnanga (1, ANSP); Atjeh, Tretet (1, ANSP); Palembang (1, USNM); Indragiri River (1, USNM); Palo Babi (1, USNM); Pulo Tuangku (1, USNM); Simalur Island (2, USNM); Rhio Archipelago, Pulo Ungar (2, USNM); no specific locality (1, ANSP). BALI—Soember Klampok (2, AMNH); Banjoe Wedan (1, AMNH).

Sus scrofa zeylonensis Blyth
Sri Lankan Wild Boar

Sus zeylonensis Blyth. 1851. J. Asiat. Soc. Bengal, 20:173.
Sus zeylanensis Gray. 1868. Proc. Zool. Soc. London, p. 24 (lapsus calumi).
Sus ceylonensis Rolleston. 1879. Trans. Linn. Soc., ser. 2, 1:266.
Sus vittatus zeylanensis Major. 1883. Zool. Anz., 6:296.
Sus cristatus cristatus Phillips. 1935. Manual Mamm. Ceylon, p. 349.

Holotype: In the Indian Museum, Calcutta, India; skull; collected by E. F. Kelaart in 1850 from the type locality.

Type locality: Sri Lanka.

Range: Sri Lanka.

Morphological characters: The adult coat coloration is blackish gray with some light bristle tips. The younger animals are browner in color, while the older animals are grayer (Phillips, 1935). The juvenile animals are striped (Phillips, 1935; Eisenberg and Lockhart, 1972). No underfur is present (Phillips, 1935).

The means and maxima (in parentheses) for the external body measurements (in mm) and masses (in kg) of five adult males and three adult females were: head-body length, males = 1,553 (1,702)/females = 1,325 (1,372); tail length, males = 275 (292)/females = 283 (318); hind foot length, males = 288 (330)/females = 279 (280); ear length, males = 156 (165)/females = 130 (140); shoulder height, males = 828 (889)/females = 604 (610); body mass, males = 101 (127)/females (n = 2) = 54 (57) (Phillips, 1935).

The means and ranges of skull measurements (in mm) of three adult males were: condylobasal length = 335.5 (333–338); nasal length = 186.3 (182–187); zygomatic breadth = 149.5 (147–152); mandibular length = 266.0 (262–269); lower 3d molar length = 44.0 (42.7–45.2); width of first cusp row of lower 3d molar = 18.8 (17.1–19.8). The mean of the lacrimal indices was 1.08, and the range was 0.90–1.21. All three specimens had convex cranial profiles. Blyth (1851) stated that the *Sus zeylonensis* skull was longer than and third molars were larger than those of the Indian wild boar.

Comments: Groves (1981) synonymized this subspecies under *Sus scrofa affinis.*

Specimens examined (3): SRI LANKA—no specific locality (1, USNM; 2, MCZ).

Appendix C

*Use of Scientific Nomenclature
for Domestic and Feral Swine*

Domestic Swine

Although domestic swine are invariably included in the species *Sus scrofa*, the validity of a subspecies designation, and to some extent the choice of a species designation, for this domestic mammal have been controversial. In the tenth edition of *Systema Naturae* (1758), Linnaeus included all the common domestic animals among the numerous organisms he named and described. At present, however, some researchers are uncertain what to call some domestic animals when a scientific name is required, as illustrated by the inconsistent usage of scientific names for some of these animals (Groves, 1971; Clutton-Brock, 1981). Because domestic animals do not constitute naturally evolved populations and can produce fertile offspring when crossbred with their wild ancestors, it has even been suggested that these forms should not be covered by zoological nomenclature (Clutton-Brock et al., 1976; Clutton-Brock, 1981). The concept that an artificial situation such as domestication does not warrant a separate specific or subspecific taxonomic designation for the resulting forms is accepted by some systematic zoologists (Groves, 1971; R. M. Wetzel, pers. comm.). This problem is further complicated by the lack of a specific set of nomenclatural rules for determining the scientific names of domestic animals (Ride et al., 1985).

Since the late 1950s, there has been an ongoing debate regarding the use of scientific names for domestic animals.

Bohlken (1958) began the controversy by stating that binomial names that had been applied to domestic forms first should never be used for wild ancestors. He used the example of the yak, referred to in Ellerman and Morrison-Scott (1951) as *Bos grunniens mutus*. The wild yak had been described originally as *Bos mutus* Przewalski, and the domestic yak as *Bos grunniens* Linnaeus. Bohlken (1958) felt this was an unrealistic nomenclatural relationship because the wild yak did not evolve from the domestic yak; instead, the domestic yak should be referred to as *Bos mutus grunniens,* since the domestic form had been derived from the wild one. This line of reasoning was later considered by Groves (1971) to be in violation of the rules of nomenclature.

Trumler (1961) carried Bohlken's views a step further, applying them to domestic animals of multiple species origin. As an example Trumler used the domestic donkey, which was derived from two wild species he referred to as *Asinus africanus* and *Asinus taeniopus*. Linnaeus had originally named the domestic donkey *Equus asinus*. By Trumler's scheme, the domestic donkey should then be referred to as *Asinus africanus asinus* or *Asinus taeniopus asinus,* depending on which breed was being referred to.

Bohlken (1961) later modified his view, stating that domestic animals were "ecological races" of their wild ancestors. He recommended that the domestic be considered a form, or "forma," of the wild animal, denoting these domestics with an *f.* before the subspecific name. Using Bohlken's (1958) earlier example, the scientific name of the domestic yak would become *Bos mutus* f. *grunniens*. This is the nomenclature method favored by archaeo-zoologists (Clutton-Brock, 1981).

Dennler de la Tour (1968) proposed the use of the word *"familiaris"* as a standard name for domestic forms. For domestic animals that have a known wild ancestor, the species name of the wild form is followed by the key word *"familiaris,"* and then by the name given to the domesticate. The domestic yak, therefore, would become *Bos mutus "familiaris" grunniens*. Domestic animals of uncertain or disputed origin would be referred to by their original name, which would then be followed by the key word, for example, *Equus asinus "familiaris."*

Because of the problems with using scientific names for domestic animals, Groves (1971) proposed the removal from zoological nomenclature of the forms and the names originally assigned to them. He argued that domesticated animals were already excluded from the International Code of Zoological Nomenclature as products of artificial selection, with the phrases "in nature" and "hypothetical concepts" in Article 1 of the code restricting the use of zoological nomenclature from these forms. Groves (1971) proposed amending Article 1 to explicitly omit domestics. When one wanted to refer specifically to a domestic form, it could be qualified in the vernacular; for example, domestic cattle could be referred to as *Bos primigenius* (domestic form); *Bos primigenius,* Jersey cow; *Bos primigenius* dom.; or *Bos primigenius taurus*. But the italicized identifications such as *Bos taurus* or *Bos primigenius taurus* would not be permitted to refer specifically to domestic forms. After all, cultivated plants are excluded from botanical nomenclature and there is a separate International Code of Nomenclature for Cultivated Plants. Groves did not believe that there was a need for such a separate code for domestic animals; however, he did note that there was no reason why, in principle, long-established and easily recognized feral animal populations should not be classified with suitable scientific names.

Holthuis and Husson (1971) disagreed with Groves's "drastic solution" to the problem. They claimed that amending the code to include a section on domestic animals would only lead to endless controversies because of the lack of a good definition for domestic animals and the many gradations between truly domestic and truly wild animals. They failed to see what objection there could be to using the oldest name for both. If wild and domestic animals were not proved to be one species, Holthuis and Husson (1971) felt there was no choice but to treat these animals as distinct species. They admitted that there were problems with nomenclature for domestic animals, but adding a species clause in the code for domestic animals would only result in numerous difficulties.

Lemche (1971), responding to Groves (1971), proposed that rather than changing the code and gradually "spoiling it" through time, an interpretive declaration should be issued at the next revision of the code, stating that: " 'races' produced through the domestication of animals are to be considered as infra-subspecies status" and would therefore not fall under the rules of the code.

Eisenmann (1972) objected to Groves's 1971 proposal because it would upset well-established usage and raise many difficult problems. With the exception of a few breeds believed to be of hybrid origin, domestic animals should belong to the same taxonomic unit, at least at the generic and specific level, as the living or extinct wild ancestors. To categorically prohibit those names given to domestic animals would cause many problems, involving homonymy as well as synonymy, that probably would, in some instances, involve generic and familial group names (Eisenmann, 1972).

Groves (1977) reviewed the comments to his proposal and agreed with the various respondents with regard to using nomenclature for domestic animals, but he did not feel that any of their solutions were sufficient. He presented two solutions that would avoid "spoiling the code," although he admitted that neither was really satisfactory. The first solution, as presented in his original proposal, would treat domestic forms as hypothetical concepts and so exclude them under Article 1. The other would exclude them under Article 24(c) as probable hybrids, since "most domestic breeds are likely to have received an occasional injection of genes from wild stock in the vicinity, even if they are not the product of subspecific or even specific crossing in the first place" (Groves, 1977). In conclusion, Groves stated that nomenclature was supposed to be an aid to taxonomy, but where domestic animals were concerned it was only contributing confusion. He again insisted that the only solution was to remove the source of confusion, since such classifications do not reflect biological reality (Groves, 1977).

Melville (1977) commented that Groves's proposal should be granted, but only after the problem had been more clearly defined. Specialists within each group should present the commission with lists of available names of domesticates so that these could be excluded from zoological nomenclature. In addition, he suggested that certain names could be stabilized by the designation of neotypes from wild populations if the exclusion of such names would create too many problems. Melville further felt that dictionary definitions of "domestic" animals were too vague for nomenclatural or biological purposes and proposed the following definition: "Domestic animal. Any animal of which the living conditions and breeding are controlled by man for his use or pleasure, other than individuals taken in the wild for purposes of conservation or research and their progeny" (1977).

Van Gelder (1979) felt that the problem was not as deplorable as Groves found it to be. The exclusion of the names of domestic animals from zoological nomenclature would create more problems than it would solve because Groves's system would exclude names up through the familial level. This same complaint had been stated earlier by Eisenmann (1972). Van Gelder (1979) contended that zoological nomenclature does not imply, suggest, or concern itself with ancestor-descendant relationships. In addition, he pointed out that many of Linnaeus's taxa were applicable to both the domestic and the wild forms (e.g., *Felis catus, Equus asinus, Bos bubalus, Bos taurus,* and *Capra hircus*). Van Gelder also returned to the problem of the definition of a domestic animal. He distinguished domestic animals (a population) from domesticated animals (individuals). Groves's solution of excluding domestic animals as either hypothetical concepts or hybrids lacked merit because domestic animals exist as identifiable genetic populations, and the concept of domestication is far from hypothetical. Furthermore, the definition of a hybrid in the code glossary was "the product of the crossing of two species." Van Gelder (1979) contended that Groves confused the influx of genes of wild subspecies into domestic animals as hybridization, and he concluded by stating that an available zoological name indicating that the animals in question were domestic populations would serve a useful purpose. He suggested that the commission recommended that only one name within a species be permitted to represent all domestic forms of that taxon, which would avoid a proliferation of names based on domesticates in different regions.

Groves (1979) responded to Van Gelder's (1979) comments, specifically disputing several points, but his strongest argument for excluding domestic animals was that most specialists who work with domestic animals have already chosen to ignore the rules of the code by giving scientific names (e.g., *Bos primigenius* f. *taurus*) with no regard for priority. Whether the names for domestics were banned or another solution adopted, Groves felt that something had to be done to regulate the use of zoological nomenclature for domestic animals.

The scientific name for domestic swine is a good example of the problem resulting from a lack of a set of nomenclatural guidelines. Linnaeus (1758) first classified domestic swine as a separate species, *Sus domesticus*, distinct from the wild-living Eurasian wild boar, *Sus scrofa*. Erxleben (1777) was the first authority to reduce this domestic form to a racial designation, referring to it as *Sus scrofa domesticus*. In addition to other subsequent subspecific designations, several new genera and species were assigned to various breeds of domestic swine before the late 1800s (Gray, 1862; Fitzinger, 1864; Gray, 1868). By 1900, however, most authorities had agreed on *Sus* as the generic designation. Unfortunately, the specific designation applied to domestic swine has remained unresolved. It is usually given as *Sus scrofa*, but *Sus domesticus* and *Sus domestica* are also used. This was further complicated earlier in the 1900s by the species identification of the Eurasian wild boar (see Appendix B). Before the mid 1900s, Eurasian wild boar were usually separated into two species, *Sus scrofa* and *Sus cristatus*, with the species *Sus leucomystax* and *Sus vittatus* occasionally added to the list. This would have invalidated the use of zoological nomenclature for this mammal. However, since all Eurasian wild boar are now synonymized under *Sus scrofa*, it is agreed that domestic swine were derived from this one wild species (Epstein, 1971; Clutton-Brock, 1981). At present, *Sus scrofa domesticus* is the most frequently used trinominal designation for domestic swine, with *Sus scrofa domestica* occasionally cited. No subspecific designations are used with either *Sus domesticus* or *Sus domestica*.

The following is the synonymy for domestic swine:

Sus scrofa domesticus Linnaeus

Sus domesticus Linnaeus, 1758. Syst. Nat., 10th ed. 1:49.
Sus scrofa domesticus Erxleben, 1777. Linn. Syst. Nat., p. 217.
Sus leucomystax sinensis Fitzinger, 1858. Sitz. Ak. Wiss. Wien, 29:367.
Sus scrofa crispa turcica Fitzinger, 1858. Sitz. Ak. Wiss. Wien, 29:368.
Sus scrofa crispa hugarica Fitzinger, 1858. Sitz. Ak. Wiss. Wien, 29:369.
Sus scrofa crispa syrmiensis Fitzinger, 1858. Sitz. Ak. Wiss. Wien, 29:370.
Sus scrofa crispa anatolica Fitzinger, 1858. Sitz. Ak. Wiss. Wien, 29:371.
Sus scrofa crispa polonica Fitzinger, 1858. Sitz. Ak. Wiss. Wien, 29:372.
Sus scrofa crispa nana Fitzinger, 1858. Sitz. Ak. Wiss. Wien, 29:372.
Sus scrofa macrotis vulgaris Fitzinger, 1858. Sitz. Ak. Wiss. Wien, 29:373.
Sus scrofa macrotis moravica Fitzinger, 1858. Sitz. Ak. Wiss. Wien, 29:374.
Sus scrofa macrotis italica Fitzinger, 1858. Sitz. Ak. Wiss. Wien, 29:375.

Sus scrofa macrotis germanica Fitzinger, 1858. Sitz. Ak. Wiss. Wien, 29:375.
Sus scrofa macrotis hispida Fitzinger, 1858. Sitz. Ak. Wiss. Wien, 29:376.
Sus scrofa macrotis bavarica Fitzinger, 1858. Sitz. Ak. Wiss. Wien, 29:377.
Sus scrofa macrotis suevica Fitzinger, 1858. Sitz. Ak. Wiss. Wien, 29:377.
Sus scrofa macrotis jutica Fitzinger, 1858. Sitz. Ak. Wiss. Wien, 29:378.
Sus scrofa macrotis zeelandica Fitzinger, 1858. Sitz. Ak. Wiss. Wien, 29:379.
Sus scrofa macrotis gallica Fitzinger, 1858. Sitz. Ak. Wiss. Wien, 29:380.
Sus scrofa macrotis boloniensis Fitzinger, 1858. Sitz. Ak. Wiss. Wien, 29:381.
Sus scrofa macrotis normanna Fitzinger, 1858. Sitz. Ak. Wiss. Wien, 29:382.
Sus scrofa macrotis ardennica Fitzinger, 1858. Sitz. Ak. Wiss. Wien, 29:383.
Sus scrofa macrotis campaniensis Fitzinger, 1858. Sitz. Ak. Wiss. Wien, 29:383.
Sus scrofa macrotis pictaviensis Fitzinger, 1858. Sitz. Ak. Wiss. Wien, 29:384.
Sus scrofa macrotis caroliensis Fitzinger, 1858. Sitz. Ak. Wiss. Wien, 29:385.
Sus scrofa macrotis velauniensis Fitzinger, 1858. Sitz. Ak. Wiss. Wien, 29:386.
Sus scrofa macrotis petroviensis Fitzinger, 1858. Sitz. Ak. Wiss. Wien, 29:386.
Sus scrofa macrotis variegata Fitzinger, 1858. Sitz. Ak. Wiss. Wien, 29:387.
Sus scrofa macrotis credoniensis Fitzinger, 1858. Sitz. Ak. Wiss. Wien, 29:388.
Sus scrofa macrotis anglica Fitzinger, 1858. Sitz. Ak. Wiss. Wien, 29:389.
Sus scrofa macrotis rudvicensis Fitzinger, 1858. Sitz. Ak. Wiss. Wien, 29:390.
Sus scrofa macrotis cestriensis Fitzinger, 1858. Sitz. Ak. Wiss. Wien, 29:390.
Sus scrofa macrotis glocestriensis Fitzinger, 1858. Sitz. Ak. Wiss. Wien, 29:391.
Sus scrofa macrotis antoniensis Fitzinger, 1858. Sitz. Ak. Wiss. Wien, 29:391.
Sus scrofa macrotis cornubica Fitzinger, 1858. Sitz. Ak. Wiss. Wien, 29:392.
Sus scrofa macrotis licestriensis Fitzinger, 1858. Sitz. Ak. Wiss. Wien, 29:392.
Sus scrofa macrotis eboracensis Fitzinger, 1858. Sitz. Ak. Wiss. Wien, 29:393.
Sus scrofa macrotis lincoloniensis Fitzinger, 1858. Sitz. Ak. Wiss. Wien, 29:394.
Sus scrofa macrotis suffolciensis Fitzinger, 1858. Sitz. Ak. Wiss. Wien, 29:394.
Sus scrofa macrotis befortiensis Fitzinger, 1858. Sitz. Ak. Wiss. Wien, 29:395.
Sus scrofa macrotis norfolciensis Fitzinger, 1858. Sitz. Ak. Wiss. Wien, 29:395.
Sus scrofa macrotis barcheriensis Fitzinger, 1858. Sitz. Ak. Wiss. Wien, 29:396.
Sus scrofa macrotis essexiensis Fitzinger, 1858. Sitz. Ak. Wiss. Wien, 29:397.
Sus scrofa macrotis westernii Fitzinger, 1858. Sitz. Ak. Wiss. Wien, 29:398.
Sus scrofa macrotis sussexiensis Fitzinger, 1858. Sitz. Ak. Wiss. Wien, 29:399.
Sus scrofa macrotis hibernica Fitzinger, 1858. Sitz. Ak. Wiss. Wien, 29:399.
Sus scrofa macrotis caesareensis Fitzinger, 1858. Sitz. Ak. Wiss. Wien, 29:400.
Sus scrofa macrotis scotica Fitzinger, 1858. Sitz. Ak. Wiss. Wien, 29:401.
Sus scrofa macrotis wiltoniensis Fitzinger, 1858. Sitz. Ak. Wiss. Wien, 29:403.
Sus scrofa macrotis northantoniensis Fitzinger, 1858. Sitz. Ak. Wiss. Wien, 29:404.
Sus scrofa macrotis salopiensis Fitzinger, 1858. Sitz. Ak. Wiss. Wien, 29:404.

Sus scrofa macrotis herfordiensis Fitzinger, 1858. Sitz. Ak. Wiss. Wien, 29:404.
Sus scrofa macrotis derbicensis Fitzinger, 1858. Sitz. Ak. Wiss. Wien, 29:405.
Sus scrofa macrotis cortwrightii Fitzinger, 1858. Sitz. Ak. Wiss. Wien, 29:405.
Sus scrofa macrotis suecica Fitzinger, 1858. Sitz. Ak. Wiss. Wien, 29:407.
Sus scrofa macrotis sibirica Fitzinger, 1858. Sitz. Ak. Wiss. Wien, 29:408.
Sus pliciceps Gray, 1862. Proc. Zool. Soc. London, p. 13.
Ptychoerus plicifroms Fitzinger, 1864. Sitz. Ak. Wiss. Wien, 50:409.
Centurious pliciceps Gray, 1868. Proc. Zool. Soc. London, p. 41.
Sus scrofa var. *domesticus* Rolleston. 1879. Trans. Linn. Soc., ser. 2, 1:273.

Feral Swine

The scientific designation of feral forms carries the nomenclatural controversy a step further. In contrast to domestic populations, feral populations
once again operate under natural selection. A number of geographically
distinct populations of feral swine from Southeast Asia and the East Indies
have been classified as separate species and subspecies. Because of the
possibility of the continued additional introduction of domestic stock into
these populations, they may not truly represent distinct closed gene pools. In
addition, since the morphological differences between domestic and feral
swine may be only the result of different environmental factors, these two
morphotypes may in actuality be ecotypes. Nevertheless, some forms have
been given different names. The following is a listing of the synonymies of
feral populations that have been given either specific or subspecific
designations:

Andaman Islands

Sus andamanensis Blyth, 1858. J. Asiat. Soc. Bengal, 27:267.
Sus vittatus andamanensis Major, 1883. Zool. Anz., 6:296.
Sus cristatus andamanensis Lydekker, 1900. Great and Small Game of India,
    p. 265.
Sus scrofa andamanensis Ellerman and Morrison-Scott, 1966. Checklist
    Palearctic & Indian Mamm., p. 346.

Nicobar Islands

Sus nicobaricus Miller, 1902. Proc. U.S. Natl. Mus., 24:735.
Sus vittatus nicobaricus Lydekker, 1900. Great and Small Game of India,
    p. 284.
Sus scrofa nicobaricus Ellerman and Morrison-Scott, 1966. Checklist
    Palearctic & Indian Mamm., p. 347.

Timor[a]

Sus timoriensis Müller and Schlegel, 1842. Wilde Zwijnin van den Ind. Arch., 1:42.
Sus scrofa var. Giebel, 1855. Säugethiere, p. 225.
Sus timorensis Gerrard, 1862. Cat. Bones Mamm. British Mus., p. 278 (lapsus calumi).
Sus vittatus timoriensis Major, 1883. Zool. Anz., 6:296.

New Guinea[b]

Sus papuensis Lesson and Garnot, 1826. Bull. Sci. Nat., 7:80.
Sus ternatensis Meyer, 1877. Trans. Linn. Soc., ser. 2, 1:276.
Sus arnensis Rosenberg, 1878. Malay Archip., p. 362.
Sus niger Finsch, 1886. Proc. Zool. Soc. London, p. 217.
Procula papuensis Fitzinger, 1864. Sitz. Ak. Wiss. Wien, 50:23.
Sus vittatus papuensis Major, 1883. Zool. Anz., 6:296.
Sus vittatus ternatensis Major, 1883. Zool. Anz., 6:296.

Flores

Sus florensianus Jentink, 1905. Notes Leyden Mus., 26:178.
Sus vittatus florensianus Lydekker, 1915. Cat. Ungul. British Mus., p. 325.

---

a Groves (1981) argued that wild pigs on Timor were feral *Sus celebensis* and not *Sus scrofa*.
b Groves (1981) reported the wild pigs on Papua New Guinea to be hybrids between *Sus celebensis* and *Sus scrofa*.

Appendix D

*Survey of the Status of
Wild Pigs in the United States
in 1988*

The following pages reproduce questions from the questionnaire sent to appropriate state agencies in 1988 in order to determine the current status of populations of wild pigs in the United States. The results indicated were based on returns submitted by the states of Alabama, Arizona, Arkansas, California, Florida, Georgia, Hawaii, Iowa, Kentucky, Louisiana, Mississippi, Missouri, New Hampshire, New Mexico, Oklahoma, Oregon, South Carolina, Tennessee, Texas, Virginia, Washington, and West Virginia. Although no survey form was returned by the state of North Carolina, other information submitted by their personnel was used to update the information on the status of wild pig populations in that state, as presented in the text.

Tabulated Responses to a 1988 Questionnaire to Determine the Status of Wild Pigs in the United States.

---

1) Are the wild pigs in your state
   a) feral hogs (wild-living domestic stock)?   68.1%
   b) Eurasian wild boar?   36.4% (*n* = 22)[a]
   c) feral hog/wild boar hybrids?   50.0%
   [Please check all appropriate answer(s)]

2) Are the wild pigs in your state considered to be game animals? (i.e., is the harvest of these animals regulated by state game laws)

   yes 28.6%
   no 71.4%   (*n* = 21)

3) If these animals are considered to be game animals, are they classified as
   a) big game?   50.0%
   b) unspecified?   20.0%   (*n* = 10)[b]
   c) other (please specify)?   30.0%

4) If these animals are regulated by state game laws,
   a) is there a specific hunting season?

   yes 40.0%   (please specify)
   no 60.0%   (*n* = 15)[b]

   b) is there a daily or seasonal bag limit?

   yes 33.3%   (please specify)
   no 66.7%   (*n* = 15)[b]

5) To hunt these animals in your state, are hunting licenses required to be purchased by
   a) a state resident hunting on his own land?

   yes 40.0%
   no 60.0%   (*n* = 20)[c]

   b) a nonresident?

   yes 65.0%
   no 35.0%   (*n* = 20)[c]

6) Are any special tags or permits required to hunt these animals in your state?

   yes 20.0%   (please specify)
   no 80.0%   (*n* = 20)[c]

7) Are there any weapons restrictions for persons hunting these animals in your state?

> yes <u>35.0%</u> (please specify)
> no <u>65.0%</u> $(n = 20)^c$

8) Are there any special regulations governing the use of dogs when hunting these animals in your state?

> yes <u>35.0%</u> (please specify)
> no <u>65.0%</u> $(n = 20)^c$

9) Is it legal to live-trap these animals in your state?

> yes <u>65.0%</u>
> no <u>35.0%</u> $(n = 20)^c$

10) If the answer to the previous question was yes, is a trapping license required to be purchased by
    a) a state resident trapping on his own land?

> yes <u>15.4%</u>
> no <u>84.6%</u> $(n = 13)$

    b) a state resident trapping on land other than his own?

> yes <u>15.4%</u>
> no <u>84.6%</u> $(n = 13)$

    c) a nonresident?

> yes <u>23.1%</u>
> no <u>76.9%</u> $(n = 13)$

11) Have any projects (past or present) been undertaken to study these animals in your state?

> yes <u>42.9%</u>
> no <u>57.1%</u> $(n = 21)^d$

12) Are any such projects planned for the future?

> yes <u>25.0%</u>
> no <u>75.0%</u> $(n = 20)^d$

13) Have any extirpation or extermination programs on wild pigs been implemented in your state?

> yes <u>38.1%</u>
> no <u>61.9%</u> $(n = 21)^d$

14) Are any such programs planned for the future?

> yes <u>17.6%</u>
> no <u>82.4%</u> $(n = 17)^d$

15) Does "open range" (i.e., releasing domestic livestock into unfenced areas as a rearing practice) legally exist in your state?

yes <u>26.3%</u>

no <u>73.7%</u> $(n = 19)^d$

16) If your state has regulations governing the annual hunter harvest of these animals, in what year did your state initiate these regulations?

(Please specify)

1957 (Calif.)	1927 (Haw.)	1966 (La.)	1979 (W.V.)
1955 (Fla.)	1988 (Ky.)	1936 (Tenn.)	

17) If your state has no regulations governing the annual hunter harvest of these animals, are any such regulations planned for the future?

yes _____

no <u>100%</u> $(n = 13)^d$

18) If the answer to the previous question was yes, can you give any details as to when, harvest season, bag limits, etc.?

(See text for responses.)

19) How abundant are these animals in your state? (Check one)
    a) Very abundant/easily seen   <u>12.5%</u>
    b) Common/occasionally seen   <u>41.7%</u> $(n = 24)^{ce}$
    c) Rare/seldom seem   <u>45.8%</u>

20) If there is any other information which you can give us that you feel may be of interest or benefit to our survey, please include it below or on the back of this page:

(See text for responses.)

---

a Since each respondent could reply to more than one category, the total of the percentages exceeds 100%.

b Although only 6 respondents indicated that wild pig harvest was regulated by state game laws (question 2), 10 and 15 individuals replied to questions 3 and 4, respectively.

c Iowa and Missouri did not respond to a number of questions because there are no longer any wild pigs in their states.

d Some states were unable to respond to certain questions because of ambiguous conditions that prevented them from providing a simple yes or no answer.

e Some respondents indicated more than one category.

Appendix E

Distribution Maps for
Wild Pig Populations
in the United States

The following pages reproduce portions of a map showing the distribution of feral/wild swine populations in the United States in 1988. The map was originally published, at a scale of approximately 1 inch = 100 miles, by the Southeastern Cooperative Wildlife Disease Study (SCWDS, 1988). As originally published, distribution of swine was divided into three categories: (1) more than 10 swine per square mile, (2) 10 or less swine per square mile, and (3) swine rare or absent. These coded areas are reproduced here exactly as presented on the original map and without regard to whether or not the details of these distributions agree with the findings of the present study. Disagreements between the two, however, have been noted and discussed in the text.

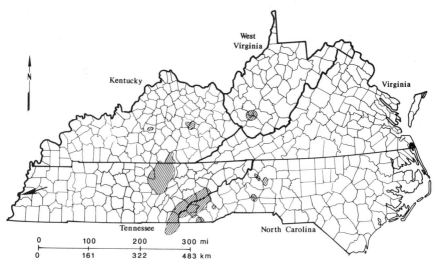

Appendix E-1. Distribution of wild pig populations in Virginia, West Virginia, Kentucky, Tennessee, and North Carolina. Blackened areas represent densities reported as more than 10 per square mile, shaded areas represent densities of less than 10 per square mile, and no shading represents areas where wild pigs are reported as rare or absent. Arrow indicates population in the Anderson-Tully Wildlife Management Area in western Lauderdale County. Redrawn from SCWDS (1988).

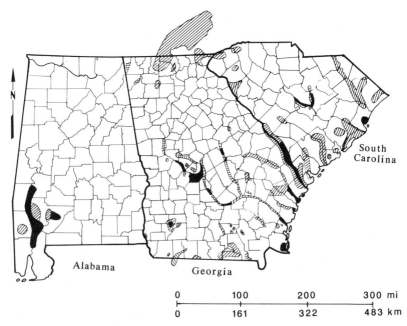

Appendix E-2. Distribution of wild pig populations in Alabama, Georgia, and South Carolina. Blackened areas represent densities reported as more than 10 per square mile, shaded areas represent densities of less than 10 per square mile, and no shading represents areas where wild pigs are reported as rare or absent. Redrawn from SCWDS (1988).

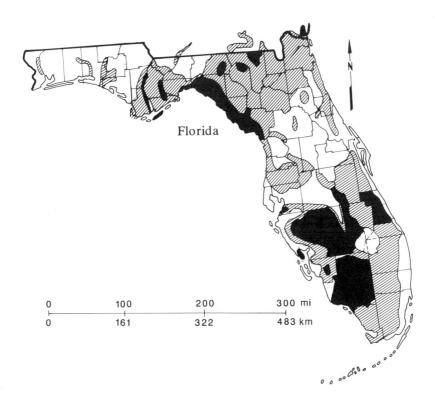

Appendix E-3. Distribution of wild pig populations in Florida. Blackened areas represent densities reported as more than 10 per square mile, shaded areas represent densities of less than 10 per square mile, and no shading represents areas where wild pigs are reported as rare or absent. Redrawn from SCWDS (1988).

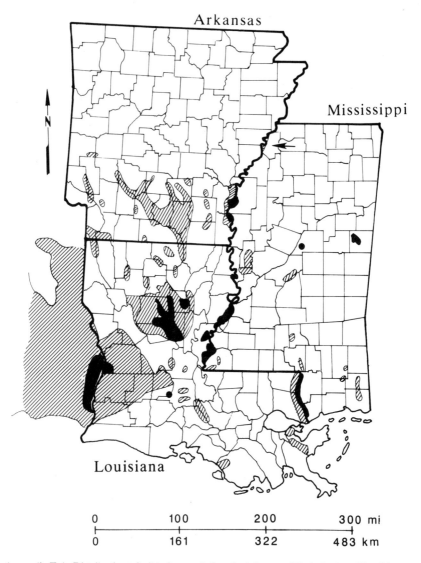

Appendix E-4. Distribution of wild pig populations in Arkansas, Mississippi, and Louisiana. Blackened areas represent densities reported as more than 10 per square mile, shaded areas represent densities of less than 10 per square mile, and no shading represents areas where wild pigs are reported as rare or absent. Redrawn from SCWDS (1988).

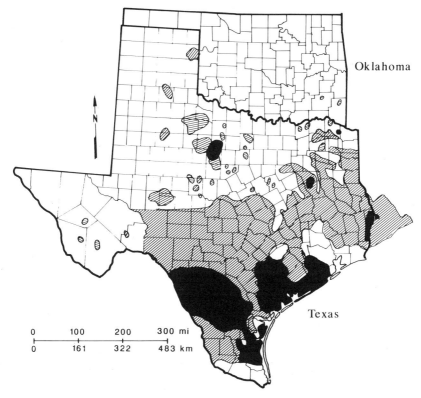

Appendix E-5. Distribution of wild pig populations in Oklahoma and Texas. Blackened areas represent densities reported as more than 10 per square mile, shaded areas represent densities of less than 10 per square mile, and no shading represents areas where wild pigs are reported as rare or absent. Redrawn from SCWDS (1988).

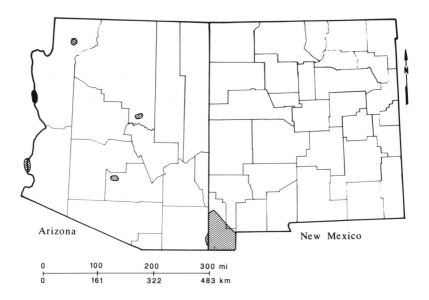

Appendix E-6. Distribution of wild pig populations in Arizona and New Mexico. Blackened areas represent densities reported as more than 10 per square mile, shaded areas represent densities of less than 10 per square mile, and no shading represents areas where wild pigs are reported as rare or absent. Redrawn from SCWDS (1988).

California

| 0 | 100 | 200 | 300 | mi |
| 0 | 161 | 322 | 483 | km |

Appendix E-7. Distribution of wild pig populations in California. Blackened areas represent densities reported as more than 10 per square mile, shaded areas represent densities of less than 10 per square mile, and no shading represents areas where wild pigs are reported as rare or absent. Redrawn from SCWDS (1988).

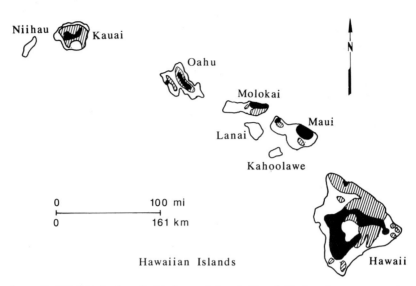

Niihau   Kauai

Oahu

Molokai

Lanai   Maui

Kahoolawe

0   100 mi

0   161 km

Hawaiian Islands   Hawaii

Appendix E-8. Distribution of wild pig populations in Hawaii. Blackened areas represent densities reported as more than 10 per square mile, shaded areas represent densities of less than 10 per square mile, and no shading represents areas where wild pigs are reported as rare or absent. Redrawn from SCWDS (1988).

286

Literature Cited

Abdulali, H. 1962. The wild pigs of the Andamans. J. Bombay Nat. Hist. Soc., 59:281-283.

Adlerberg, G. 1930. Preliminary synopsis of Russian and Mongolian wild boars. Compt. Rend. Acad. Sci. USSR, no. 4:91-96.

Alexander, J. H. 1967. Sportsman's Hawaii: Part I. Sports Afield, 157:43-45, 106-116.

Allen, G. M. 1930. Pigs and deer from the Asiatic expeditions. Amer. Mus. Novitates, 430:1-5.

———. 1940. The Mammals of China and Mongolia. Amer. Mus. Nat. Hist., New York.

———. 1954. A Checklist of African Mammals. Spaulding-Moss Co., Boston.

Allen, R. E., G. Robinson, M. J. Parsons, R. A. Merkel, and W. T. Magee. 1982. Age-related changes in the effects of serum from lean and obese pigs on myogenic cell proliferation in vitro. J. Anim. Sci., 54:763-768.

Amato, C. 1976. The wild boar. Explorer's J., 54(2):50-57.

Amon, R. 1938. Abstammung, Arten und Rassen der Wildschweine Eurasiens. Z. Tierz. Zuchtbiol., 40:49-88.

Anderson, J., and W. E. de Winton. 1902. Zoology of Egypt: Mammalia. Hugh Rees, London.

Anderson, S., and J. K. Jones, Jr. 1967. Recent Mammals of the World. Ronald Press Co., New York.

Andrzejewski, R. 1974. Spotty mutation of the wild boar, *Sus scrofa* Linnaeus 1758. Acta Theriol., 19(11):159-163.

Angwin, P. 1955. What comes naturally. Outdoor Life, 115(5):28.

Anon. 1949. The wild razorback. Louisiana Conserv., 1(5):9.

Anon. 1956. Tusker tug of war. Florida Wildl., 11(4):34.

Anon. 1967. The outdoor world. Georgia Game & Fish, 2(11):14.

Anon. 1968. Exotic Big Game in California. California Dept. Fish & Game Pamphlet 5/68.

Anon. 1970. Feral Animals in California. California Dept. Fish & Game Pamphlet 6/12/70-300.

Anon. 1972. Wild boars reappearing in Adirondack Mountains. Bennington Banner, June 20, 5.

Anon. 1976. Slater-Marietta. Greenville County Monitor, Oct. 22-29, 8.

Anon. 1979. Russian boar killed in Boerne. Boerne Star, Dec. 10, 8.

Anon. 1980a. Feral animal removal part of settlement involving San Clemente Island. End. Spp. Tech. Bull., 5(2):6, 8.

Anon. 1980b. Hog mania alive and well in east Texas. Texas Parks and Wildl., 38(2):20.

Anon. 1985. California goes hog wild. Newsweek, 106:29.

Anon. 1987. Great red waddle hogs—only a few in the entire world. Rare Breeds J., 1(1):6.

Ansell, W. F. H. 1971. Part 15: Order Artiodactyla. In: Meester, J., and H. W. Setzer, eds. The Mammals of Africa: An Identification Manual. Smithsonian Inst. Press, Washington, D.C.

Antonius, O. 1922. Grundzuge einer Stammesgeschichte der Haustiere. Verlag von Gustav Fischer, Jena.

Asahi, M. 1971. A naked wild boar, *Sus scrofa leucomystax,* hunted in Kyoto Prefecture. J. Mammal. Soc. Japan, 5:108.

Baber, D. W., and B. E. Coblentz. 1986. Density, home range, habitat use, and reproduction in feral pigs on Santa Catalina Island. J. Mamm., 67:512-525.

——. 1987. Diet, nutrition and conception in feral pigs on Santa Catalina Island. J. Wildl. Mgmt., 51:306-317.

Baker, J. 1984. The pigs of Ossabaw Island. Sedgwick County Zoo Members' Zooletter. March/April, 1-2.

Baker, J. K. 1976. The feral pig in Hawaii Volcanoes National Park. In: Abstracts, First Conf. Sci. Res. Natl. Parks, New Orleans, La.

Bangs, O. 1901. Notes on a small collection of mammals from the Liukiu Islands. Amer. Nat., 35:561-562.

Banoglu, N. A. 1952. Turkey: A Sportsman's Paradise. The Press, Broadcasting, and Tourism Department, Ankara, Turkey.

Baron, J. 1982. Effects of feral hogs (*Sus scrofa*) on the vegetation of Horn Island, Mississippi. Amer. Midland Nat., 107:202-205.

Baron, J. S. 1979. Vegetation damage by feral hogs on Horn Island, Gulf Islands National Seashore, Mississippi. M.S. thesis, Univ. Wisconsin, Madison.

Barrett, R. H. 1970. Management of wild hogs on private lands. Trans. California-Nevada Sec. Wildl. Soc., 17:71-78.

——. 1971. Ecology of the Feral Hog in Tehama County, California. Ph.D. thesis. Univ. California, Berkeley.

——. 1977. Wild pigs in California. In: Wood, G. W., ed. Research and Management of Wild Hog Populations. Belle Baruch Forest Sci. Inst., Clemson Univ., Georgetown, S.C.

——. 1978. The feral hog at Dye Creek Ranch, California. Hilgardia, 46:283-356.

Barrett, R. H., and D. L. Pine. 1980. History and status of wild pigs, *Sus scrofa*, in San Benito County, California. California Fish & Game, 67:105-117.

Barrows, P. L., H. M. Smith, A. K. Prestwood, and J. Brown. 1981. Prevalence and distribution of *Sarcocystis* sp. among wild swine in southeastern United States. J. Amer. Vet. Med. Assoc., 179:1117-1118.

Bauer, E. A. 1970a. Wild boars and red sheep. Outdoor Life, 146(1):44-47, 90-92.

——. 1970b. Christmas trip to Texas. Outdoor Life, 146(6):35-37, 112-114.

Baumann, F. 1949. Die Freilebenden Säugetiere der Schweiz. Verlag Hans Huber, Bern.

Baynes, E. H. 1931. Wildlife in the Blue Mountain Forest. Macmillan Co., New York.

Bee, J. W. 1957. Personal field notes, Aug. 24. Deposited at University of Kansas Mus. Nat. Hist., Lawrence.

Belden, R. C. 1989. History and biology of feral swine. In: Black, N., ed. Proc. Feral Pig Symposium, April 27-29, Orlando, Fla. Livestock Conservation Institute, Madison, Wisc.

Belden, R. C., and W. B. Frankenberger. 1977. Management of feral hogs in Florida—past, present and future. In: Wood, G. W., ed. Research and Management of Wild Hog Populations. Belle Baruch Forest Sci. Inst., Clemson Univ., Georgetown, S.C.

——. 1979. Job completion report —Brunswick hog study. Florida Game & Fresh Water Fish Commission, Gainesville, Fla.

Benke, A. 1973. The ugliest Texans. Texas Parks & Wildl., 31(1):6-9.

Berry, R. J. 1969. The genetic implications of domestication in animals. In: Ucko, P. J., and G. W. Dimbleby, eds. The Domestication and Exploitation of Plants and Animals. Aldine Publishing Co., Chicago.

Bish, T. 1967. Ruger .44 Mannlicher. Gun World, 7(1):20-24.

Blanford, W. T. 1875. List of Mammalia collected by the late Dr. Stoliczka in Kashmir, Ladak, Eastern Turkestan, and Wakhan; with descriptions of new species. J. Asiat. Soc. Bengal, 44(2):105-112.

——. 1888. The Fauna of British India Including Ceylon and Burma. Taylor & Francis, London.

Blennerhasset, A. 1907. Wildcat and boar. Forest & Stream, 68(24):935.

Blyth, E. 1851. [No title available.] J. Asiat. Soc. Bengal, 20:173.

——. 1858. Report of the curator, zoology department (*Sus* from Andaman and Nicobar islands). J. Asiat. Soc. Bengal, 27:267-269.

——. 1860. Report from Captain Hodge, commanding the guard-ship "Sesostris" at Port Blair. Two additional collections of sundries at that locality. The list of Andamanese mammalia is now extended to five species. J. Asiat. Soc. Bengal, 29:102-115.

Boback, A. W. 1957. Das Schwarzwild. Biologie und Jagd. Neumann Verlag, Leipzig.

Bobrinskii, N. A. 1944. Ungulata. In: Bobrinskii, N. A., B. A. Kuznetzov, and A. P. Kuzyakin. Key to the Mammals of the USSR. Gov. Publ. Office Sovetzkaya Nauka, Moscow.

Boddaert, A. P. 1785. Elenchus Animalium, sistens quadrupedia hie usque notata, eorumque varietas ad ductum Naturee, quantum fieri potuit disposita. C. R. Hake, Rotterdam, 1:1-174.

Bohlken, H. 1958. Zur Nomenklatur der Haustiere. Zool. Anz., 160:167-168.

——. 1961. Haustiere und zoologische Systematik. Z. Tierz. Zuchtbiol., 76:107-113.

Bokonyi, S. 1974. History of domestic mammals in central and eastern Europe. Akademiai Kiado, Budapest.

Bolton, H. E. 1916. Spanish Exploration in the Southwest, 1542-1706. C. Scribner's Sons, New York.

Bosma, A. A. 1976. Chromosomal polymorphism and G-banding patterns in the wild boar (*Sus scrofa* L.) from the Netherlands. Genetica, 46:391-399.

Bratton, S. P. 1975. The effects of the European wild boar, *Sus scrofa*, on gray beech forest in the Great Smoky Mountains. Ecology, 56:1356-1366.

——. 1977. Wild hogs in the United States—Origin and Nomenclature. In: Wood, G. W., ed. Research and Management of Wild Hog Populations. Belle Baruch Forest Sci. Inst., Clemson Univ., Georgetown, S.C.

Brickell, J. 1968. The Natural History of North Carolina. Johnson Publishing Co., Murfreesboro, N.C.

Briedermann, L. 1970. Zum Korper- und Organwachstum des Wildschweines in der Deutschen Demokratischen Republik. Arch. Forstwes., 19:401-420.

Brisbin, I. L., Jr. 1989. Feral animals and zoological parks: Conservation concerns for a neglected component of the world's biodiversity. Proc. Southeastern Regional Conf. Amer. Assoc. Zool. Parks & Aquar., 523-530.

Brisbin, I. L., Jr., R. A. Geiger, H. B. Graves, J. E. Pinder III, J. M. Sweeney, and J. R. Sweeney. 1977a. Morphological characteristics of two populations of feral swine. Acta Theriol., 22(4):75-85.

Brisbin, I. L., Jr., M. W. Smith, and M. H. Smith. 1977b. Feral swine studies at the Savannah River Ecology Laboratory: An overview of program goals and design. In: Wood, G. W., ed. Research and Management of Wild Hog Populations. Belle Baruch Forest Sci. Inst., Clemson Univ., Georgetown, S.C.

Britt, K. 1978. The joy of pigs. Natl. Geogr., 153(3):398-415.

Brown, V., and G. Lawrence. 1965. The Californian Wildlife Region. Naturegraph Publ., Healdsburg, Calif.

Brown, W. L. 1930. Personal field notes, Dec. Deposited at Nat. Mus. Nat. Hist., Washington, D.C.

Bruce, J. 1941. Mostly about the habits of the wild boar. California Conserv., 6(5):14, 21.

Brugh, M., Jr., J. W. Foster, and F. A. Hayes. 1964. Studies of comparative susceptibility of wild European and domestic swine to hog cholera. Amer. J. Vet. Res., 25:1124-1127.

Bryan, L. W. 1937. Wild pigs in Hawaii. Paradise Pacific, 49(12):31-32.

Buhlinger, C. A., P. J. Wangsness, R. J. Martin, and J. H. Ziegler. 1978. Body composition, *in vitro* lipid metabolism and skeletal muscle characteristics in fast-growing, lean and in slow-growing, obese pigs at equal age and weight. Growth, 42:225-236.

Bump, G. 1941. The introduction and transplantation of game birds and mammals into the state of New York. Trans. N. Amer. Wildl. Conf., 5:409-420.

Bustad, L. K., V. G. Horstman, and D. C. England. 1966. Development of Hanford miniature swine. In: Bustad, L. D., and R. O. McClellan, eds. Swine in Biomedical Research. Battelle Memorial Inst., Pacific N.W. Lab., Richland, Wash.

Cabon, K. 1958. Untersuchungen uber die Schadelvariabilitat des Wildschweines, *Sus scrofa* L. aus Nord-ost Polen. Acta Theriol., 2(6):107-140.

Cabrera, A. 1914. Fauna Iberica Mamíferos. Museo Nacional de Ciencias Naturales, Madrid.

Caire, W., J. D. Tyler, B. P. Glass, and M. A. Mares. 1989. Mammals of Oklahoma. Univ. Oklahoma Press, Norman.

Carter, H. A. 1943. Letter to Mr. James Silver concerning wild hogs on the Georgia coastal islands. On file at Natl. Mus. Nat. Hist., Washington, D.C.

Clutton-Brock, J. 1981. Domesticated Animals from Early Times. Univ. Texas Press, Austin.

Clutton-Brock, J., G. B. Corbet, and M. Hillo. 1976. A review of the family Canidae, with a classification by numerical methods. Bull. British Mus. Natl. Hist., Zool., 29:119-199.

Cockerham, J. 1985. Pork chops the hard way. Louisiana Conserv., 37(3): 18-21.

Coleman, S. 1984. Control efforts in Great Smoky Mountains National Park since 1978. In: Tate, J., ed. Techniques for controlling wild hogs in Great Smoky Mountains National Park. Proc. of a workshop, November 29-30, 1983. U.S. Dept. Int., Natl. Park Serv., Research/Resources Mgmt. Rpt. SER-72.

Conley, R. H., V. G. Henry, and G. H. Matschke. 1972. Final Report for the European Hog Research Project W-34. Tennessee Game & Fish Comm., Nashville.

Cook, H. B. S., and A. F. Wilkinson. 1978. Suidae and Tayassuidae. In: Maglio, V. J., and H. B. S. Cook, eds. Evolution of African Mammals. Harvard Univ. Press, Cambridge, Mass.

Coombs, D. W., and M. D. Springer. 1974. Parasites of feral pig × European wild boar hybrids in southern Texas. J. Wildl. Dis., 10:436-441.

Cooney, J. 1976. Nowhere but straight up. Archery World, 25(3):30-31, 52.

Cope, E. D. 1870. Fourth contribution to the history of the fauna of the Miocene and Eocene periods in the United States. Proc. Amer. Philos. Soc. Philadelphia, 11(83):285-294.

Corbet, G. B. 1978. The Mammals of the Palearctic Region: A Taxonomic Review. Cornell Univ. Press, Ithaca, N.Y.

Cornefert-Jensen, F., W. C. D. Hare, and D. A. Abt. 1968. Identification of the sex chromosomes of the domestic pig. J. Hered., 59:251-255.

Corruccini, R. S., and R. M. Beecher. 1982. Occlusal variation related to soft diet in a non-human primate. Science, 218:74-76.

Cote, P. J., P. J. Wangsness, H. Varela-Alvarez, L. C. Griel, Jr., and J. F. Kavanaugh. 1982. Glucose turnover in fast-growing, lean and in slow-growing, obese swine. J. Anim. Sci., 54:89-94.

Cottam, C. 1956. The problem of wildlife introductions—its successes and failures. Proc. 46th Conv. Internat. Assoc. Game, Fish & Conserv. Comm., 94-111.

Cox, J. 1981. More than a match for hunters. Texas Parks & Wildl., 39(1):8-10.

Crandall, L. S. 1964. Management of wild mammals in captivity. Univ. Chicago Press, Chicago.

Crossman, J. 1965. Sportsman's bonus from the military. Sports Afield, 153(4):44-45, 95-99.

Crouch, L. C. 1983. Movements of and habitat utilization by feral hogs at the Savannah River Plant, South Carolina. M.S. thesis, Clemson Univ., Clemson, S.C.

Dalrymple, B. W. 1970. Complete Guide to Hunting Across North America. Harper & Row, New York.

Darwin, C. A. 1867. Variation of Animals and Plants Under Domestication. D. Appleton & Co., London. 2 vols.

Davis, T. F. 1935. Juan Ponce de Leon's voyages to Florida. Quart. Per. Florida Hist. Soc., 14:1-70.

Day, G. E. 1915. Productive Swine Husbandry. J. B. Lippincott Co., Philadelphia.

Day, N. R. 1979. Going to market. Natl. Parks Conserv. Mag., 53(11):15-16.

De Beaux, O. 1928. Risultati zoologici della Missione inviata dalla R. Societa Geografica Italiana per l'esplorazione dell oasi di Giarabub. Ann. Mus. Civ. Shr. Nat., 53:39-76.

De Beaux, O., and E. Festa. 1927. La Ricomparsa del Cinghiale Nell'Italia Settentrionale-Occidentale. Mem. Soc. Ital. Sci. Nat. Mus. Civ. Stor. Nat. Milano, 9(3):265-324.

Decker, E. 1978. Exotics. In: Schmidt, J. L., and D. L. Gilbert, eds. Big Game of North America: Ecology and Management. Stackpole Books, Harrisburg, Pa.

Dennler de la Tour, G. 1968. Zur Frage der Haustier-Nomenklatur. Säugetierk. Mitt., 16:1-20.

De Vos, A., R. H. Manville, and R. G. Van Gelder. 1956. Introduced mammals and their influence on native biota. Zoologica, 41:163-194.

Diehl, B., Jr. 1965. The ghosts of Jekyll. In: Baker, E., ed. The Jekyll Island Story. Jekyll Island, Ga.

Diong, C. H. 1973. Studies of the Malayan wild pig in Perak and Johore. Malayan Nat. J., 26(3-4):120-151.

Dixon, W. J., and M. B. Brown. 1979. BMDP Biomedical Computer Programs, P-Series, Univ. California Press, Berkeley.

Donkin, R. A. 1985. The peccary—with observations on the introduction of pigs to the New World. Trans. Amer. Philos. Soc., 75:1-152.

Doran, B. B. 1988. De Soto introduces swine to the southeast. Florida Wildl., 42(6):10-12.

Dotson, T. L. 1986. Wild boar. In: 1986 Big Game Bull. West Virginia Dept. Nat. Res., Wildl. Res. Div., Charleston, W.Va.

———. 1987. Wild boar. Enclosure to a form letter sent to licensed boar hunters in West Virginia. Photocopy.

Doutt, J. K. 1959. Personal field notes, May 1-2. Deposited at Carnegie Mus. Nat. Hist., Pittsburgh.

———. 1962. Personal field notes, Jan. 1-2. Deposited at Carnegie Mus. Nat. Hist., Pittsburgh.

Duever, M. J., et al. 1986. The Big Cypress National Preserve. Res. Rept. no. 8, National Audubon Society, New York.

Duff, J. F. 1934. Wild pig hunting in the Kuling Foothills. China J., 20:204-205.

Duncan, B., and B. Schwab. 1986. Virginia's bay of pigs. Virginia Wildl., 47:16-21.

Durov, V. V., and V. N. Alexandrov. 1968. Measurements and weights of wild boars in the Caucasus Sanctuary. Trudi Kaukazkovo Gosubarstuewoyu Zapodvednika, 10:294-301.

Eisenberg, J. F., and M. Lockhart. 1972. An ecological reconnaissance of Wilpattu National Park, Ceylon. Smithsonian Contrib. Zool., 101:1-118.

Eisenmann, E. 1972. Comment on proposal to exclude from zoological nomenclature names for domestic animals. Bull. Zool. Nomencl., 29(3):108.

Elder, W. H., and C. M. Hayden. 1977. Use of discriminant function in taxonomic determination of canids from Missouri. J. Mamm., 58:17-24.

Ellerman, J. R., and T. C. S. Morrision-Scott. 1951. Checklist of Palearctic and Indian Mammals 1758 to 1946. British Mus. Nat. Hist., London.

———. 1966. Checklist of Palearctic and Indian Mammals 1758 to 1946. 2d ed. British Mus. Nat. Hist., London.

Elman, R., ed. 1974. The Hunter's Field Guide. Alfred A. Knopf, New York.

Ensminger, M. E. 1961. Swine Science. The Interstate Printers & Publishers, Danville, Ill.

Epstein, H. 1971. The Origin of the Domestic Animals of Africa. Africana Publishing Corp., New York. 2 vol.

Erxleben, J. C. P. 1777. Systemae Regni Animalis per Classes, Ordines, Genera, Species, Varietates, cum Synonymia et Historia Animalium. Classis I. Mammalia. Weygandianus, Lipsiae.

Ezekwe, M. O., and R. J. Martin. 1975. Cellular characteristics of skeletal muscle in selected strains of pigs and mice and the unselected controls. Growth, 39:95-106.

Feder, F. 1978. Vergleichende Untersuchungen und Haaren von Schweinen. Säugetierk. Mitt., 26:199-206.

Findley, J. S., A. H. Harris, D. E. Wilson, and C. Jones. 1975. Mammals of New Mexico. Univ. New Mexico Press., Albuquerque.

Finsch, O. 1886. On a new species of wild pig from New Guinea. Proc. Zool. Soc. London, 1886:217-218.

Fitzinger, L. J. 1853. Versuch einer Geschichte der Menagerien des Österreichisch-Kaiserlichen Hofes. Sitz. Akad. Wiss. Wien, 10:300-403.

——. 1858. Die Racen des zahmen oder Hausschweines. Sitz. Akad. Wiss. Wien, 29:361-408.

——. 1864. Revision der big jetz bekannt gewordenen Arten der Familie der Borstentiere oder Schweine (Setigera). Sitz. Akad. Wiss. Wien, 50:383-434.

Foley, C. W., R. W. Seerley, W. J. Hansen, and S. E. Curtis. 1971. Thermoregulatory responses to cold environment by neonatal wild and domestic piglets. J. Anim. Sci., 32:926-929.

Foreyt, W. J., A. C. Todd, and K. Foreyt. 1975. *Fascioloides magna* (Bassi, 1875) in feral swine from southern Texas. J. Wildl. Dis., 11:554-559.

Frankenberger, W. B., and R. C. Belden. 1976. Distribution, relative abundance and management needs of feral hogs in Florida. Proc. Southeastern Game & Fish Comm., 30:641-644.

French, H. 1974. Wild boar hunt. Guns & Ammo, 18(11)38-39.

Gaffrey, G. 1961. Merkmale der Wildlebenden Säugetiere Mittel Europas. Akademische Verlagsgesellschaft, Leipzig.

Garga, D. P. 1945. A large wild boar (*Sus cristatus*). J. Bombay Nat. Hist. Soc., 46:398-399.

Genov, P. 1981. Die Verbreitung des Schwarzwildes (*Sus scrofa* L.) in Eurasien und seine Anpassung an die Hahrungsverhaltnisse. Z. Jagdwiss., 27:221-231.

Gerrard. 1862. Catalog of the Bones of the British Museum. British Mus. Nat. Hist., London.

Giannoni, M. A., and I. Ferrari. 1977. Estudo biometrico do Cariotipo da especie *Sus scrofa*. Cientifica, 5:199-208.

Giebel, C. G. A. 1855. Die Säugethiere in Zoologischer, Anatomischer und Palontologischer Beziehung Umfassend Dargestellt. A. Abel, Leipzig.

Gilbertson-Beadling, S. K., R. Vasilatos-Younken, and D. R. Hagen. 1988. Placental and fetal development in straightbred and reciprocal crosses of Yorkshire and Ossabaw swine. Growth, Devel. & Aging, 52:97-101.

Gillelan, G. H. 1967. First Aransas deer hunt. Outdoor Life, 139(4)106-110.

Gimenez-Martin, G., J. F. Lopez-Saez, and E. G. Monge. 1962. Somatic chromosomes of the pig. J. Hered., 53:281-290.

Gmelin, J. F. 1788. Caroli a Linne . . . Systema naturae per regua tria naturae, secundum classes, ordines, genera, species, cum characteribus, differentiis, synonymis, locis. 13th ed. Impensis Georg. Emanuel. Beer, Lipsiae, 1:1-500.

Godin, A. J. 1977. Wild Mammals of New England. Johns Hopkins Univ. Press, Baltimore.

Golley, F. B. 1964. Mammals of Georgia. Univ. Georgia Press, Athens.

——. 1966. South Carolina Mammals. The Charleston Museum, Charleston, S.C.

Gray, J. E. 1843. List of the Specimens of Mammalia in the Collections of the British Museum. British Mus. Nat. Hist., London.

——. 1847. Catalog of the Osteological Specimens in the British Museum. British Mus. Nat. Hist., London.

——. 1862. On the skull of the Japanese pig (*Sus pliciceps*). Proc. Zool. Soc. London, 1316.

——. 1868. Synopsis of the species of pigs (Suidae) in the British Museum. Proc. Zool. Soc. London, 17-49.

——. 1873a. Observations on pigs (*Sus* Linn., *Sertifera* Illiger) and their skulls, with the description of a new species. Ann. Mag. Nat. Hist., 11(66):431-439.

——. 1873b. Hand-List of the Thick-Skinned Mammals in the British Museum. British Mus. Nat. Hist., London.

Gropp, A., D. Giers, and U. Tettenborn. 1969. Das chromosomen Komplement des Wildschwein (*Sus scrofa* L.). Experientia, 25:778.

Groves, C. P. 1971. Request for a declaration modifying Article 1 so as to exclude names proposed for domestic animals from zoological nomenclature. Bull. Zool. Nomencl., 27:269-272.

——. 1977. Comments on request for a declaration modifying Article 1 so as to exclude names proposed for domesticated animals from zoological nomenclature. Z.N. (S.) 1935. Bull. Zool. Nomencl., 34:137-139.

——. 1979. Reply by Colin P. Groves. Bull. Zool. Nomencl., 36:9-10.

——. 1981. Ancestors for the pigs: Taxonomy and phylogeny of the genus *Sus*. Dept. of Prehistory, Res. School of Pacific Studies, Australian Natl. Univ., Tech. Bull. no 3:1-96.

Guilday, J. E. 1977. Sabertooth cat, *Smilodon floridanus* (Leidy), and associated fauna from a Tennessee cave (40 DV 40), the First American Bank Site. J. Tennessee Acad. Sci., 52:84-94.

Gunchak, N. S. 1978. Morphological characteristics of *Sus scrofa* in Ukrainian Carpathians. Zool. Zhurnal, 57:1870-1877.

Gustavsson, I., M. Hageltorn, L. Ech, and S. Reiland. 1973. Identification of the chromosomes in a centric fusion/fission polymorphic system of the pig (*Sus scrofa* L.). Hereditas, 75:153-155.

Hagen, D. R., and K. B. Kephart. 1980. Reproduction in domestic and feral swine. I. Comparison of ovulatory rate and litter size. Biol. Reprod., 22:550-552.

Hagen, D. R., and K. B. Kephart, and P. J. Wangness. 1980. Reproduction in domestic and feral swine. II. Interrelationships between fetal size and spacing and litter size. Biol. Reprod., 23:929-934.

Halloran, A. F. 1943. Management of deer and cattle on the Aransas National Wildlife Refuge, Texas. J. Wildl. Mgmt., 7:203-216.

Halloran, A. F., and J. A. Howard. 1956. Aransas Refuge wildlife introductions. J. Wildl. Mgmt., 20:460-461.

Haltenorth, T., and W. Trense. 1956. Das Grosswild der Erde und Seine Trophaen. Bayerischer Landwirtschaftsverlag, Bonn.

Hammond, J. 1962. Some changes in the form of sheep and pigs under domestication. Z. Tiez. Zuchtbiol., 77:156-158.

Hamnett, W. L., and D. C. Thornton. 1953. Tar Heel Wildlife. North Carolina Wildl. Resources Comm., Raleigh.

Handley, C. O., Jr. 1979. Mammals of the Dismal Swamp: A historical account. In: Kirk, P. W., Jr., ed. The Great Dismal Swamp. Univ. Press Virginia, Charlottesville.

Hansen, W. J., C. W. Foley, R. W. Seerley, and S. E. Curtis. 1972. Pelage traits in neonatal wild, domestic and "crossbred" piglets. J. Anim. Sci., 34:100-102.

Hansen-Melander, E., and Y. Melander. 1974. The karyotype of the pig. Hereditas, 77:149-158.

Hanson, R. P., and L. Karstad. 1959. Feral swine in the southeastern United States. J. Wildl. Mgmt., 23:64-75.

Harrington, F. A. 1977. A Guide to the Mammals of Iran. Dept. Environment, Tehran, Iran.

Harrison, D. L. 1968. The Mammals of Arabia. Vol. II. Carnivora, Artiodactyla, Hyracoidea. Ernest Benn, London.

Harrison, J. 1974. An Introduction to Mammals of Singapore and Malaya. Singapore National Printers, Singapore.

Hatt, R. T. 1959. The Mammals of Iraq. Misc. Publ. no. 106, Mus. Zoology, Univ. Michigan, Ann Arbor.

Hausman, G. J., D. R. Campion, and G. B. Thomas. 1983. Semitendinosis muscle development in several strains of fetal and perinatal pigs. J. Anim. Sci., 57:1608-1617.

Hausman, G. J., and R. J. Martin. 1981. Subcutaneous adipose tissue development in Yorkshire (lean) and Ossabaw (obese) pigs. J. Anim. Sci., 52:1442-1449.

(Hediger, H. 1966. Diet of animals in captivity. Internat. Zool. Yearb., 6:37-58.)

Heim de Balsac, H. 1937. Diagnoses de Mammifères Nord-Africains. Bull. Soc. Zool. France, 62:329-334.

Heise, L., and C. Christman. 1989. American Minor Breeds Notebook. Publ. Amer. Minor Breeds Conserv., Pittsboro, N.C.

Hemmer, H. 1978. Geographische Variation der Hirngrosse in *Sus scrofa* und *Sus verrucosus* Kreis. Spixana, 3:309-320.

Henderson, J. 1979. The great Texas hog hunt. Southern Outdoors, 28(1):64, 66-68.

Henry, V. G. 1969. Detecting the presence of European wild hogs. J. Tennessee Acad. Sci., 44(4):103-104.

——. 1970. Weights and body measurements of European wild hogs in Tennessee. J. Tennessee Acad. Sci., 45:20-23.

Heptner, V. G., A. A. Nasimovic, and A. G. Bannikov. 1966. Die Säugetiere der Sowjetunion. Band I. Paarhufer aund Unpaarhufer. VEB Gustav Fisher Verlag, Jena.

Herring, S. W. 1972. The role of canine morphology in the evolutionary divergence of pigs and peccaries. J. Mamm., 53:500-512.

Hetzer, H. O. 1945. Inheritance of coat color in swine. J. Hered., 36:121-128.

Heude, P. M. 1888. Etudes sur les ruminants de l'Asie orientale—cerfs des Philippines et de l'Indo-Chine, et études sur les suilliens. Mem. concern. l'Hist. Nat. de l'Emp. Chinois, 2:1-64.

——. 1892. Une étude novelle sur les systèmes dentaire des herbivores. Mem. concern. l'Hist. Nat. de l'Emp. Chinois, 2:65-240.

——. 1897. Etudes odontologiques: carnivores. insectivores. marsupiaux.—cheiroptêres et rousettes.—suilliens: sangliers chinois. capricornes du Se-tchouan. Mem. concern. l'Hist. Nat. de l'Emp. Chinois, 3:157-198.

——. 1899. Etudes sur les suilliens: sangliers chinois. Mem. concern. l'Hist. Nat. de l'Emp. Chinois, 4:132.

Heyerdahl, T. 1979. Early Man and the Ocean: A Search for the Beginnings of Navigation and Seaborne Civilizations. Doubleday & Co., New York.

Hines, R. 1988. Kentucky's unwanted immigrant "wild boar." Kentucky's Happy Hunting Ground, 44(2):21-23.

Hodgson, B. H. 1841. Classified catalogue of mammals of Nepal. Corrected to end of 1841; first printed in 1932. J. Asiat. Soc. Bengal, 10:907-916.

Hoffman, E. C., P. J. Wangsness, D. R. Hagen, and T. D. Etherton. 1983. Fetuses of lean and obese swine in late gestation: Body composition, plasma hormones and muscle development. J. Anim. Sci., 57:609-620.

Hogg, J. E. 1920. Boar hunting on Santa Cruz. Outer's Recreation, 63:194-196, 238-240.

Holland, F. R., Jr. 1962. Santa Rosa Island: An archaeological and historical study. J. West, 1(1):45-62.

Hollister, N. 1917. Some effects of environment and habitat on captive lions. Proc. U.S. Natl. Mus., 53(2196):117-193.

Holman, R. L. 1941. Hunting wild boar. Amer. Forests, 47:512-514.

Holt, D. 1970. Pig hunting—Carolina style provides plenty of thrills. South Carolina Wildl., 17:2-3, 9.

Holthuis, L. B., and A. M. Husson. 1971. Comments on Dr. C. P. Groves's request for a declaration modifying Article 1 so as to exclude names proposed for domestic animals. Bull. Zool. Nomencl., 28:77-78.

Hoornbeek, B. M. 1980. Don't Forget the Pig. Paper presented at the Soc. Hist. Arch. meeting, Albuquerque, N.M. Photocopy.

Howe, T. D., F. J. Singer, and B. B. Ackerman. 1981. Forage relationships of European wild boar invading northern hardwood forest. J. Wildl. Mgmt., 45:748-754.

Hudson, E. M. 1978. The raccoon (*Procyon lotor*) on St. Catherines Island, Georgia. Part 2. Relative abundance in different forest types as a function of population density. Amer. Mus. Novitates, no. 2648:1-15.

Hunn, M. 1961. Wild hog safari. Florida Wildl., 15(6):24-28.

Husson, A. M. 1960. Mammals of the Netherlands Antilles. Werkgroep Nederlandse Antillen, Curaçao.

Hutchinson, C. B., ed. 1946. California Agriculture. Univ. California Press, Berkeley.

Igo, W. K. 1977. Wild boars thriving. Wonderful West Virginia, 40(12):12, 14.

Igo, W. K., T. J. Allen, and E. D. Michael. 1979. Observations on European wild boars released in southern West Virginia. Proc. Southeastern Assoc. Game & Fish Comm., 33:313-317.

Imaizumi, Y. 1960. Coloured Illustrations of the Mammals of Japan. Hoikuska I-Chrome Uehonmachi, Higashi-ku. Osaka, Japan.

———. 1973. Taxonomic study of the wild boar from the Ryukyu Islands, Japan. Mem. Nat. Sci. Mus. Tokyo, 6:113-129.

Ingels, L. G. 1954. Mammals of California and Its Coastal Waters. Stanford Univ. Press, Stanford, Calif.

Jackson, A. 1964. Texotics. Texas Game & Fish, 22(4):7-11.

Jackson, H. H. T. 1944. Big-game resources of the United States, 1937-1942. U.S. Dept. Int. Res. Rept. no. 8.

Jenkins, J. H., and E. E. Provost. 1964. The Population Status of the Larger Vertebrates on the Atomic Energy Commission Savannah River Plant Site. Publ. Office Tech. Serv., Dept. Commerce, Washington, D.C.

Jentink, F. A. 1905. *Sus*-studies in the Leyden Museum. Notes Leyden Mus., 26:155-195.

Joesting, E. 1972. Hawaii, an Uncommon History. W. W. Norton & Co., New York.

Johnson, A. S., H. O. Hillestad, S. F. Schanholtzer, and G. F. Schanholtzer. 1974. An Ecological Survey of the Coastal Region of Georgia. Natl. Park Serv. Sci. Monog. Ser. no. 3, Washington, D.C.

Jones, G. F. 1987. The 1780 siege of Charleston as experienced by a Hessian officer. South Carolina Historical Mag., 88:23-33.

Jones, G. S. 1975. Catalog of the type specimens of the mammals of Taiwan. Quart. J. Taiwan Mus., 28:183-217.

Jones, P. 1959. The European Wild Boar in North Carolina. Publ. North Carolina Wildl. Resources Comm., Raleigh.

Keller, C. 1902. Die Abstammung der Altesten Hausetiere. Verlag von Fritz Amberger Vorm. David Burkli, Zurich.

Kellogg, R. 1939. Annotated listed of Tennessee mammals. Proc. U.S. Natl. Mus., 86(3051): 245-303.

Kelm, H. 1938. Die postembryonale Schadelentwicklung des Wild- und Berkshire-Schweine. Z. Anat. Entwicklungsgesch., 108:499-559.

———. 1939. Zur Systematic der Wildschweine. Z. Tierz. Zuchtgsbiol., 43(3):362-369.

King. S. T., and J. R. Schrock. 1985. Controlled Wildlife. Vol. III. State Wildlife Regulations. Assoc. Syst. Coll., Lawrence, Kans.

Kozlo, P. G. 1975. Dikij Kaban. Izd. Urodzaj, Minsk, 33 SS.

Kramer, R. J. 1971. Hawaiian Land Mammals. Charles E. Tuttle Co., Rutland, Vt.

Kuroda, N. 1924. On New Mammals from the Riu Kiu Islands and the Vicinity. Published by the author, Tokyo, 8:1-14.

———. 1935. Formosan mammals preserved in the collection of Marquis Yamashina. J. Mamm., 16:277-291.

Kurten, B. 1968. Pleistocene Mammals of Europe. Weidenfeld & Nicolson, London.

Kurz, J. C. 1971. A Study of Feral Hog Movements and Ecology on the Savannah River Plant, South Carolina. M.S. thesis, Univ. Georgia, Athens.

Lacki, M. J., and R. A. Lancia. 1986. Effects of wild pigs on beech growth in Great Smoky Mountains National Park. J. Wildl. Mgmt., 50:655-659.

Lamb, S. H. 1938. Wildlife problems in Hawaii National Park. Trans. N. Amer. Wildl. Conf., 3:597-602.

Lawrence, B., and W. H. Bossert. 1967. Multiple character analysis of *Canis lupis, latrans*, and *familiaris*, with a discussion of the relationship of *Canis niger*. Amer. Zool., 7:223-232.

———. 1969. The cranial evidence for hybridization of New England *Canis*. Brevoria, 330:1-13.

Lawrence, H. L. 1982. Time to think "Tennessee boar." Field & Stream, 87(3):86-87.

Lawson, J. 1937. Lawson's History of North Carolina. Garrett & Massie, Publishers, Richmond, Va.

Laycock, G. 1966. The Alien Animals. The Natural History Press, Garden City, N.Y.

Layne, J. N. 1974. The land mammals of south Florida. In: Gleason, P. J., ed. Environments of South Florida: Present and Past. Mem. 2, Miami Geol. Soc., Miami.

Lekagul, B., and J. A. McNeely. 1977. Mammals of Thailand. Publ. Assoc. Conserv. Wildl., Bangkok, Thailand.

Lemche, H. 1971. Comment on the request to modify Article 1 so as to exclude names proposed for domestic animals. Bull. Zool. Nomencl., 28:140.

Lesson, R. P., and P. Garnot. 1826. Mammifères nouveaux ou pen connus, dacrits et figures dans l'Atlas Zoologique du voyage autour du monde de la corvette "La Coquille." Ferussac's Bull. Sci. Nat. Geol., 8.

Lewinsohn, R. 1964. Animals, Men and Myths. Fawcett Publications, Greenwich, Conn.

Lewis, J. C. 1966. Observations of pen-reared European hogs (*Sus scrofa*) released for stocking J. Wildl. Mgmt., 30:832-835.

Lewis, J. C., R. Murray, and G. Matschke. 1965. Hog subcommittee report to the chairman of the forest game committee. Southeastern Sec. Wildl. Soc.

Lewis, T. H. 1907. The narrative of the expedition of Hernando de Soto by the Gentleman of Elvas. In: Jameson, J. F., ed. Spanish Explorers in the Southern United States, 1528-1543. Charles Scribner's Sons, New York.

Lineaweaver, T. H. 1955. This Is New Hampshire. Outdoor Life, 166(5):54-55, 85-88.

Linnaeus, C. V. 1758. Systema naturae per regua tria naturae, secundum classes, ordines, genera, species, cum characteribus, differentiis, synonymis, locis. 10th ed. Laurentii Salvii, Stockholm, 1:1-824.

Linzey, A. V., and D. W. Linzey. 1971. The Mammals of Great Smoky Mountains National Park. Univ. Tennessee Press, Knoxville.

Loche, V. 1867. Histoire Naturelle des Mammifères. Expl. Sci. de l'Algeria Pendant les Années 1840, 1841, 1842. Imprimiere Nationale, Paris.

Lowe, V. P. W., and A. S. Gardiner. 1974. A re-examination of the subspecies of red deer (*Cervus elaphus*) with particular reference to the stocks in Britain. J. Zool., 174:185-201.

Lowery, W. 1959. The Spanish Settlements Within the Present Limits of the United States: Florida, 1562-1574. Russell & Russell, New York.

Lydekker, R. 1900. The Great and Small Game of India. Rowland Ward, London.

———. 1901. The Great and Small Game of Europe, Western and Northern Asia and America, their Distribution, Habits, and Structure. Rowland Ward, London.

———. 1907. The Game Animals of India, Burma, Malaya and Tibet. Rowland Ward, London.

———. 1908a. The Game Animals of Africa. Rowland Ward, London.

———. 1980b. A Guide to the Domesticated Animals (Other than Horses) Exhibited in the Central and North Halls of the British Museum (Natural History). British Mus. Nat. Hist., London.

——. 1914. Ward's Records of Big Game. 7th ed. Rowland Ward, London.

——. 1915. Catalog of the Ungulate Mammals of the British Museum (Natural History). British Mus. Nat. Hist., London, 4:1-438.

Lynn, J. 1978. Problem pigs, cool owls and smiling kids. Nat. Conserv. News, 28(6):4-7.

Major, C. J. F. 1882. L'origine della fauna della nostre isole. Atti Soc. Tosc. Sci. Nat., Proc. Verb., 3:113-133.

——. 1883. Studien zur Geschichte der Wildschweine (Gen. *Sus*). Zool. Anz., Leipzig, 6(140):295-300.

Maly, F. 1976. Outdoor Column. San Antonio Light, Jan. 9, 6-F.

Mansfield, T. M. 1978. Wild pig management on a California public hunting area. Trans. California-Nevada Sec. Wildl. Soc., 25:187-201.

Manville, R. H. 1964. History of Corbin Preserve. Smithsonian Inst. Misc. Publ. no. 4581, 427-446.

Marchinton, R. L., R. B. Aiken, and V. G. Henry. 1974. Split guard hairs in both domestic and European wild swine. J. Wildl. Mgmt., 38:361-362.

Marczynska, B., and H. Pigon. 1973. Somatic chromosomes of the North African pig. Cytologia, 38:111-116.

Marshall, P. 1967. Wild Mammals of Hong Kong. Oxford Univ. Press, Hong Kong.

Martin, J. 1945. Snaggle tushes. Outdoor Life, 95(1):20-22, 89.

Martin. J. S. 1970. 23,000 acres inside a fence! Yankee, 34(12):90-95, 207-208, 211.

Martin, P. S., and H. E. Wright, Jr. 1967. Pleistocene Extinctions: The Search for a Cause. Yale Univ. Press, New Haven, Conn.

Martin, R. J., J. L. Hartsock, T. H. Gobble, H. B. Graves, and J. H. Ziegler. 1973. Characterization of an obese syndrome in the pig. Proc. Soc. Exp. Biol. Med., 143:198-203.

Martin, R. J., and J. Herbein. 1976. A comparison of the enzyme levels and *in vitro* utilization of various substrates for lipogenesis in pair-fed lean and obese pigs. Proc. Soc. Exp. Biol. Med., 151:231-235.

Mason, I. L. 1969. A World Dictionary of Livestock Breeds, Types and Varieties. Publ. Commonwealth Agricultural Bureaux, Bucks, England.

Matschie, P. 1918. Das Wildschwein von Naliboki im Weissrussland. Sitzb. der Gesell. der Naturf. Freun. Berlin, 8:300-306.

Matschke, G. H. 1967. Aging European wild hogs by dentition. J. Wildl. Mgmt., 31: 109-113.

Mauget, R. 1980. Regulations Ecologiques, Comportementales et Physiologiques (Fonction de Reproduction), de l'Adaptation du Sanglier, *Sus scrofa* L., an Milien. Ph.D. dissert., Academie de Tours, Orleans, France.

Mayer, J. J. 1984. Morphological analyses detect the presence of wild boar phenotypes in the SRP wild pig population. In: SREL 1984 Annual Report, Savannah River Ecology Laboratory, Aiken, S.C.

Mayer, J. J., and P. N. Brandt. 1982. Identity, distribution, and natural history of the peccaries, Tayassuidae. Chapter 22. In: Mares, M. A., and H. H. Genoways, eds. Mammalian Biology in South America. Special Publ. Ser. vol. 6. Pymatuning Laboratory of Ecology, Univ. of Pittsburgh, Pa.

Mayer, J. J., and I. L. Brisbin, Jr. 1988. Sex identification of *Sus scrofa* based on canine morphology. J. Mamm., 69:408-417.

Mayer, J. J., I. L. Brisbin, Jr., and J. M. Sweeney. 1989. Temporal dynamics of color phenotypes in an isolated population of feral swine. Acta Theriol., 34:243-248.

Mayer, J. J., and R. M. Wetzel. 1986. *Catagonus wagneri*. Mamm. Species, 259:1-5.

——. 1987. *Tayassu pecari*. Mamm. Species, 292:1-7.

Maynard, C. J. 1872. Catalogue of the mammals of Florida. Bull. Essex Inst., 4(9-10):1-16).

Maynard, T. 1930. De Soto and the Conquistadores. Longmans, Green & Co., New York.

McConnell, J., N. S. Fechheimer, and L. O. Gilmore. 1963. Somatic chromosomes of the domestic pig. J. Anim. Sci., 22:374-379.

McCrary, P. F. 1979. Letter to Ms. Nancy R. Day concerning feral hogs on Cumberland Island National Seashore. On file at the Cumberland Island National Seashore Headquarters, St. Marys, Ga.

McFee, A. F., M. W. Banner, and J. M. Rary. 1966. Variation in chromosome number among European wild pigs. Cytogenetics, 5:75-81.

McKnight, T. 1964. Feral Livestock in Anglo-America. Univ. California Publ. Geol. no. 16.

——. 1976. Friendly Vermin: A Survey of Feral Livestock in Australia. Univ. California Publ. Geol. no. 21.

McMeekan, C. P. 1940a. Growth and development in the pig with special reference to carcass quality characters. Internat. J. Agric. Sci., 30:276-336.

——. 1940b. Growth and development in the pig with special reference to carcass quality characters. Part II. The influence of the plane of nutrition on growth and development. Internat. J. Agric. Sci., 30:387-427.

——. 1940c. Growth and development in the pig with special reference to carcass quality characters. Part III. Effect of the plane of nutrition on the form and composition of the bacon pig. Internat. J. Agric. Sci., 30:511-569.

McNally, T. 1955. Too tough to squeal. Outdoor Life, 115(5):50-51, 111-115.

Mearns, E. A. 1907. Mammals of the Mexican Boundary of the United States. Smithsonian Inst., Washington, D.C., U.S. Natl. Mus. Bull. 56, pt. I.

Melville, R. V. 1977. Comments on request for a declaration modifying Article 1 so as to exclude names proposed for domesticated animals from Zoological Nomenclature. Z.N. (S.) 1935. Bull. Zool. Nomencl., 34:139-140.

Meyer, A. B. 1877. Letter to Dr. George Rolleston, as cited by Rolleston (1879).

Miller, G. S. 1902. Mammals of the Andaman and Nicobar Islands. Proc. U.S. Natl. Mus., 24:751-795.

——. 1903. Mammals collected by Dr. W. L. Abbott on the coast and islands of northwest Sumatra. Proc. U.S. Natl. Mus., 26(1317):437-484.

——. 1906. Notes on Malayan pigs. Proc. U.S. Natl. Mus., 30:737-758.

——. 1912. Catalogue of the Mammals of Western Europe. British Mus. Nat. Hist., London.

Milne-Edwards, M. A. 1872. Rapport presente a l'Assemblee de Mm. Les Professeurs-Administrateurs du Museum d'Histoire Naturelle; 21 Nov. 1871; sur l'Etat actuel des collections dependante de la chaire des mollusques annelides, vers et zoophytes par M. Deshayes. Nouvelles Arch. Mus. l'Hist. Nat. Paris, 7:67-100.

Moore, J. C. 1946. Mammals from Welaka, Putnam County, Florida. J. Mamm., 27:49-59.

Mori, T. 1922. On some new mammals from Korea and Manchuria. Ann. Mag. Nat. Hist., London, 10:607-614.

Mount, L. E. 1967. The Climatic Physiology of the Pig. Monographs of the Physiological Society, no. 18. Williams & Wilkins Publishers, Cambridge, England.

Müller and Schlegel. 1842. Wilde Zwijnen van den Indischen Archipel. In: Temminck, C. J., ed. Verhandelingen over de natuurlijke Geschledenis der Nederlandsche overzeesche bezittingen door de lenden der natuurkundige commissie in Indie en andere schrijvers. 1839-1844. Zoogdieren, Leiden.

Muramoto, J., S. Makino, T. Ishikawa, and H. Kanagawa. 1965. On the chromosomes of the wild boar and the boar-pig hybrids. Proc. Japan Acad., 41:236-239.

Nehring, A. 1885. Uber das japanische Wildschwein (*Sus leucomystax* Temm.) Zool. Gart. Frankfurt Jahrg., 26(11):325-336.

——. 1889. Uber Säugethiere von Wladiwostock in Sudost-Sibirien. Sitz. Berl. Ges. Nat. Freunde Berlin, 7:141-144.

Nettles, V. F. 1989. Disease of wild swine. In: N. Black, ed. Proc. Feral Pig Symp., April 27-29, Orlando, Fla. Livestock Conserv. Inst., Madison Wisc.

Nettles, V. F., J. L. Corn, G. A. Erickson, and D. A. Jessup. 1989. A survey of wild swine in the United States for evidence of hog cholera. J. Wildl. Dis., 25:61-65.

Nichols, L., Jr. 1962a. Ecology of the wild pig. Hawaii Div. Fish & Game. Project W-5-R-13.

———. 1962b. Big Game Management in Hawaii. Hawaii Div. Fish & Game. Mimeo.

———. 1964. Ecology of the wild pig. Hawaii Div. Fish & Game. Project W-5-R-15.

Nizza, P. F. 1966. Some characteristics of the Corsican swine. In: Bustad, L. K., and R. O. McClellan, eds. Swine in Biomedical Research. Battelle Memorial Inst., Pacific N.W. Lab., Richland, Wash.

Noback, C. R. 1951. Morphology and phylogeny of hair. Ann. New York Acad. Sci., 53:476-492.

Osborn, D. J., and I. Helmy. 1980. The contemporary land mammals of Egypt (including Sinai). Fieldiana-Zool., Chicago Mus. Nat. Hist., 1309:1-579.

Pack, J. W., P. K. Srivastava, and J. F. Lasley. 1975. G-banding patterns of swine chromosomes. J. Hered., 66:344-348.

Page, W. 1954. Pigs in the Garden of Eden. Field & Stream, 59(5):37-39, 98-100.

Palmer, R. S. 1954. The Mammal Guide. Doubleday & Co., Garden City, N.Y.

Park, E. 1982. Unwanted wild pigs released in Washington. Amer. Hunter, 10(2):54.

Pelton, M. R. 1975. The mammals of Kiawah Island. In: Environmental Inventory of Kiawah Island. A report for Coastal Shores, Inc., by Environmental Research Center, Columbia, S.C.

———. 1984. Biological implications for control of European wild hogs: Some reflections of the past and present. In: Tate, J., ed. Techniques for controlling wild hogs in Great Smoky Mountains National Park. Proc. of a workshop, November 29-30, 1983. U.S. Dept. Int., Natl. Park Serv., Research/Resources Mgmt. Rept. SER-72.

Penney, J. T. 1950. Distribution and bibliography of the mammals of South Carolina. J. Mamm., 31:81-89.

Pfeffer, P. 1960. L'Ecologie du Sanglier en Asie Centrale d'apres les Recherches d'A. A. Sloudsky. Terre et la Vie, 108:368-372.

Phillips, W. W. A. 1935. Manual of the Mammals of Ceylon. Dulau & Co., London.

Pine, D. S., and G. L. Gerdes. 1969. Wild pig study in Monterey County. California Dept. Fish & Game Wildl. Mgmt. Rept.

———. 1973. Wild pigs in Monterey County. California Fish & Game, 59(2):126-137.

Pira, A. 1909. Studien zur Geschichte der Schweinrassen insbesondere derjenigen Schwedens. Zool. Jahrb. (Syst), Suppl. 10(II):233-426.

Poliquin, G. 1982. Chief blasts boar in Bath. Union Leader, Jan. 20, 15.

Pope, G. W. 1934. Determining the age of farm animals by their teeth. U.S.D.A. Farmer's Bull. no. 1721.

Popescu, C. P., J. P. Quere and P. Franceschi. 1980. Observations chromosomiques chez le sanglier français (*Sus scrofa scrofa*). Ann. Genet. Sel. Anim., 12:395-400.

Prater, S. H. 1965. The Book of Indian Mammals. Bombay Nat. Hist. Soc., Bombay.

Prentice, R. 1987. Red waddle hogs or red wattle hogs? Rare Breeds J., 1(2):22-23.

Presnall, C. C. 1958. The present status of exotic mammals in the United States. J. Wildl. Mgmt., 22:45-50.

Prestwood, A. K., F. E. Kellogg, S. R. Pursglove, and F. A. Hayes. 1975. Helminth parasitisms among intermingling insular populations of white-tailed deer, feral cattle, and feral swine. J. Amer. Vet. Med. Assoc., 166:787-789.

Pullar, E. M. 1953. The wild (feral) pigs of Australia: Their origin, distribution and economic importance. Mem. Natl. Mus. Melbourne, 18:7-23.

Quinn, J. H. 1970a. Occurrence of *Sus* in North America. Geol. Soc. Amer. Abstr., 2(4):298.

———. 1970b. Special note. Soc. Vert. Paleontol. News Bull., 8:33.

Rachford, C. E. 1939. Big-game inventory of the United States. U.S. Fish & Wildl. Serv., Wildl. Leafl. 175.

Randi, E., M. Apollonio, and S. Toso. 1989. The systematics of some Italian populations of wild boar (*Sus scrofa* L.): a craniometric and electrophoretic analysis. Z. Sangetierkunde, 54:40-56.

Rary, J. M., V. G. Henry, G. H. Matschke, and R. L. Murphree. 1968. The cytogenetics of swine in the Tellico Wildlife Management Area, Tennessee. J. Hered., 59:201-204.

Reitz, E. J. 1980. Vertebrate remains from Santa Elena, South Carolina, 1979 excavations. Appendix B, in: South, S., ed. The discovery of Santa Elena. Research Manuscript Ser. no. 165. Inst. Archeol. & Anthropol., Univ. South Carolina, Columbia.

Rempel, W. E., and A. E. Dettmers. 1966. Minnesota's miniature pigs. In: Bustad, L. K., and R. O. McClellan, eds. Swine in Biomedical Research. Battelle Memorial Inst., Pacific N.W. Lab., Richland, Wash.

Ride, W. D. L., C. W. Sabrosky, G. Bernardi, R. V. Melville, J. O. Corliss, J. Forest, K. H. L. Key, and C. W. Wright. 1985. International Code of Zoological Nomenclature. 3d ed. Internat. Trust Zool. Nomencl., London.

Ridgeway, R. A. 1912. Color Standards and Color Nomenclature. U.S. Natl. Mus., Washington, D.C.

Rittmannsperger, C. 1971. Chromosomen untersuchungen bei Wild- und Hausschweinen. Ann. Genet. Sel. Anim., 3:105.

Roberts, T. J. 1977. The Mammals of Pakistan. Ernest Benn, London.

Robertson, P. 1978. Refuge for wildlife, retreat for man. South Carolina Wildl., 25(5)48-55.

Rolleston, G. 1879. On the domestic pig of prehistoric times in Britain, and on the mutual relations of this variety of pig and *Sus scrofa ferus, Sus cristatus, Sus andamanensis,* and *Sus barbatus.* Trans. Linnaean Soc. London, Ser. 2, 1:251-286.

Romic, S. 1975. Body dimensions of wild boar. Conspectus Agriculturae Scientificus, 34(44):13-24.

Roots, C. 1976. Animal Invaders. Universe Books, New York.

Rosenberg, C. B. H. 1878. Der Malayische Archipelago. G. Weigel. Leipzig.

Ruch, J. 1966. All-season big game. Field & Stream, 72(4):48-49, 81-82.

Rue, L. L. 1969. Sportsman's Guide to Game Animals. Harper & Row, New York.

Russell, W. C. 1947. Personal field notes, Oct. 10. Deposited at Mus. Vert. Zool., Univ. California, Berkeley.

Rutimeyer, L. 1878. Einige weitere Beitrage uber das zahme Schwein und das Hausrind. Verhandlungen Naturf. Gesell. 6:464-498.

Rutledge, A. 1965. Demons of the delta. Sports Afield, 153(5):68-69, 167-170.

Salley, A. S., Jr. 1911. Narratives of early Carolina: 1650-1708. Charles Scribner's Sons, New York.

Scheffer, V. B. 1941. Management studies on transplanted beavers in the Pacific Northwest. Trans. N. Amer. Wildl. Conf., 6:320-326.

Schell, R. F. 1966. DeSoto Didn't Land at Tampa. Island Press. Ft. Myers Beach, Fla.

Schoning, W. M. 1978. Wildlife sanctuary at missile base. Catalyst, 15:14-15.

Schultz, V., E. Legler, Jr., W. A. Griffin, G. A. Webb W. M. Weaver, Jr., and J. A. Fox. 1954. Statewide Wildlife Survey of Tennessee. Tennessee Game & Fish Comm., Nashville.

Sclater, P. L. 1860. Note on additions to the Society's Menagerie. Proc. Zool. Soc. London, 443.

Scott, P. P. 1968. The special features of nutrition of cats, with observations on wild Felidae nutrition in the London Zoo. In: Crawford, M. A., ed. Comparative Nutrition of Wild Animals. Symp. Zool. Soc. London, no. 21.

Scott, R. A., S. G. Cornelius, and H. J. Mersmann. 1981a. Effects of age on lipogenesis and lipolysis in lean and obese swine. J. Anim. Sci., 52:505-511.

——. 1981b. Fatty acid composition of adipose tissue from lean and obese swine. J. Anim. Sci., 53:977-981.

SCWDS. See Southeastern Cooperative Wildlife Disease Study.

Sealander, J. A. 1979. A Guide to Arkansas Mammals. River Road Press, Conway, Ark.

Searle, A. G. 1968. Comparative genetics of coat color in mammals. Academic Press, London.

Setzer, H. W. 1956. Mammals of the Anglo-Egyptian Sudan. Proc. U.S. Natl. Mus., 106(3377):447-587.

Seymour, G. 1970. Wild pig: Sportsmen show increased interest in the European wild pig. California Dept. Fish & Game Wildl. Leafl.

Sharp, A. 1956. Ancient Voyages in the Pacific. Avery Press, Wellington, New Zealand.

Shaw, A. C. 1941. The European wild hog in America. Trans. N. Amer. Wildl. Conf., 5:436-441.

Siegler, H. R. 1962. New Hampshire Nature Notes. Equity Publishing Corp., Oxford, N.H.

Silver, H. 1957. A History of New Hampshire Game and Furbearers. Evans Printing Co., Concord, N.H.

Simoons, F. J. 1961. Eat Not this Flesh. Univ. Wisconsin Press, Madison.

Singer, F. J. 1976. The European Wild Boar in the Great Smoky Mountains National Park: Problem Analysis and Proposed Research. Uplands Field Res. Lab., Gatlinburg, Tenn. Photocopy.

———. 1981. Wild pig populations in the national parks. Environ. Mgmt., 5:263-270.

Singer, F. J., and D. Stoneburner. 1977. Feral Pig Management: Memorandum to Superintendent, Cumberland Island National Seashore.

Singer, F. J., W. T. Swank, and E. E. C. Clebsch. 1984. Effects of wild pig rooting in a deciduous forest. J. Wildl. Mgmt., 48:464-473.

Sisson, S., and J. D. Grossman. 1938. The Anatomy of the Domestic Animals. W. B. Saunders Co., Philadelphia.

Sjarmidi, A., and J. F. Gerard. 1988. Autour de la Systematique et la Distribution des Suides. Monitore Zool. Italia, 22:415-448.

Sloudsky, A. A. 1956. The Wild Boar: Ecology and Economic Importance. Ed. Acad. Sci. Kazakhstan SSR (Alma-Ata).

Smith, C. R., J. Giles, M. E. Richmond, J. Nagel, and D. W. Yambert. 1974. The mammals of northeastern Tennessee. J. Tennessee Acad. Sci., 49:88-94.

Smith, F. B. 1974. Naturalist's Color Guide and Supplement. Amer. Mus. Nat. Hist., New York.

Smith, M. W., M. H. Smith, and I. L. Brisbin, Jr. 1980. Genetic variability and domestication in swine. J. Mamm., 61:39-45.

Snethlage, K. 1950. Das Schwarzwild. Verlag Paul Parey, Berlin.

Sokolov, I. I. 1959. Fauna of SSSR—Mammals, Tome 1, Bnd. 3. Hooved Animals (Perissodactyla and Artiodactyla). Acad. Sci., Moscow.

Southeastern Cooperative Wildlife Disease Study. 1988. Feral/Wild Swine Populations 1988. A map prepared in cooperation with the Emergency Programs, Veterinary Services, Animal and Plant Health Inspection Service, U.S. Dept. Agric., through Cooperative Agreement no. 12-16-93-032. Univ. Georgia, Athens.

Sowerby, A. de C. 1917. On Heude's collection pigs, sika, serows and gorals, in the Sikawei Museum, Shanghai. Proc. Zool. Soc. London, 7-26.

———. 1918. Sport and science on the Sino-Mongolian frontier. 8 vol., London.

———. 1923. Big game in Shansi. China J. Sci. and Arts, 1:269-271.

———. 1930. The wild boar in China. China J., 12(1):49-50.

———. 1943. The mammals of the Japanese Islands. Musée Heude, Notes de Mamm. 1:1-66.

Spears, J. R. 1893. The Corbin Game Park. Annual Rept. Smithsonian Inst. for 1891.

Spellman, C. W. 1948. The agriculture of the early north Florida indians. Florida Anthropologist, 1:37-48.

Springer M. D. 1975. Food Habits of Wild Hogs on the Texas Gulf Coast. M.S. thesis. Texas A&M Univ., College Station.

——. 1977. Ecologic and economic aspects of wild hogs in Texas. In: Wood, G. W., ed. Research and Management of Wild Hog Populations. Belle Baruch Forest Sci. Inst., Clemson Univ., Georgetown, S.C.

Staffe, A. 1922. Uber den Schadel und das Haarkleid von *Sus leucomystax sibiricus* eine neue sud sibirische Wildschweinform. Arb. Lehr. f. Tierz. Hochs. f. Bodenkultur, Wien, 1:51-98.

Stansbury, C. F. 1925. The Lake of the Great Dismal. Albert & Charles Boni, New York.

Stegeman, L. 1938. The European wild boar in the Cherokee National Forest, Tennessee. J. Mamm., 19:279-290.

Stephenson, R. L. 1979. Introduction. In: South, S. The search for Santa Elena on Parris Island, South Carolina. Res. MS Ser. no. 150. Inst. Archeol. & Anthropol., Univ. South Carolina, Columbia.

Stewart, W. C. 1989. Eradication. In: N. Black, ed. Proc. Feral Pig Symp., April 27-29, Orlando, Fla. Livestock Conserv. Inst., Madison, Wisc.

Stone, C. P. 1985. Feral pig (*Sus scrofa*) research and management in Hawaii National Park. U.S. Natl. Park Serv., Hawaii Nat. Park, Hawaii.

Stone, C. R. 1987. Forty years in the Everglades. Atlantic Publishing Co., Tabor City, N.C.

Stone, W. 1973. Wild boars on exhibit at Bronx Zoo. Anim. Kingdom, 76(2):28.

Stribling, H. L., I. L. Brisbin, Jr., J. R. Sweeney, and L. A. Stribling. 1984. Body fat reserves and their prediction in two populations of feral swine. J. Wildl. Mgmt., 48:635-639.

Strobel, P. 1882. Studio Comparativo sul teschio del Porco della Mariere. Atti Soc. Tosc. Sci. Nat. Pisa, Proc. Verb., 25:21-85.

Stubbe, M., and N. Chotolchu. 1968. The mammalian fauna of Mongolia. Mitt. Zool. Mus. Berlin, 44(1):5-121.

Sweeney, J. M. 1970. Preliminary Investigation of a Feral Hog (*Sus scrofa*) Population on the Savannah River Plant, South Carolina. M.S. thesis, Univ. Georgia, Athens.

Swift, L. W. 1957. Letter to Mr. Perry Jones concerning wild boar in the U.S. national forests.

Swinhoe, R. 1863. On the mammals of the island of Formosa (China). Proc. Zool. Soc. London, 1862:360-361.

——. 1864. Letters from Mr. Swinhoe. Proc. Zool. Soc. London, 381-383.

——. 1870. Catalogue of the mammals of China (south of the river Yangtzse) and of the island of Formosa. Proc. Zool. Soc. London, 615-653.

Tate, G. H. H. 1947. Mammals of Eastern Asia. Macmillan Co., New York.

Tate, J., ed. 1984. Techniques for controlling wild hogs in Great Smoky Mountains National Park. Proc. of a workshop, November 29-30. U.S. Dept. Int., Natl. Park Serv., Research/Resources Mgmt. Rept. SER-72.

Taylor, S. 1980. Wild boar: Still a few around. New Hampshire Times, 10(2):8.

Temminck, C. J. 1842. Aperçu general et specifique sur les mammifères qui habitent le Japon et les îles qui en dependent. Siebold's Fauna Japonica. Lugduni Batavorum.

Thomas, O. 1912. The races of the European wild swine. Proc. Zool. Soc. London, 390-393.

Tikhonov, V. N., and A. I. Troshina. 1974. Identification of chromosomes and their aberrations in karyotypes of subspecies of *Sus scrofa* L. by differential staining. Doklady Akademii Nauk, SSR, 214(4):932-935.

Tinsley, R. 1968. Wild hogs. Fur-Fish-Game, 64(4):12-13, 43-44.

Tisdell, C. A. 1979. Feral Pigs (and Other Feral or Introduced Mammals) in National Parks and Nature Reserves in New South Wales: A Report on a Survey Covering Damages and Control. Dept. Econ., Univ. Newcastle, N.S.W., Australia.

——. 1982. Wild Pigs: Environmental Pest or Economic Resource. Pergamon Press, New York.

Tomich, P. Q. 1969. Mammals of Hawaii. Bishop Mus. Press, Honolulu, Hawaii.

Towne, C. W., and E. N. Wentworth. 1950. Pigs from Cave to Cornbelt. Univ. Oklahoma Press, Norman.

Tristram, H. B. 1866. Report on the mammals of Palestine. Proc. Zool. Soc. London, 84–93.

Trouessart, E. 1905. La faune des mammifères de l'Algerie, du Maroc et de la Tunisie. Causeries Scientifiques Soc. Zool. France, 1(10):353–410.

Trumbull, P. F., and C. A. Reed. 1974. The fauna from the terminal Pleistocene of Palegawra Cave; A Zarzian occupation site in northeastern Iraq. Fieldiana-Anthropol., Chicago Mus. Nat. Hist., 63:81–146.

Trumler, E. 1961. Entwurf einer Systematik der rezenter Equiden und ihrer fossilen Verwandten. Säugetierk. Mitt., 9:109–125.

Tye, M. 1971. Wild hog hunt. Georgia Game & Fish, 6(12):8–10.

U.S. Department of the Interior. 1969. Environmental Impact of the Big Cypress Jetport. U.S. Dept. Int., Washington, D.C.

USDI. See U.S. Department of the Interior.

Van den Brink, F. H. 1968. A Field Guide to the Mammals of Britain and Europe. Houghton Mifflin Co., Boston.

van der Leek, M. 1989. Wild swine—resource or risk? In: Black, N., ed. Proc. Feral Pig Symp., April 27–29, Orlando, Fla. Livestock Conserv. Inst., Madison, Wisc.

Van Gelder, R. G. 1979. Comments on request for a declaration modifying Article 1 so as to exclude names proposed for domestic animals from zoological nomenclature. Z.N. (S.) 1935. Bull. Zool. Nomencl., 36:5–9.

Van Peenan, P. F. D., P. F. Ryan, and R. H. Light. 1969. Preliminary Identification Manual for Mammals of South Viet Nam. Smithsonian Inst., Washington, D.C.

Varner, J. G., and J. J. Varner, eds. 1951. The Florida of the Inca. Univ. Texas Press, Austin.

———. 1983. Dogs of the conquest. Univ. Oklahoma Press, Norman.

Vereshchagin, N. K. 1967. The Mammals of the Caucasus: A History of the Evolution of the Fauna. Acad. Sci. USSR, Zool. Inst. Translation from Russian by Israel Program Sci. Transl.; U.S. Dept. Commerce, Washington, D.C.

Vinson, C. 1946. Wild Rooshians of Tennessee. Nature Mag., 39:405–406.

von Bloeker, J. C., Jr. 1938. The Mammals of Monterey County, California. M.S. thesis, Univ. California, Berkeley.

Von Nathusius, H. 1864. Vorstudien fur Geschichte und Zucht der Haustiere zunachst am Schweinschadel. Weigandt und Hempel, Berlin.

Wagner, J. A. 1839. [No title available.] Akad. Wiss. Munch. Gelehrt. Anz., 9:433–440.

———. 1844. Die Säugetiere in Abbildungen nach der Natur, mit Beschreibungen (J. C. D. Schreber), supplementband, Leipzig, 4:1–523.

Wahlenberg, W. G. 1946. The Longleaf Pine. C. L. Pack Forestry Foundation, Washington, D.C.

Wakely, P. C. 1954. Planting the Southern Pine. U.S.D.A. Forest Serv. Agric. Monogr. no. 18.

Walker, E. P. 1975. Mammals of the World. 3d ed. Johns Hopkins Press, Baltimore, Md. 2 vol.

Wangsness, P. J., R. J. Martin, and J. H. Gahagan. 1977. Insulin and growth hormone in lean and obese pigs. Amer. J. Physiol., 233:E104–E108.

Warwick, E. J. 1962. New breeds and types. In: Stefferud, A., ed. After a Hundred Years: The Yearbook in Agriculture. U.S. Gov. Printing Office, Washington, D.C.

Waters, J. H., and C. J. J. Rivard. 1962. Terrestrial and Marine Mammals of Massachusetts and Other New England States. Standard-Modern Printing Co., Brockton, Mass.

Weaver, M. E., and D. L. Ingram. 1969. Morphological changes in swine associated with environmental temperatures. Ecology, 50:710–712.

Webb, S. D., ed. 1974. Pleistocene Mammals of Florida. Univ. Presses Florida, Gainesville.

Webster, W. D. 1988. The mammals of Nags Head Woods Ecological Preserve and surrounding areas. Assoc. Southeastern Biol. Bull., 35:223–229.

Weddle, R. S. 1973. Wilderness Manhunt: The Spanish Search for La Salle. Univ. Texas Press, Austin.

Welch, L. C., and M. J. Twiehaus. 1966. Nebraska miniature swine development. In: Bustad, L. K., and R. O. McClellan, eds. Swine in Biomedical Research. Battelle Memorial Inst., Pacific N.W. Lab., Richland, Wash.

Wheeler, S. A. 1944. California's little known Channel Islands. U.S. Naval Inst. Proc., 70(493):257-270.

Whitaker, J. O., Jr. 1980. The Audubon Society Field Guide to North American Mammals. Alfred A. Knopf, New York.

Wodzicki, K. A. 1950. Introduced mammals of New Zealand. Dept. Sci. and Industry Res., Bull. no. 98.

Wood, G. W., and R. H. Barrett. 1979. Status of wild pigs in the United States. Wildl. Soc. Bull., 7:237-246.

Wood, G. W., and R. E. Brenneman. 1977. Research and management of feral hogs on Hobcaw Barony. In: Wood, G. W., ed. Research and Management of Wild Hog Populations. Belle Baruch Forest Sci. Inst., Clemson Univ., Georgetown, S.C.

Wood, G. W., and T. E. Lynn, Jr. 1977. Wild hogs in southern forests. So. J. Appl. Forestry, 1(2):12-17.

Wooters, J. 1973. Hunting wild hogs. Guns & Ammo, 17(7):44-45.

——. 1975. The Czar of Swine. In: Elman, R., and G. Peper, eds. Hunting America's Game Animals and Birds. Winchester Press, New York.

Young, S. P. 1958. The Bobcat of North America. Stackpole Co., Harrisburg, Pa.

Zeuner, F. E. 1963. A History of Domesticated Animals. Harper & Row, New York.

Zivkovic, S., V. Joranovic, I. Isakovic, and M. Milosevia. 1971. Chromosome complement of the European wild pig (*Sus scrofa* L.). Experientia, 27:224-226.

Zurowski, W., H. Sindowa, and B. Galka. 1970. Effect of a single pig's blood addition on the local wild boar (*Sus scrofa*) population. Internat. Congr. Game Biol., 9:235-238.

Index